A Colour Atlas of FOOD QUALITY CONTROL

Jane P. Sutherland
BSc, PhD, AIFST
Senior Microbiologist
J. Sainsbury plc, London

A. H. Varnam
BSc, PhD
Consultant Microbiologist
Southern Biological, Reading

M. G. Evans
Photographer

A Wolfe Science Book

Copyright © J. P. Sutherland, A. H. Varnam, M. G. Evans, 1986
Published by Wolfe Publishing Ltd, 1986
Printed by Royal Smeets Offset b.v., Weert, The Netherlands
ISBN 0 7234 0815 7

This title is one of a rapidly expanding science series.
For a full list of Atlases in the series, plus forthcoming titles
and details of our other Atlases, please write to:
Wolfe Publishing Ltd, Wolfe House, 3 Conway Street, London W1P 6HE.

Contents

Acknowledgements

The authors would like to thank their colleagues in the food industry, particularly members of the technical staffs of J. Sainsbury plc; G. Ruddle & Co. plc and North Farm Poultry Ltd, who have helped in the production of this Atlas both by providing examples for photography and by providing technical information. We would also like to thank other acquaintances who have offered material for photography.

In addition we are especially grateful to Ms Debbie Catlow who produced the colour drawings on pages 151, 239, 240 and 241 and the black and white drawings on pages 77, 81, 214, 221, 228, 245 and 249; to Ms Soo Downs for preparation of the typescript and to Mr Nigel Nixon of Fotokiné (Reading) Ltd for his help and advice concerning some of the technical aspects of photography.

Preface

It is intended that A Colour Atlas of Food Quality Control should serve as a guide not only to persons actively involved in Food Quality Control but also to students and trainees as well as to non-technical food industry personnel who wish to enhance their product knowledge.

The selection of illustrations has been based on those quality defects most commonly encountered at retail or final inspection level, together with less common defects which illustrate a point of particular significance. The scope of the Atlas is such that it is not possible to illustrate all potential conditions, and some defects which have been described in detail elsewhere, such as spectacular colour changes in milk occurring as a result of microbial activity, are not included as they are now of no commercial significance. At the same time actual spoilage or visible quality deterioration of some shelf-stable products is rare and in such cases illustrations of good quality material are provided to serve as a reference point. Particular attention in this respect is paid to 'exotic' imported goods such as Oriental fermented products the nature of which may be unfamiliar to many persons involved in Food Inspection.

In discussing food 'quality' it is necessary to state that the Atlas is concerned primarily with technical aspects of quality and that subjective judgements based on hedonistic properties as in the comparison of wines of different vintage cannot be considered in detail. However where technical quality is concerned it is the intention that where a visual fault is identified the Atlas should, where practical, assist in identifying the cause. To this end the faults illustrated are related, where possible, to the manufacturing technology, using in many cases flow charts based on the Hazard Analysis: Critical Control Point Technique. In addition examples of laboratory tests which may be of value in confirming visual diagnoses are included.

Food poisoning agents, whether microbial or chemical in nature, cannot usually be detected by visual examination, but anything containing food poisoning agents must *a priori* be considered to be of poor quality, and thus for each type of food specific problems of a Public Health nature are briefly discussed.

The basic plan of the Atlas is to devote a chapter to each commodity group (e.g. Fresh Meat) with two non-commodity chapters concerned with Precepts of Food Quality Control and Foreign Bodies and Infestations. There are inevitably foods which could, with justification, be placed in more than one chapter and in such cases full cross-references are included.

Laboratory Tests

Introductory Notes

Examples of laboratory tests, where appropriate, are given at the end of each commodity chapter (or in some cases each commodity group). It should be appreciated that these are examples only of commonly applied tests to finished products and that more specialist, or alternative tests, may be applied according to circumstance. No details of methodology are given and these should be obtained from specialist publications such as ICMSF (1978), and Harrigan and McCance (1976) for general microbiology; Egan *et al* (1981) for classical chemical and Bergmeyer (1974) for enzymatic methods of analysis. Specialist techniques for isolation of food poisoning organisms and toxins are discussed in detail in Corry *et al* (1982) while those interested in alternative approaches to microbiology may find Sharpe (1980) of interest. Identification of insects and other foreign bodies is best made with the assistance of illustrated guides, references to which may be found at the end of Chapter 17, Foreign Bodies and Infestations.

Considerable technical information is also available from suppliers of laboratory materials such as (in the U.K.) Oxoid Ltd, Basingstoke; Difco Laboratories, West Molesey and Becton-Dickinson (U.K.) Ltd, Wembley. Advice on the application of instrumental chemical techniques such as spectrophotometry and the various types of chromatography is available from instrument manufacturers and includes, where appropriate, operator training. Examples of major manufacturers in the U.K. include Pye-Unicam, Cambridge; Perkin-Elmer, Beaconsfield and LKB Instruments, South Croydon.

Explanation of Symbols, Formulae and Abbreviations

Microbial numbers are expressed as **10** raised to a **power**. Thus 10^6 is equivalent to **1,000,000** and 10^3 is equivalent to **1000** etc. The numbers refer to colony forming units per gram or millilitre of food; a colony forming unit being either a single micro-organism, a clump or a chain of micro-organisms which give rise to a single colony during the plate count procedure. Where referred to in the text microbial numbers appear in the following form: 10^6 colony forming units/g; 10^3 colony forming units/ml.

pH value. A measure of acidity and alkalinity based on a scale of 0–14. A pH value of 0 is extremely acid and a pH value of 14 extremely alkaline; a pH value of 7 is neutral. Examples of pH values of some common foods are:

Alkaline (pH value >7.0)
Egg white, up to 9.6

Neutral (pH value 7.0–6.5)
Milk 7.0–6.8; chicken meat ca. 6.7; luncheon meat 6.8–6.5

Low acid (pH value 6.5–5.3)
Bacon 6.0–5.6; canned vegetables 6.5–5.4; fresh meat 6.0–5.5

Medium acid (pH value 5.3–4.5)
Canned soups ca. 5.0; soft cheese ca. 4.5; pickled cucumbers 5.0–4.5

Acid (pH value 4.5–3.7)
Mayonnaise 4.1–3.0; yogurt 4.2–3.8; orange juice 4.0

High acid (pH value <3.7)
Pickles ca. 3.5; canned citrus fruits 3.7–3.5; sauerkraut 3.7–3.1
(N.B. The categories of acid foods quoted are those commonly used in the canning industry)
Corresponding pH values for growth of common food associated micro-organisms are:

Bacteria
Acetobacter spp 2.8–4.3; *Escherichia coli* 4.4–9.0; *Salmonella spp* 4.0–9.0; *Bacillus cereus* 4.9–9.3; *Clostridium botulinum* 4.7–8.5; *Lactobacillus spp* 3.8–7.0; *Staphylococcus aureus* 4.0–9.8; *Streptococcus lactis* 4.3–9.2

Yeasts
Hansenula canadensis 2.1–8.6; *Saccharomyces cerevisiae* 2.3–8.6; *Saccharomyces pastorianus* 2.1–8.8

Moulds
Aspergillus spp 1.6–9.3; *Penicillium spp* 1.6–11.1; *Fusarium spp* 1.8–11.1

Aw level: Water activity. Technically water activity is defined as the ratio of the **water vapour pressure** of the food (p) to that of pure water (p_o) at the same temperature.

$$Aw = p/p_o$$

In practical terms water activity determines the amount of water that is available to micro-organisms, thus the lower the Aw level the less susceptible the food to microbial spoilage. It is important not to confuse Aw level with water content; although the two parameters are related, foods of the same water content may have different Aw levels and conversely foods of the same Aw level may have different water contents. Examples of the Aw levels of some common foods are:

Aw level >0.98
Fresh meat and fish; milk; canned vegetables
Aw level 0.98–0.93
Evaporated milk; bread; lightly cured meats
Aw level 0.93–0.85
Semi-dry salami sausage; mature cheese; sweetened condensed milk
Aw level 0.85–0.60
Dried fruit; jams and jellies; dried salt fish
Aw level <0.60
Chocolate; dried milk; biscuits
Corresponding minimum Aw levels for growth of common food associated micro-organisms are:

Bacteria
Escherichia coli 0.95; *Salmonella spp* 0.95; *Bacillus cereus* 0.95; *Clostridium botulinum* 0.94–0.97; *Lactobacillus spp* 0.94; *Staphylococcus aureus* 0.86; *Halobacterium salinarium* 0.75

Yeasts
Debaryomyces hansenii 0.83; *Saccharomyces cerevisiae* 0.90; *Saccharomyces rouxii* 0.62

Moulds
Aspergillus niger 0.77; *Penicillium spp* 0.79–0.83; *Mucor spp* 0.93; *Rhizopus spp* 0.93
It should be noted that many micro-organisms are able to survive for long periods at Aw levels below the minimum for growth.

R.H.A.S.	Recipe Hazard Analysis System.
NaCl	Sodium chloride; common salt.
$NaNO_2$	Sodium nitrite.
$NaNO_3$	Sodium nitrate.
SO_2	Sulphur dioxide.
–S:S–	Disulphide group (Sulphur bridge): a chemical group important in the production of the desirable characteristics of bread dough.
Redox potential	The ability of a food to act as either a reducing or an oxidising agent.
$^w/_v$	Weight for volume, for example 10g solid material in 90ml liquid produces a 10%($^w/_v$) solution (or suspension).
$^v/_v$	Volume for volume, for example 10ml of liquid A in 90ml liquid B produces a 10%($^v/_v$) solution.
ml	Millilitre (0.001 litre).
g	Gram (0.001 kilogram).
mg	Milligram (0.000001 kilogram).
ppm	Parts per million (may also be expressed as mg/l). For example 0.2ppm is equivalent to 0.00002%.
lb/in^2	Pounds per square inch (pressure).
>	Greater than.
$\ngtr$	Not greater than.
<	Less than.
$\nless$	Not less than.
ca.	Approximately.
F1 hybrid	Plant varieties produced by crossing two pure breeding strains of a particular species or variety and where the progeny are both more vigorous than either parent and of very uniform appearance are known as F1 hybrids.

Guide to suitability of food for sale

It is often necessary for a retailer to decide whether or not food in his possession is suitable for sale. This decision may be difficult to make in the absence of fixed criteria and as an aid illustrations in this Atlas are colour coded as follows.

● Suitable for sale and of acceptable quality.
○ Suitable for sale (under the terms of the Food Act, 1984 in the U.K. or equivalent legislation elsewhere) but of poor quality (see note 3b)
● Unsuitable for sale (Vendor liable for prosecution under the Food Act, 1984 in the U.K. or equivalent legislation elsewhere).

It must be stressed that the colour coding is intended as a guide only and cannot be considered to be definitive. Judgement of food quality and suitability for sale is, in many cases, subjective and varies according to preferences held on both an individual and a cultural basis. Similarly the legal interpretation of suitability for sale can vary, according to circumstance, from one nation to another. Where judgement of suitability for sale is particularly subject to individual or cultural differences, or where the basis for national legislation may differ this is indicated by an asterisk placed alongside the colour code. When the Atlas is used to judge suitability of food for sale it is also necessary for its usage to be in the context of the following factors.

(1) If there is any reason, circumstantial or otherwise, to believe that the food has caused, or is capable of causing, any illness or injury or if it has been exposed to contamination or stored at incorrect (high) temperatures it is implicitly unsuitable for sale.

(2) Appearance is only one factor in judging a foods suitability for sale; smell and taste are of equal, or greater, importance. Food should **not**, however, be tasted if there is any doubt as to its safety, if it has been stored under unsuitable conditions or if it shows any sign of overt spoilage.

(3) Food which is normal in appearance, taste and smell, which presents no known hazard and which has been correctly stored is not necessarily suitable for sale. There are two main reasons for this.

(a) The food is incorrectly or falsely labelled, does not meet compositional requirements such as meat content of sausages, contains excess water or non-permitted ingredients etc. It must be stressed that in the U.K. at least the retailer ultimately bears responsibility for the food he sells even though he must often depend on the manufacturer for an assurance that it meets the appropriate legislative requirements. If the retailer has purchased from a reputable manufacturer he may well not be prosecuted in the event of complaint but even the possession of a written Warranty from the manufacturer cannot guarantee his exemption from legal proceedings (see, for example Shropshire County Council *vs* Safeway Stores Ltd, 1983).

(b) Under the Food Act, 1984 in the U.K. and equivalent legislation in some other countries poor quality may itself be sufficient justification for legal proceedings. In the case of the U.K. the terminology of the Food Act – 'Food not of the substance and quality demanded' is open to individual interpretation and inevitably creates grey areas although most prosecutions under this Section of the Act are brought for well defined reasons such as mould growth, foreign bodies, illegal preservatives etc. Prosecutions involving quality *per se* are usually linked to the way in which the food is described. Thus a sausage of the minimum legal meat content containing a high, but permitted, level of rind might attract prosecution if labelled 'Premium' or Extra-quality' but not otherwise. Similarly a young, bland cheese while perfectly satisfactory in its own right could attract prosecution if labelled as 'mature' or 'well-ripened'.

Chapter 1

Food Quality Assurance Systems

Food requires inspection at all stages from primary production to final retail distribution to ensure that required standards of quality are met. The purpose of inspection is multi-fold and may be designed to meet legislative, quality and safety standards, or a combination of these. In addition different inspection regimes will be required for different foods and to meet the varying requirements of different manufacturers and retailers; it is recognised that the majority of readers may not be concerned either with the inspection of all foods or with all of the stages from primary production onwards.

The foremost objective of this book is to illustrate food quality and spoilage but these cannot be divorced from a knowledge of the broader aspects of an entire food handling and processing system, which includes processing plant, premises etc. The principles of Food Quality Assurance systems (excluding Government systems) are therefore outlined below.

The Process

The degree of processing and the technology involved will differ from one food product to another. Where it is possible to relate quality directly to stages in the manufacturing process the processes are outlined in the Introduction to the appropriate commodity chapter. In general terms, however it is of fundamental importance that persons involved in Food Quality Assurance at both manufacturing and post-manufacturing stages should:

(a) Be aware of individual processes involved in the preparation of food(s) for which they are responsible.
(b) Be familiar with the technological objectives of each stage.
(c) Be confident that these objectives are consistently and reliably achieved.

(Technological objectives must be assumed to include parameters relating directly to safety.)

Each process should have an up-to-date specification, a proforma or flow diagram (see **Hazard Analysis Critical Control Point Technique: HACCPT** p. 13). Alterations to the process should not be made without full consideration of the effects on safety and quality of the final product; equally if the nature of the final product changes, the possibility of ill considered, unauthorised or unintentional process alterations should be investigated. It is also necessary to ensure that all process records are correctly maintained.

Quality Assurance personnel at a post-manufacturing stage are likely to be less involved in day-to-day plant operations. As indicated above a basic knowledge of the process and its objectives should be acquired and a pro-forma for the process should be available.

The Ingredients

It is necessary to ensure that ingredients meet their technological requirements with respect to flavour, texture, preservative properties etc., as well as ensuring that ingredients do not themselves adversely affect the chemical or microbiological safety and stability of the finished product.

The need to monitor the quality of major ingredient(s) is self-evident but less attention is often paid to minor ingredients which are purchased as proprietary pre-formulations with some potential loss of quality assurance by the user. It should be appreciated that small changes in minor ingredients may result in significant changes in the finished product. An extension of HACCPT known as **Recipe Hazard Analysis** (RHAS) (Zottola and Wolfe, 1981) may be found to be of value in assessing the role of ingredients with respect to food quality and safety. Although control of ingredients is primarily the responsibility of Quality Assurance personnel at the manufacturing level, post-manufacture personnel should be aware of the ingredients present in a given product and understand their purpose.

The Formulation (Recipe)

At its most basic level the objective of product formulation is to produce a food of good perceived quality with respect to the intended consumer. Economic restraints will apply according to the market for which the product is intended and there may also be constraints of ingredient availability and, in some cases, suitability of available process plant.

It is necessary to ensure that product formulations meet the required legislative standards, for example meat content of sausages, and that where preservatives such as sodium nitrite are added the levels are sufficiently high (although again meeting legislative requirements). In designing product formulation it is necessary to ensure that interaction between ingredients will not result in undesirable chemical or microbiological consequences or that the presence of a given ingredient will not prevent a technological objective from being attained during processing. Systems based on RHAS may be beneficial during formulation.

In addition to ascertaining that the product formulation inherently meets the required conditions it must be ensured that in practice the formulation is correctly produced. This is likely to involve laboratory testing of the final product and it is desirable that the required tests should be decided upon at the design stage of the formulation.

A final point must be made with respect to formulation. Food additives such as colours and preservatives may in some circumstances be added unnecessarily; sodium metabisulphite (sulphur dioxide), for example, a common ingredient of frozen comminuted meat products, can obviously have no antimicrobial role in a frozen product. Avoidance of 'redundant' additives is beneficial from an economic and often an aesthetic viewpoint.

The Premises

A knowledge that premises are suitable for their purpose and that they are adequately maintained is necessary. There are a number of specialist technical publications describing the detailed requirements for food manufacturing and storage premises which should be consulted if details are required. In the present context principles only will be discussed. The basic functions of food handling premises are as follows:
(a) To provide suitable storage space for ingredients, packaging and finished product.
(b) To protect ingredients and finished product from contamination.
(c) To provide a suitable work area for efficient production of the foodstuffs (including adequate facilities for production personnel).
In consideration of (a) it should be appreciated that productive (i.e. revenue generating) space is at a premium and that inadequate storage space is a frequent contributory cause of infestation and foreign body contamination.

Where refrigeration is required inadequate facilities can lead to loss of quality not only through too high temperatures leading to rapid microbial growth and thus spoilage, but in some cases too low temperatures leading to general quality deterioration. An example of the latter is that of dairy cream desserts which may 'collapse' if stored at temperatures below 1°C (see pp. 42–43).

The protection of ingredients and finished products (b) refers not only to protection from contamination from outside sources but also requires that the premises are constructed and maintained in order that they themselves do not become a focal point of contamination. Ease and practicability of cleaning is an important consideration and where dry powder ingredients are concerned build-up in poorly accessible areas often leads to rapid development of an infestation. The fabric of the building must be such as not to be a source of foreign bodies (see pp. 250–251).

An important point must be made at the start of the discussion of (c). Changing public tastes, economic circumstances and the mass marketing policies of large scale food retailers have placed considerable strain on some small manufacturers, and premises which meet the requirements for small scale (or even effectively cottage scale) manufacture usually do not provide a suitable work area for large scale manufacture. Equally where expansion has been self-generated quality problems may well arise unless premises have been suitably enlarged or modified. In any premises, whether old or new, the importance of providing adequate working space for production personnel together with a suitable overall environment with regard to lighting, heating, ventilation etc., cannot be over-estimated.

Peripheral facilities such as a pleasant canteen are of importance in maintaining employee morale and special consideration should be given to toilet and hand washing facilities. The provision of these is an important factor in maintenance of food hygiene standards and a senior member of staff should be responsible for ensuring that toilets are clean and well maintained and that washrooms are clean with an adequate supply of hot water, unscented soap and suitable drying facilities. 'Now wash your hands' notices are an important visual reminder.

Apart from specific points discussed above the general condition and housekeeping of premises is often indicative of general attitudes to quality and is an important factor of which Quality Assurance personnel should be aware.

Plant and Utensils

As with the discussion concerning premises there are a number of specialist technical publications such as Food Trades Journal and Food Processing (Chicago) which describe both the function of the various pieces of plant and the special requirements of design and construction to be met for use with foodstuffs. It is intended to deal only with principles.

Personnel involved with Food Quality Assurance should have at least a basic knowledge of the function of the relevant plant, principles of operation and consequences of malfunction. Where involvement is at a manufacturing level it must be ensured that equipment is used for the purpose for which it was designed, that it is kept in good order and is fitted with all necessary instrumentation. The equipment should not only be kept clean but be readily cleanable.

When designing cleaning schedules the particular problems associated with each piece of plant should be taken into account and where **cleaning-in-place (CIP)** systems are used care must be taken to ensure that the circulating fluids reach all parts of the plant. Micro-processor controlled CIP systems should not be regarded as infallible but should be checked on a regular basis.

Particular attention should be given to plant repairs. It is important that repairs should be finished to the same standard as the original and that the same, or an equivalent, material is used. Where there is contact with food it is essential to ensure that the material is of 'food grade standard'. Temporary repairs, made with pieces of string etc., are prone to becoming permanent resulting in shedding machinery parts into the foodstuffs, and should be avoided. Where alterations are made to the layout of pipework handling liquid foods care should be taken to avoid creating dead ends which serve as a focal point for build up of contamination.

Small utensils present particular problems in that they are easily overlooked during cleaning and may be used for purposes other than those intended. The use of food handling utensils to dispense cleaning fluids etc., presents a particular hazard which may be reduced by providing special coloured containers for detergents and disinfectants. The condition of utensils should be checked on a regular basis and worn or damaged items replaced. Managerial attitudes which permit spending significant sums on process plant and yet are reluctant to replace a scrubbing brush with loose bristles are to be condemned.

Production Personnel

In any given food manufacturing or handling situation there is likely to be, at some stage, a conflict between production requirements and those of quality. In such situations personnel involved in Quality Assurance are likely to find themselves in the role of Devil's Advocate, and in resolving conflicts between quality and production it is often necessary to assess the strengths and weaknesses of production personnel for it is they who are responsible for actually producing the food.

It is an old adage that quality must come up from the bottom and down from the top. Unless senior management and executive directors have a commitment to quality no Quality Assurance system is likely to be successful.

Similarly it is necessary to ensure that non-management production workers are adequately trained and have at least a basic understanding of why procedures must be followed. This is of particular importance with respect to hygiene precautions. Food handlers must appreciate that they bear an implicit responsibility with respect to food safety. Management must provide laid down procedures to deal with illness amongst staff, contact with sufferers from food-borne disease etc., and ensure the staff understand that these are not intended to criticise personal hygiene but to ensure food safety.

With respect to involvement of non-management personnel in obtaining food quality, good results are likely to be obtained where industrial relations and working conditions are good. Care should be taken to avoid a wages structure where the increment of the total wage due from bonus payments is too high since the loss of quality is likely to more than offset the benefits of increased productivity.

The management having the greatest day-to-day involvement with performance of non-management staff are of course line management of middle and junior management levels. These latter sectors are of considerable importance in other aspects of food quality in that it is at their level most immediate decisions are made, and also because it is at this level the conflict between production and quality is likely to be most acute.

Where outside inspection of a food processing unit is required it is necessary for the person responsible to be aware of the capabilities of the unit's own Quality Assurance personnel. 'Watching the watchdogs' in this manner may appear to be onerous but there is no doubt that some manufacturers 'deliberately' employ weak and/or inexperienced persons as Quality Assurance personnel, and a knowledge of the capabilities of such persons is thus no less important than a knowledge of plant, process etc., when an outside inspection of a plant must be made.

Sampling – the Statistics of Madness

Sampling schemes vary as widely as the requirements of individual food handling units vary and for each problem a number of possible approaches to a solution exist. It is thus not possible to discuss in detail the design of sampling schedules. For these the reader is referred to discussions such as that of Kilsby (1982) while the two major approaches to sampling, the attribute schemes and the variable sampling schemes, are described in detail elsewhere (ICMSF, 1974; Kilsby et al, 1979).

It is most important that before designing a sampling scheme it is clearly understood exactly what the sampling is intended to achieve and that the choice of approach, and final design of the scheme, are based on logical premises. Moreover the logic must be understood by those who are required to apply the scheme because conceptually sound sampling schemes may be rendered worthless by misunderstanding or misapplication.

Failure of sampling schemes with potentially dangerous consequences can be due to two basic causes. The first lies in a misunderstanding of the rationale behind a scheme and is illustrated by the following example: 'A production unit experienced microbiological problems leading to overt product spoilage with significant attendant financial loss. The subsequent investigation showed that the problems invariably involved product manufactured during the latter part of the working day. The intermediate cleaning schedules were adjudged to be inadequate and were revised. It was agreed to monitor the new schedules by sampling products hourly (eight times a day). The instruction to the staff responsible for sampling which merely stated "eight samples daily" was duly carried out. For a variety of reasons all eight samples were taken from the same product batch during the first hour of production. The laboratory tests made on the samples thus failed totally to meet the purpose for which they were devised. It should be noted that the plant involved was considered to be above average with respect to its Quality Control.'

The second area of failure lies quite simply in the procurement and transport of samples to the laboratory. Problems are most acute where samples are obtained at a distant site and transported to a central laboratory and where inexperienced, untrained or unsuitable personnel are involved. Failure by management to recognise these factors may be due either to ignorance of reality or to a lack of commitment to quality. Typical examples are failure to adequately mix a vat of liquid before sampling, misidentification of and mislabelling of samples, loss of samples and temperature abuse during transport. Where feasible the latter may be controlled by use of recording thermographs but this may be precluded on cost grounds.

Finally mention must be made of the **deliberately misleading** sample exemplified by specially clean samples for microbiological analysis or specially supervised formulations for chemical analysis. Such samples may be produced for intra-plant purposes, for example to avoid criticism of cleaning procedures or for external purposes such as submission of samples to a potential customer. Whether samples of this nature are produced in ignorance of the potentially serious consequences or whether knowingly intended to deceive the practice cannot be condemned too strongly.

Data Handling

Quality Assurance systems generate large quantities of information which is often under-utilised. Once again it is not intended to discuss systems and means of data handling in detail but rather to highlight the major principles in relation to Quality Assurance.

In deciding the manner in which data is to be handled and presented it is first essential to define what the individual Quality Assurance system is intended to achieve. A system, for example, may be intended to monitor the performance of a single plant, possibly by comparing individual production lines or to compare the output of a number of plants. The data handling requirements of the two examples are likely to be different. Equally the information from a given Quality Assurance operation may be used retrospectively to provide management information as to performance or on a day-to-day or even hour-to-hour basis to provide production control. Once again the data handling requirements are likely to differ.

Computers are now widely used in data handling and small but powerful systems such as the Commodore, ICL, Apple and Digital Equipment series are available with peripheral devices such as printers for relatively low costs. Computers have considerable advantages not only in reducing the physical work and avoiding the inherent errors associated with manual systems but also in vastly increasing the ways in which data can be manipulated and presented. It is necessary however to avoid the temptation to employ the power of a computer to produce superfluous statistical information which can divert the emphasis of Quality Assurance data away from its primary objective.

In planning a computer based data handling system it is important that the programmer or systems analyst responsible for software design (and possibly choice of hardware) is fully acquainted with what the Quality Assurance system is intended to achieve, what use will be made of the information and by whom. At the same time Quality Assurance personnel who will be required to operate the data handling system must be aware that the computer is not a universal panacea, a philosopher's stone to solve all problems. Before introducing a computer based system it is advisable that the users acquire a basic understanding of computer principles and applications, preferably including 'Hands-on' experience.

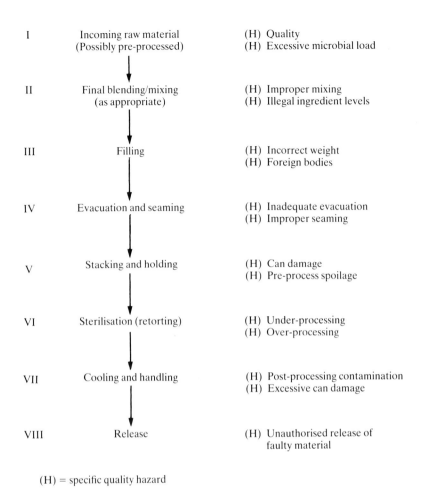

1. A flow diagram typical of a canning process. The technical details and thermal processing applied will vary according to the product being canned.

Hazard Analysis – Critical Control Point Technique (HACCPT)

HACCPT is a technique which is intended to extend the principle of **Good Manufacturing Practice (GMP)** toward zero defect manufacture. The technique, which was developed by the Pillsbury Corporation on behalf of the United States Government (Anon 1973), is being increasingly used in the food industry particularly in the U.S.A. and Japan. HACCPT incorporates all the principles discussed above but in the context of a formal framework whereby hazards are identified and (in some cases) categorised (e.g. quality 'hazard'; public health 'hazard') and the critical control points are identified.

The first stage in applying HACCPT is to prepare an audit of the process, the main feature of which is a **flow diagram** depicting the manufacturing process from receipt of raw materials to storage of the finished product. (In some cases HACCPT is extended beyond the plant to distribution and even to point of consumption.)

The flow diagram (**1**) may depict not only the manufacturing process itself but consider peripheral functions which can affect control points such as compressed air supply and process instrumentation. Hazards are indicated, identified according to type, and the control points are marked. In some cases degree of hazard is quantified. Flow diagrams based on HACCPT principles have been used, where considered appropriate, to illustrate the relationship between stages in the manufacturing process and quality. This particular example is used for canning (see p. 212).

While initial preparation of the audit and flow chart is usually made by in-house personnel it is increasingly common practice to have the audit surveyed by a 'godfather' who may be from another sector of the same organisation or an outside consultant.

HACCPT is a powerful tool in assuring the quality of foods. It is important that the technique should be recognised as a tool for use where required and that it does not represent a blanket solution to problems. The successful application of HACCPT depends on many factors including the attitude of management and overall commitment to quality and safety, and it is usually of greatest benefit when made the responsibility of one senior member of Quality Assurance staff rather than that of a committee.

Shelf Life of Food and Quality

The manufacture of perishable foods was for a long period organised on a local basis and food was sold shortly after manufacture. The development of centralised production, the demands of mass marketing and the development of effective means of extending storage life such as widespread availability of refrigeration and vacuum packing now means that perishable foods are stored, in many cases, for significant periods before consumption.

It is necessary for retailers to be able to recognise the age of foods and to ensure they are sold while still in a wholesome condition. This has led to the development of date-coding and 'sell-by' dating which defines the life of the food. It should be noted that date-coding refers only to spoilage of food. It cannot, and is not intended to, provide a guarantee of safety; the age of a food has no bearing on the presence or absence of pathogens although food bearing high microbial numbers carries a higher risk of non-specific food poisoning, and low temperature pathogens such as *Yersinia enterocolitica* may develop to significant numbers in foods stored for excessive periods at low temperatures.

The principle of date-coding has in many countries been extended to non-perishable foods. In the U.K. perishable foods with a storage life not greater than 6 weeks are coded with a 'sell-by' date whereas non-perishable foods are coded with a 'best before' date. The convention in other countries may be different. The underlying principles of 'sell-by' and 'best before' coding in the U.K. context are different in that perishable foods may be expected to spoil after the sell-by date has passed (allowing 2 to 3 additional days for customer storage and several days safety margin) whereas non-perishable foods are not expected to spoil but to show detectable organoleptic deterioration.

After a certain point foods whether perishable or non-perishable begin to lose quality and there is no doubt that the trend towards long storage lives leads to significant loss of quality within the stated product life. Indeed the ease with which overtly spoiling food within its life may be found in large supermarkets (see Table 1) illustrates that lives are unrealistically long. It should be noted that pressure for long lives comes primarily from retailers rather than manufacturers and persons concerned with food quality should be fully aware of the consequences of commercial demands in this respect. It is also necessary to consider whether products with excessively long storage lives may truthfully be described as 'fresh'.

Table 1   Survey of Overt Spoilage within Life

Food Involved and Spoilage	Days Before 'Sell-By'	Display Temperature °C
Medium/Large Town Centre Supermarkets		
Rhubarb yogurt blowing	3	6
Walnut gateau (cheese) possible gas formation	10	5
Hard cheese blowing	8	7
Coleslaw blowing	4	7
Numerous dairy	up to 15	not indicated
Black pudding mould	not coded	not indicated
Small Suburban Supermarkets		
Cottage cheese blowing	4	8
Hard cheese mould	not coded	8
None (very limited perishable display)	—	6–7
Large Suburban Supermarkets		
Pâté (pre-pack) blown	3	5
Brie spoilage mould	10	7.5
Taramosalata (pre-pack) blown	5	6
Coleslaw vinaigrette blown	6	6
Bulk salad gas formation	not coded	6
Camembert spoilage mould	10	12
Hard cheese mould	not coded	12
Very Large Edge of Town Hypermarket		
None	—	2–4
Food Hall (Self-Service) Department Stores		
Toffee yogurt blowing	2	7
Crème dessert gas formation	2	7
Roast beef slime	not coded	not indicated
Bacon discoloration	not coded	not indicated
Fresh pork slime	1	not indicated

N.B. The survey was based on a single visit to food stores in the Greater Reading area on a single day in June 1983. It is not suggested that these findings would necessarily be repeated on other occasions. It should also be noted that display temperatures were obtained from dial thermometers mounted in the cabinets and that high temperatures may be a consequence of a cabinet's de-frost cycle.

Domestic Storage and Quality

Beyond printed storage instructions on pre-packed foods it is obvious that once sold neither the retailer nor the manufacturer has any control over the subsequent handling. However it is necessary to be aware of the vagaries of domestic storage both in assessing the validity of complaints from customers and in determining realistic storage lives. In general terms domestic refrigeration is less efficient than its commercial counterpart. Domestic refrigerators frequently have large temperature gradients within the cabinet and temperatures below 7°C are rarely obtained in all places.

Microbial growth rates of psychrotrophic bacteria increase rapidly between 2°–4°C (commonly accepted nominal commercial storage temperature) and 7°–8°C, and quality may deteriorate as rapidly in 2 to 3 days at 7°–8°C as in much longer periods at a lower temperature. In addition most domestic refrigerators are unable to rapidly cool large quantities of foods from temperatures above their normal operating temperature including food which has been allowed to become warm during transit from the store. The likelihood of this happening is increased by the growth of 'one stop' shopping at very large stores. Such stores (hypermarkets) are often situated out of town and feature facilities such as restaurants and garden centres, thereby increasing the likelihood of delay in transport. Frozen foods are likely to thaw out under these circumstances and suffer quality loss as a result of re-freezing.

One stop shopping also produces a demand for longer customer life. Late night shopping on a Tuesday, for example, may mean purchase of a joint of meat for consumption on Sunday which means a customer life of 5 days as opposed to the 2 to 3 days usually planned for in shelf life calculations.

It is finally to be noted that a significant proportion of the population still do not own a domestic refrigerator. These are likely to be in low income categories including the elderly who, with the demise of neighbourhood food shops, are often unable, particularly in rural areas, to shop for perishable food on a daily basis.

Quality Circles

Quality circles represent the ultimate commitment to production of high quality products. The concept originated in the Japanese optical and electronic industries as a means of realising the ideal of **zero defect manufacture**. It is also widely used in the U.S.A. where Quality Circles have been introduced in some food plants but applications in the U.K. have, up to the present, been limited and as far as is known are not used in the food industry.

The focal point of the system is the Quality Circle itself. This consists of a number of members of production staff drawn from all levels who meet on a regular basis to discuss the plant or unit's performance with respect to quality and, if necessary, approaches to improvement. All relevant Quality Control data is available to members of the Circle who meet on an equal basis irrespective of position in the organisation's hierarchy. In addition to the Quality Circle itself a commitment to quality must be obtained from all members of staff. In many cases this takes the form of a signed declaration. Summarised Quality Control data is displayed in the production area and staff are encouraged to develop a 'pass it on' or 'tell it like it is' approach to quality whereby sub-standard product may be removed by all personnel rather than relying on specific stages and procedures for inspection. (These procedures are, of course, still necessary.)

In operating a Quality Circle based production unit successfully, attaining a high quality standard, it is usual to give financial rewards to employees. These vary in nature from a simple bonus system based on defect rate to competitions between units of the same organisation and bonus prizes such as consumer goods and holidays. The importance of rewards is greatest in the setting up and early operating stages of a Quality Circle.

In addition to a genuine commitment to quality from both management and employees the successful operation of a Quality Circle based system requires good management:employee relations and above all an acceptance that the worth of comment and suggestion is not in direct proportion to the position the person making it holds in the organisation. In the U.K. context a major reason for the non-acceptance of the Quality Circle concept must lie with the retributive management style current in 'born-again' managers of all levels.

Dairy

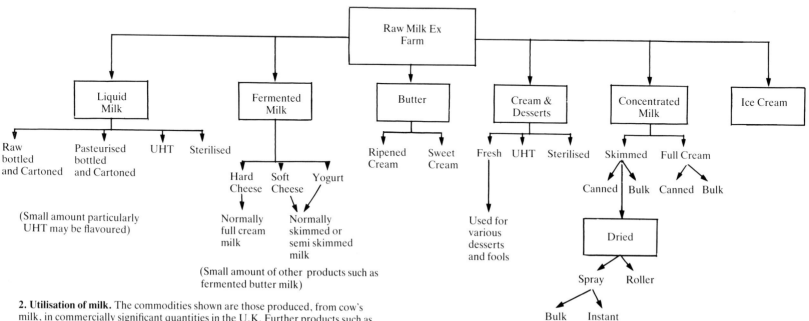

Raw Milk Ex Farm

Liquid Milk — Raw bottled and Cartoned — Pasteurised bottled and Cartoned — UHT — Sterilised

(Small amount particularly UHT may be flavoured)

Fermented Milk — Hard Cheese — Soft Cheese — Yogurt

Normally full cream milk

Normally skimmed or semi skimmed milk

(Small amount of other products such as fermented butter milk)

Butter — Ripened Cream — Sweet Cream

Cream & Desserts — Fresh — UHT — Sterilised

Used for various desserts and fools

Concentrated Milk — Skimmed — Full Cream

Canned | Bulk — Canned | Bulk

Dried — Spray — Roller

Bulk — Instant

Ice Cream

2. Utilisation of milk. The commodities shown are those produced, from cow's milk, in commercially significant quantities in the U.K. Further products such as the mare's milk based kumiss in the U.S.S.R. are produced in large quantities elsewhere.

Introduction

The dairy industry is unique in producing from a single raw material a complex range of products in many cases of an apparently diverse nature.

Liquid Milk (3–20)

The term liquid milk is used to describe sales of unprocessed (apart from pasteurisation or other heat treatment) milk to the retail market. In the present context it is broadened slightly to include retail sales of skimmed or part-skimmed (fat-reduced) and flavoured milks. Liquid milk accounts for ca. 45% of milk consumption in the U.K. and in most countries precedence is given to the liquid milk market in time of shortage. The U.K. is unique in that a high proportion of sales are made by direct delivery to the consumer.

Direct delivery sales are normally in returnable (re-used) glass bottles which present a contamination problem if improperly cleaned, while milk standing outside a house on a warm day undoubtedly undergoes quality deterioration including loss of Vitamin C. Direct delivery is under threat due to increasing sales

A simplified flow diagram (2) illustrates the ways in which raw milk may be processed. In recent years the dairy industry has undergone change in that the fall in consumption of such products as sweetened condensed and evaporated milk has been matched by vastly increased consumption of yogurt and soft cheeses and the introduction of new products such as desserts. At the same time the wide scale introduction of the UHT (ultra heat treatment) process has permitted the introduction of a large range of room-temperature stable high-quality dairy products.

In overall terms the dairy industry in most industrialised countries of Europe and the U.S.A. is dominated by large organisations, either private enterprise or co-operatives. Thus with the exception of small specialist plants the tendency is to large centralised milk processing units dealing with milk from more than one farm. The liquid nature of milk means that in most cases a high level of automation is possible if not actually applied.

It is not pertinent to discuss in length topics such as 'the butter mountain' or the protectionist policies of various Governments towards their respective dairy industry. It should however be noted that the dairy industry is a highly political one and that it may be affected directly by decisions of a political nature rather than by economic logic.

of marginally cheaper milk through supermarkets, cheap UHT milk and the high cost of delivery. It is however an emotive topic in the U.K. and in an attempt to increase revenue many dairies now deliver other foodstuffs with milk including, controversially, wine.

Milk on receipt from the farm is subject to a variety of checks to ensure compositional and bacteriological standards are met. Where milk is delivered to the dairy in small units, e.g. churns containing ca. 10 gallons, the influence of a single churn on overall composition or quality is likely to be minimal. Exceptions are some types of feed taint and the presence of penicillin. Where milk is delivered from bulk farm tanks containing several hundred gallons the influence on overall quality is greater. Poor quality raw milk may affect the quality of retail milk even after heat processing.

Pasteurised Milk

The majority of liquid milk is pasteurised although UHT is of increasing importance. Pasteurisation is achieved by heating milk either in vats at **62.8°C** for **30 minutes (low temperature; long time)** or, more commonly, at **71.7°C** for **15 seconds (high temperature; short time)**. In the latter case a plate heat exchanger fitted with a metering pump and a holding tube is used. It is a legislative requirement that a by-pass device must prevent improperly heated milk from passing the pasteurisation plant and a recording thermograph must be fitted to provide a permanent record of pasteurisation, which is sufficient to kill the majority of vegetative micro-organisms including pathogens but not microbial endospores.

After pasteurisation milk is cooled and stored in refrigerated tanks prior to filling either into glass bottles, plastic bottles or a variety of cardboard or cardboard/plastic cartons. The majority of spoilage problems derive from post-pasteurisation contamination usually by Gram-negative bacteria although surviving spores may outgrow and cause spoilage if subsequent storage temperatures are high.

Ultra Heat Treated Milk (UHT)

UHT milk is of increasing importance. While in the U.K. it accounts for only ca. 1% of liquid milk sales, in Western Europe the equivalent figure is ca. 35% while in urban areas of European U.S.S.R. virtually 100% of milk is ultra heat treated. The UHT process consists of heating milk to a temperature of **132°C** for **not less than one second**. This is achieved either using a plate heat exchanger (APV system) or by injecting live steam into the milk (de Laval, Stork). In the latter case it is necessary to evaporate off the water added as a result of steam injection. The treated milk is then aseptically packaged in plastic coated cardboard cartons, usually using the Tetra-Brik system.

Ultra Heat Treatment destroys all vegetative micro-organisms and all but the most heat resistant endospores. Organoleptic changes are less than in conventionally pasteurised milk and UHT milk has a shelf life at room temperature of several months. Although spoilage can be caused by outgrowth of surviving endospores this is very rare in practice and spoilage is normally due to contamination at the packing stage or is of a physico-chemical nature due to the pre-existing state of the raw milk. Skimmed, semi-skimmed and flavoured milks are also UHT processed.

Sterilised Milk

Traditional means of producing sterilised milk involved heating in sealed bottles in either continuous or batch retorts at a temperature of **not less than 100°C**. More modern processes involve UHT processing followed by either aseptic filling into sterile bottles or a secondary in-bottle heat treatment of ca. **123°C** for a total (i.e. including warm up) time of **20 minutes**. Sterilised milk, particularly when produced by traditional means, has a distinctive flavour due to caramelisation during heating. In the U.K. most sterilised milk is consumed in industrial areas of the North and Midlands where many older members of the population prefer the taste to that of pasteurised milk.

Lactoperoxidase System and Preservation of Milk

The lactoperoxidase system (LPS) is a naturally occurring defence mechanism against bacterial infections in man and other animals. LPS, which is thought to derive its inhibitory effect by damaging the inner membrane of bacteria, consists of three components; **lactoperoxidase, thiocyanate** and **hydrogen peroxide** (H_2O_2). All of these three components must be present for the LPS to be inhibitory. Lactoperoxidase is present in ample quantities in cows' milk and thiocyanate, although often low as a consequence of feeding practice may readily be increased by incorporating clover or cruciferous crops into the diet (Reiter, 1981). H_2O_2 however is present only in trace amounts in freshly drawn milk and thus under most conditions lack of sufficient H_2O_2 is the limiting factor for inhibition of bacteria by the LPS.

Stimulation of the LPS by addition of small quantities of H_2O_2 is therefore an effective means of reducing the rate of microbial growth in milk. However the effect is short lived and while the addition is permitted in developing countries such as Kenya where high concentrations of H_2O_2 are added as a preservative *per se* it is unlikely that legislation permitting the addition of even the small quantities required to activate the LPS would be introduced in western nations. The use of **xanthine oxidase** and **glucose oxidase** for endogenous production of H_2O_2 has been proposed to overcome these objections and a system first described by Björck and Rosen (1976) avoids addition of enzyme and substrate to milk by using firstly β-**galactosidase** immobilised on glass beads to produce glucose from lactose and secondly immobilised glucose oxidase to generate H_2O_2 from glucose. Law and Mabbitt (1983) commented that packing the mixed enzyme-associated beads into a column formed the basis of a continuous flow **cold-sterilisation** unit.

Raw Milk

A small amount of milk is sold in the U.K. without pasteurisation and in less developed parts of the world the percentage is higher. Raw milk is not only of markedly poorer keeping quality than heat treated but also carries the risk of microbial pathogens. Although tuberculosis has been almost completely eradicated from the national herds in the U.K., Continental Europe and the U.S.A., other milk associated pathogens such as *Brucella* (brucellosis) and *Coxiella burnetii* (Q fever) are still likely to be present. **Q fever** is worthy of special note in that it is the only known **rickettsial** disease transmitted to man by means other than a tick-bite. In addition the chance of other microbial pathogens being present is obviously much greater than in pasteurised milk.

Homogenisation

A final note is perhaps pertinent concerning homogenisation of milk. Homogenisation is a process whereby milk under pressure is forced through a narrow orifice. Fat globules are dispersed and the milk no longer shows a distinct layer of cream at the top. All sterilised and UHT milk is homogenised as is most of the pasteurised milk, packed in opaque cartons. Bottled milk sold in London is homogenised but that sold in the provinces is not. One of the difficulties in ascribing quality attributes is illustrated by the responses to homogenised milk *vs* non-homogenised. Consumers accustomed to non-homogenised dislike the absence of a cream line, considering it to be watery and insipid, whereas many consumers of homogenised milk react strongly against the cream line to the extent of finding it actively disgusting.

Examples of laboratory tests are shown in Table 2, p. 20.

3. Milk is a good substrate for microbial growth and souring can be rapid. The appearance of the milk after souring depends on the micro-organism responsible. Souring by lactic streptococci (in this example *Streptococcus lactis*) is illustrated in sample A. Such souring is the usual form of raw milk spoilage although the sample shown was, in fact, pasteurised. Sample B illustrates souring by Gram-negative rod-shaped bacteria (Family *Enterobacteriaceae*). Such bacteria are the most common post-pasteurisation contaminants of milk and low temperature strains are able to grow well at temperatures of below 8°C.

4 and 5. Spoilage of milk by *Bacillus cereus*. In cream *B.cereus* produces classical 'bitty' cream (Billing and Cuthbert, 1958) but in homogenised milk where the fat is dispersed the 'bits' are less discrete (**4**). Ultimately coalescence occurs as spoilage continues (**5**). Although *B.cereus* is recognised as a causative organism of mild food poisoning it is rarely implicated in dairy products.

6. Yeast spoilage of milk is distinguished from other types of spoilage by the strong smell of alcohol and large quantities of gas. The yeast involved are lactose fermenting genera such as *Candida spp*. Yeast spoilage is rare in commercially bottled milk but occurs more frequently when milk is stored in re-usable containers such as vacuum flasks in the home.

7 and 8. Clotting of UHT milk. The milk was sterile therefore clotting was not due to underprocessing or post-process contamination (**7**). In such an example clotting may be due to poor quality (low pH value) milk being processed or, more probably, to the action of heat-stable proteolytic enzymes. These latter are derived from psychrotrophic bacteria which develop in the bulk farm tanks prior to delivery to the dairy. The bacteria are non-heat resistant and thus readily destroyed by processing. The enzymes, however, survive and cause progressive clotting ultimately producing total solidification. (See also Cream p. 42.) Good quality UHT milk is shown in contrast (**8**).

9. Yellow coloration of UHT milk. The same plant used a yellow food dye in production of a flavoured milk and contamination, either accidental or deliberate, has occurred.

10 and 11. Flavoured milks. A recent feature of the liquid milk market in the U.S.A., U.K., Western Europe and elsewhere has been a growth in the range of flavoured milks. Where fruit juices or fruit derivatives are added it is necessary to ensure that the **pH value** of the products is not lowered below the **iso-electric point of the milk protein** otherwise **coagulation** will occur. When viewed from above (**10**) the appearance of the products is normal. If poured out, however, the coagulation becomes apparent (**11**). The addition of a **stabiliser** (citrate) to prevent the pH value falling below the iso-electric point of its proteins overcomes problems of this nature.

12, 13 and 14. Swelling of a carton of UHT flavoured milk (12). Note that gas production is sufficient to force the sealed ends of the pack away from the sides (A). The gas was produced by a heterofermentative *Lactobacillus sp*. Such bacteria are not endospore forming and would not survive the UHT process. Contamination therefore was post-process and probably at the filling stage. As the milk leaving the UHT process is effectively sterile growth of any single contaminating organism is likely to be particularly rapid in the absence of competition. The contents of the pack were clotted, clots being shown in close-up (**13**). Such clots are due to acid formation. A comparison of spoilt and unspoilt milk is shown (**14**).

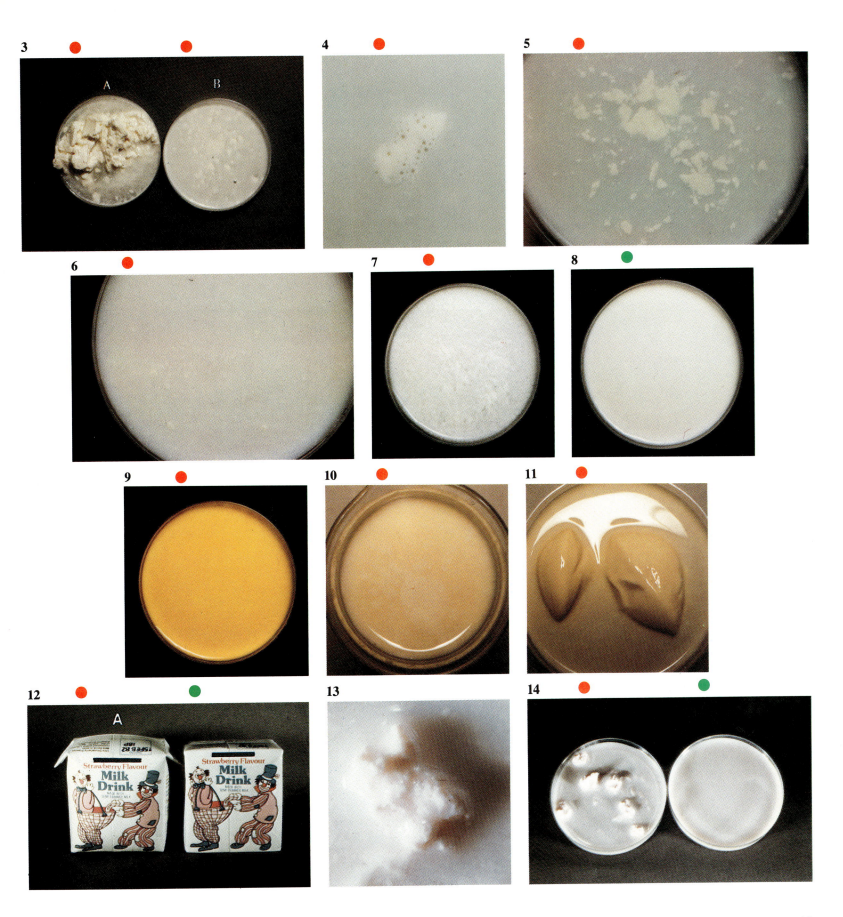

19

Table 2 Laboratory Tests Dairy Milk

I. Microbiological
Total aerobes
Enterobacteriaceae
E. coli
Sterility (UHT and sterilised)
Turbidity (sterilised only)

II. Chemical/physical
Fat

Foreign body/infestation examination not usually required.

Notes:
(1) These tests are examples only and are not inclusive.
(2) Examples are for finished products only.
(3) Techniques are detailed in specialist publications such as ICMSF (1978) – Microbiology: Egan *et al* (1981) – Chemistry.
(4) Instrumental techniques such as Impediometry may be substituted for some classical microbiological methods.

15. Foreign bodies in milk. Milk handling in modern dairies is such that foreign bodies are relatively rare although problems remain where re-usable (returnable) bottles are employed. This illustration shows 'grit' from a carton of pasteurised milk. The material is almost certainly derived from the **rubber O-rings** of the pipe joints used for conveying milk within the dairy.

16. Debris in pasteurised milk. In a single case such as this it is not possible to identify the source but a large number of such complaints would indicate a serious breakdown in plant hygiene.

17. Slivers of dried paint found in a bottle of pasteurised milk. Returnable bottles are often used as containers for a variety of household chemicals. Paint in particular may remain through the bottle washer and be missed at the inspection stage, which in most dairies consists of visual inspection of bottles passing on a conveyor in front of an illuminated screen.

18 and 19. A damaged silverfish (*Lepisma saccharina*) found in a cardboard carton of pasteurised milk (**18**). The insect is damaged but recognisable and was probably present in the packaging before filling. It is contrasted with an alleged 'silverfish' from UHT milk (**19**). Close examination shows this silverfish to be a small fraction of the milk carton.

20. Sterilised milk produced by in-bottle sterilisation. The milk has a distinctive caramelised flavour which is popular in the West Midland and North West regions of the U.K. Elsewhere consumption is limited. Most modern milk sterilisation processes are designed to attain full commercial sterility but some older proces-

Fermented Milk Products

For the purpose of the present discussion fermented milk products are defined as those products where the fundamental nature of the product is derived from souring by a culture of lactic acid bacteria. For practical purposes this means **hard and soft cheeses, yogurt and related products**. Ripened cream butter, although produced from cream soured by a culture of lactic acid bacteria, is not considered to be a fermented milk product *per se* and is discussed elsewhere (see p. 44).

Fermentation of milk is probably the earliest form of food preservation (see Scott, 1981) and similar procedures are applied to all types of milk in all parts of the world. The economic importance is considerable. In 1976 cheese and other fermented milks accounted for 21.6% of world output of fermented products (alcoholic drinks account for 72.6%) at a value of £7,500 million. Despite the huge variety of fermented milks the basic technological procedures fall into a small number of groups.

The technological description of each group is intended only to relate possible faults to the manufacturing process and not as a detailed description of that process. If full details are required use should be made of specialist books such as Cheesemaking Practice (Scott, 1981). Similarly where starter organisms are named such names are given as examples only and specialised information on starters and their action may be obtained by reference to Law and Sharpe (1978), Cox, Stanley and Lewis (1978) or to original research publications.

ses permit the survival of a small number of more highly heat resistant endospores (usually *Bacillus spp*) which ultimately cause spoilage.

Yogurt (21–33)

A flow diagram for yogurt manufacture is shown in **21**. Heating is necessary not only to remove micro-organisms of potential public health significance but also to produce favourable conditions for growth of starter organisms. In extreme cases of over-heating a caramelised flavour may be produced although this is normally masked by the acidic flavour produced after fermentation.

From a viewpoint of intrinsic quality the behaviour of the starter culture is of greatest importance. Partial or complete starter failure may occur with yogurt as with all fermented products. Yogurt starters are relatively free of problems with bacteriophage, and failure is usually due to trace amounts of antibiotics or to cleaning agents such as hypochlorite and quaternary ammonium compounds. Partial failure may lead to problems of poor texture while in extreme cases acidification by *Bacillus spp*, which have survived heat treatment, may occur. In the latter case the product formed is unlike yogurt and would normally be rejected before packing. From a viewpoint of contamination yeasts are of greatest importance in yogurt particularly in fruit containing types. Davis (1973) comments that it appears in practice impossible to prevent contamination of fruit yogurt with yeasts (and moulds) and lists measures which should be taken to reduce the incidence of contamination.

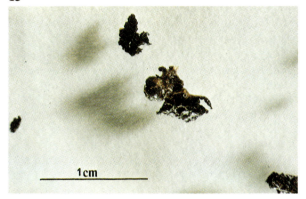

15

1cm

16 ●

17 ●

18

1cm

19

1cm

20 ●

Strains of starter cultures in use today have been selected on the basis of past performance. Major improvements in performance are likely to be made in the near future as a result of the application of genetic engineering techniques to starter strains of lactic acid bacteria (Chessy, 1984).

Yogurt and Soft Cheeses

Yogurt and unrenneted soft and curd cheeses are considered together since even with ripened soft cheese yogurt production and the first stage of soft cheese production are very similar while curd and unripened soft cheese (e.g. cottage) may be regarded as the same product as yogurt but with the loss of some whey to reduce the moisture content to 70%–80%. Davis (1973) points out that in hot countries yogurts and related products may merge with curd and soft cheese and there may be no clear demarcation line between them.

Soft, Unripened Cheese (34–43)

This includes curd, cottage and the various types of cream cheese. The basic technology is as for yogurt although *Str. cremoris* is commonly used as a starter and rennet (a mixture of the enzymes pepsin and rennin) may be used to assist curd formation. In cream cheese, for example, rennet is added shortly after the starter culture and its action assists in giving cream cheese its characteristic creamy texture. Soft cheeses are subject to starter failure although bacteriophage activity is more important than with yogurt. Failure to form a curd may also be due to the casein having been damaged by the proteolytic enzymes of psychrotrophic bacteria growing in bulk tanks before pasteurisation. On farm heat treatment (**thermisation**; Zall, 1980) and the use of LIP treated milk have been proposed to overcome this problem.

After fermentation the curd if separated from the whey either by simple filtration through muslin or after heating as in cottage cheese.

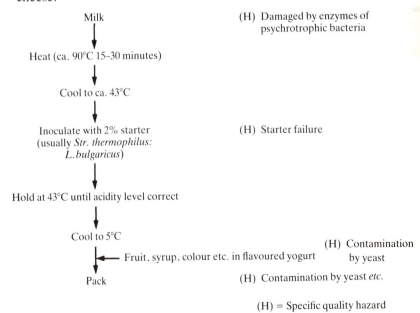

```
           Milk                         (H) Damaged by enzymes of
            ↓                                psychrotrophic bacteria
   Heat (ca. 90°C 15–30 minutes)
            ↓
      Cool to ca. 43°C
            ↓
  Inoculate with 2% starter            (H) Starter failure
  (usually Str. thermophilus:
       L.bulgaricus)
            ↓
 Hold at 43°C until acidity level correct
            ↓
       Cool to 5°C
            ↓ ← Fruit, syrup, colour etc. in flavoured yogurt   (H) Contamination
            ↓                                                        by yeast
          Pack                         (H) Contamination by yeast etc.

                                       (H) = Specific quality hazard
```

21. A flow diagram for the manufacture of stirred yogurt. Note that set yogurt is packed before incubation at 43°C. Yogurt may also receive additional post-incubation treatment such as pasteurisation.

Soft cheeses are prone to spoilage by yeast, mould and bacteria. A common source of bacteria is **wash water** (where applicable) and cottage cheese appears to be particularly prone to spoilage from this source.

The increased popularity of skim milk cheese such as cottage cheese as part of a dietary regimen has resulted in the introduction of a wide range of cottage cheese with additions. Such products must be treated with particular care because additives may not only be a source of spoilage organisms but may produce micro-habitats in which spoilage can readily occur.

Ripened Soft Cheese (44–78)

The fermentation technology of ripened soft cheese is the same as that of unripened, and indeed the division between the two is, in some cases, arbitrary. There are two types of ripened soft cheese: in the first case a significant part of the ripening is due to non-starter micro-organisms, usually moulds, whereas in the second case ripening is due to the residual biochemical activity of the starter organisms, the milk derived enzymes and, possibly, the activities of minor non-starter micro-organisms.

The first group of cheese is typified by **Camembert** and **Roquefort** and the species of mould, *Penicillium camembertii* and *Penicillium roquefortii*, responsible for ripening, are named for the cheese. The cheeses may be mould inoculated either adventitiously from the environment or using a suspension of spores. The latter method is most common today although the use of inoculating

wires dipped in a spore suspension for veined cheese is long established. The changes brought about during ripening by moulds are superimposed on the intrinsic ripening processes.

The second group of cheese is typified by **Saint Paulin**. The ripening processes are fundamentally the same as those discussed for hard cheese (see p. 30 and 31) although the residual activity of the starter organisms is probably relatively greater. The role of non-starter micro-organisms which have gained adventitious entry to the cheese is difficult to assess. Many soft, non-mould ripened cheeses contain large numbers of the biochemically active *Entero-bacteriaceae* (see below for discussion of their public health significance) which may have a role in flavour development. It should be noted that structural changes due to protein degradation are often allowed to continue to a far greater degree in ripened soft cheeses than would be desirable in hard cheese.

Enteropathogenic *Escherichia coli* in Soft Cheese

Some types of soft cheese commonly contain significant numbers of the *Enterobacteriaceae* including known enteropathogenic strains of *E. coli* (EEC). Outbreaks of food poisoning due to EEC contaminated soft cheese have occurred in the U.S.A. (Fantasia *et al*, 1975) and elsewhere. The problem is greatest with 'farmhouse' cheeses made with unpasteurised milk but the presence of *E. coli* (and other members of the *Enterobacteriaceae*) is a continuing problem in many types of soft cheese.

22, 23 and **24.   Good quality unflavoured yogurt** presents a plain surface in the pot when viewed from above (**22**). A small amount of shrinkage away from the sides is normal and a small amount of free liquid (whey) is often present on the surface due to syneresis. Viewed in depth (**23**) the body of the yogurt is seen to be of even consistency. Similar criteria apply to fruit flavoured yogurt (**24**), the example shown having a smooth texture and even colour distribution.

25 and **26.   Two packs of honey flavoured yogurt.** Pack A (ar-rowed) is swollen due to gas production by yeasts (**25**). The gas bubbles are easily seen in the yogurt of the affected pack shown alongside a similar but good quality sample (**26**).

27.   On subsequent and prolonged storage the swollen yogurt pack collapsed inwards. This is due to CO_2 produced by the contaminating yeast being re-absorbed into the yogurt producing a partial vacuum.

28.   A peach segment (arrowed) contaminating black cherry yogurt (fruit has been filtered and rinsed). This fault is due to failure to properly clean through the fruit preserve dispensing line after changing from one flavour to another.

29 to **33.   A technological fault** affecting organoleptic quality of fruit flavoured (black cherry) yogurt. Although there is little

visible difference between the four pots in the 'four pack' (**29**) the variation in amount of black cherry in each individual pot is shown in the sequence **30**, **31**, **32** and **33**. The approximately correct amount is present in **31** with excess in **30** and too little in **32** and **33**.

Three good quality examples of unripened soft cheese. In each case storage life is no more than 10 days under refrigeration.

34.   Cottage cheese. A low fat cheese which has gained popularity as part of weight control diets. This is often used as a base for admixture with other ingredients such as chives, prawns, ham etc. and may be mixed with coleslaw to produce a salad.

35.   Curd cheese. A moist low fat cheese made entirely without rennet. This is less suitable for mixing with other ingredients and lacks the popularity of cottage cheese.

36.   Fetta cheese may be made from cow or, more commonly, goat's milk. The example illustrated was produced in Bulgaria but cheeses of this nature are common throughout the Balkan States and Greece. Similar cheeses are produced in the Middle East from mare's milk. The cheese is loosely pressed in moulds, the fissures seen derive from the structure of the curd and are not defects.

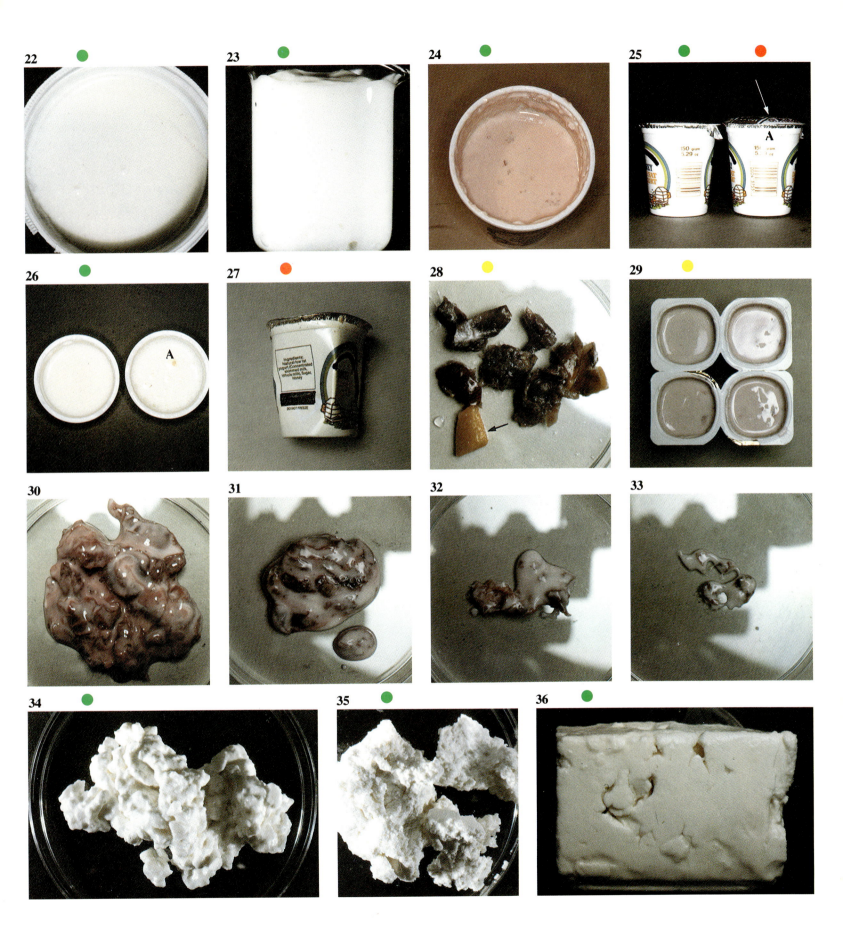

Mycotoxins in Soft Cheeses

The use of moulds to ripen soft cheese has inevitably raised the question of mycotoxin production in cheeses and considerable alarm has been raised by popular press reports such as that of the *Rhein-Zeitung* (Anon, 1973) which cast doubts on the safety of **Roquefort**, **Gorgonzola** and **Stilton** cheeses. While this report was based on misinterpretation of a scientific text (Wei *et al*, 1973), Jarvis (1976) concedes that there is no evidence that some mycotoxins may not be produced in mould ripened cheeses. The solution to the problem, as with other mould ripened products, is to ensure that the commercially used strains of moulds do not produce mycotoxins.

Examples of laboratory tests are shown in Table 3.

Table 3   Laboratory Tests Dairy Fermented Milk and Milk Products

I. Microbiological
Total aerobes
(Starter organisms excluded)
Yeast
Enterobacteriaceae including *E. coli*
Staphylococcus aureus

II. Chemical/physical
Fat
Moisture

Foreign body/infestation examination not usually required.

Notes:
(1) These tests are examples only and are not inclusive.
(2) Examples are for finished products only.
(3) Techniques are detailed in specialist publications such as ICMSF (1978) – Microbiology: Egan *et al* (1981) – Chemistry.
(4) Instrumental techniques such as Impediometry may be substituted for some classical microbiological methods.

37.   Cottage cheese is highly perishable and subject to spoilage by yeast, moulds and bacteria. The sample shown has deteriorated in quality without overt spoilage. The excess whey seen in the centre is due to temperature abuse and enhances subsequent microbial growth.

38 and 39.   The labile nature of unripened soft cheese (in this example Quark) is shown. **38** shows good quality Quark with a close, even texture and of a creamy appearance. Quark spoiled as a result of poor temperature control is illustrated (**39**). The cheese has darkened in colour and growth of yeast (A) and moulds (B) are seen. There is also an accumulation of whey (C).

40.   Soft cheese packed in bulk for sale from a delicatessen counter to the customers' requirements. Close inspection shows yeast colonies growing on the surface of the cheese (arrowed). An examination by inexperienced staff could result in the colonies being overlooked. The colonies are likely to have developed as a consequence of inadequate storage conditions since many cheeses of this type carry an intrinsic inoculum of yeast which can develop rapidly into visible colonies if held at too high a temperature (>8°C).

41, 42 and 43.   Areas of a pink/orange coloration on full fat soft cheese with chives is seen in general view (**41**), and growing into a crack in the cheese (**42**). Such discoloration is due either to growth of film yeasts or to *Brevibacterium linens*. Such growth is desirable in some types of surface ripened cheese. Some strains of *Geotrichum candidum (Oidium lactis)*, the 'dairy mould', also produce orange, pink and red discoloration but in such cases the mycelium would be visible in close-up (**43**). (N.B. Some authors continue to refer to *Geotrichum* as a yeast rather than a mould.)

44 to 47.   Control of ripening of Bucheron, a goat's milk cheese to produce two sub-varieties. **44** shows a cross-section of Bucheron cheese. The white mould may be seen on the surface (arrowed A) and the zone of ripening as the proteolytic enzymes diffuse inwards (arrowed B). In the second type of Bucheron mould growth is delayed by coating the surface of the cheese with charcoal (**45**) and no zone of ripening is seen. Examination of the surface of the cheese shows that where mould development is full the surface becomes softened (**46**), the grooves across the cheese having been made by the cheese sinking into a support grid during display. The degree of inhibition of the mould by charcoal is illustrated (**47**) and only a limited degree of softening has occurred.

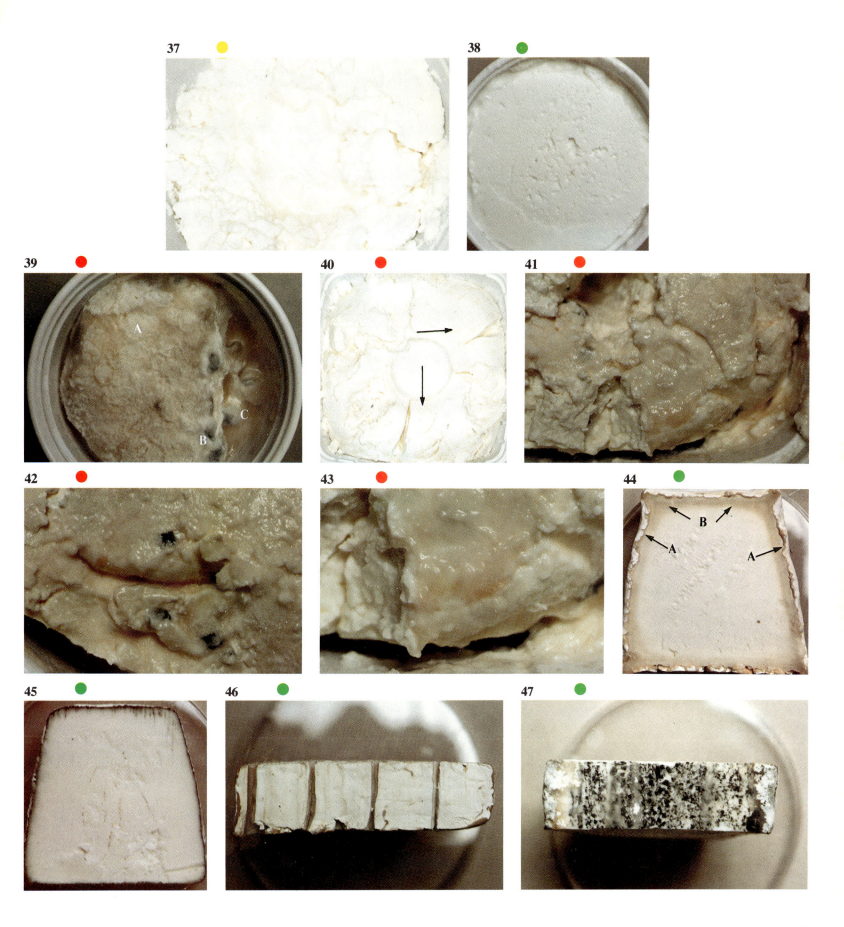

48. A very ripe wedge of Camembert cheese showing the mould usually present on the surface *(P. camembertii)* spreading across the cut surface. To some extent this 'defect' is accepted by retailers since the mould is naturally associated with the cheese, although where the mould growth is as extensive as is shown here the cheese would not normally be put on display. Such judgements are, however, highly subjective.

49. Dehydrated Camembert. Despite the adequate surface mould growth, normal ripening of this Camembert cheese has failed to take place and the cheese has merely dehydrated. The cause may be storage at too **low** a temperature markedly reducing the activity of the proteolytic enzymes.

50 and **51.** **Roquefort** is a mould ripened *(P. roquefortii)* goat's milk cheese. A general view of an over-ripe piece of Roquefort is shown (**50**). Such a cheese would probably be too strong for the majority of consumers in markets such as the U.K. The close-up (**51**) shows splitting of the cheese brought about by proteolytic enzymes from the mould.

52, 53 and **54.** **A general view of Pipo Crème**, a French mould ripened cheese (**52**). The mould *(Penicillium sp)* is seen growing along the lines of the inoculum produced by needles dipped in a spore suspension. Ripening proceeds from the interior of the cheese rather than merely from the surface in cheeses such as Camembert. In close-up the fruiting bodies of the mould may be seen as white dots (arrowed) (**53**) against the overall blue green. Mould growth also occurs on the surface (**54**).

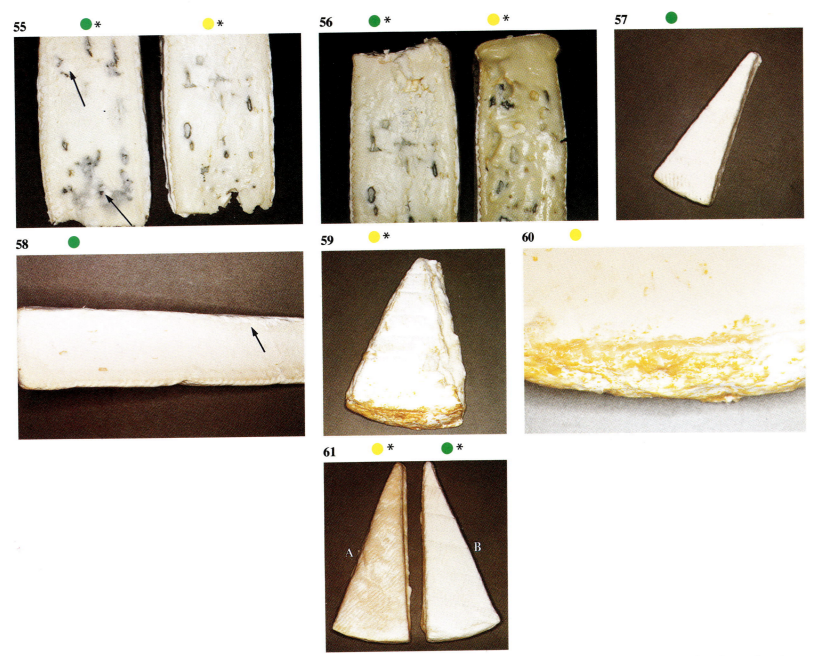

55 and 56. Lymeswold (Westminster) Blue is a new British cheese intended to compete with the better known European mould ripened cheeses. The left hand cheese shows (arrowed) a white spoilage mould overgrowing the blue inoculated *Penicillium* (**55**). This cheese also has a buttery texture due to over-ripening which is shown more clearly (**56**). The right hand sample shows breakdown due to excessive proteolytic activity resulting in a poor appearance together with strong odour and taste.

57 and 58. An immature Brie. The top view (**57**) shows the chalky white appearance of the rind while the side view (**58**) shows (arrowed) a zone just below the crust where ripening has commenced.

59. An over-ripe Brie where the body of the cheese has been rendered semi-liquid by the action of proteolytic enzymes. The associated release of ammonia gives the cheese a pungency disliked by most consumers but favoured by some connoisseurs. Note that the chalky appearance of the rind has disappeared and the body of the cheese is now a yellow/brown.

60. An over-ripe Brie in close-up which is becoming deteriorated showing in detail the brown discoloration of the exudate.

61. A comparison of two pieces of pre-packed Brie. Block A has been stored at high temperatures which has led to syneresis and collection of fluid in the base of the pack while Block B is ageing correctly at a suitable storage temperature.

62 and 63.   Growth of spoilage mould is a problem on all cheeses and growth is particularly rapid on soft cheeses particularly where cutting for pre-packing is involved. Development of mould on the cut surface of **Brie with Peppers** is shown (arrowed) (**62**) and a close-up of the mould (**63**).

64.   Progressive invasion of the cut surface of soft cheese (Brie) by white mould. Invasion is virtually total in Block A, partial in Block B and only minor in Block C. Note the shrinking of Block A. This is partially due to the dehydrating effects of mould and partially due to enzymic breakdown of the structure of the cheese.

65 and 66.   Brie cheese with peppers is a typical example of developments in the cheese market where the range of both hard and soft cheeses is extended by additions to more conventional

products. The relatively small scale of production can lead to quality problems. The even coating of peppers on the first sample (**65**) contrasts markedly with the sparse coating on the second (**66**).

67, 68 and 69.   Progressive mould spoilage of a cheese known as **walnut gateau**. This is a circular soft French cheese which has two layers of walnuts through the middle in the horizontal plane and a further layer on the top and sides. **67** shows the mould beginning to develop around the nuts on the top of the cheese. The mould probably originated from the nuts rather than from the cheese. At an advanced stage (**68**) the mould is developing around the nuts in the centre of the cheese and across the cut surface. The nuts themselves are relatively unaffected by mould growth although some large colonies are developing (**69**).

70, 71 and **72.** **A further variant of 'novelty' cheeses** is Roulé, a soft cheese with a topping such as herbs rolled and sliced (**70**). The degree of additional handling involved leads to an enhanced risk of contamination and mould development on the exterior of the Roulé is shown (**71**). Note that development of the mould was suppressed to some extent by the overwrapping film. The cheese was purchased in this condition in the display cabinet of a large London department store. Final spoilage is shown (**72**).

73 and **74.** **Bonchampi** is a soft French cheese containing slices of mushroom. Allegations of poor quality due to gas formation were made of this sample. Investigation revealed that the 'gassing' was due simply to mushrooms having become displaced when the cheese was cut for pre-packing. This is illustrated in general (**73**) and in detail in (**74**).

75. **Poor temperature control** may adversely affect the physical characteristics of soft cheese. This plate shows two whole **Bonchampi** cheeses. Cheese A on the left has been subject to poor temperature control, has become flattened and misshapen, and was rejected for pre-packing. Cheese B has been correctly stored, maintained its shape and is therefore acceptable.

76. **Bruder Basil** is a soft cheese ripened by *Propionibacterium spp.* The outer surface is coated with red wax to reduce dehydration and prevent mould growth. The pattern of small holes is characteristic of this cheese but would be considered a fault in some other cheeses ripened by *Propionibacterium*.

77 and **78.** **Soft, extensively ripened cheeses** exert an attraction to houseflies (*Musca domestica*). Unless control is adequate eggs and subsequently maggots in such cheese can be a severe problem. A general view of fly eggs in Brie is shown (**77**) and in close-up (**78**). The eggs are below the overwrapping film and contamination obviously occurred before pre-packing.

Hard Cheese (79–103)

As with fermented milks and soft cheese there is no firm demarcation line between soft and hard cheese. The latter are typically of lower water content and undergo a longer ripening period. The initial stages of production (**79**) are similar to those of soft cheese manufacture and problems may be caused either by starter culture failure or as a consequence of enzymic breakdown of the milk protein during bulk storage. After curd formation the curd is heated and, after matting, cut into slabs and turned and piled to remove whey.

Although the process illustrated (**79**) refers to **Cheddar** cheese (commercially the most important) the technology of other varieties is similar. Considerable variation in the final product is the result of minor differences in the temperature at which the curd is set, the heating it receives and the amount of moisture evaporation allowed during ripening. After the curd has been heated and turned it is salted and pressed into moulds. The amount of salt added varies according to variety and consumer preference and the degree of pressing also varies according to variety. In general mould ripened cheeses such as **Blue Stilton** have a more open texture to obtain sufficient aeration for mould development.

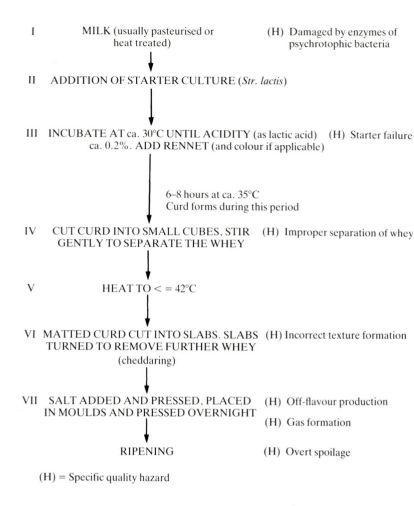

I	MILK (usually pasteurised or heat treated)	(H) Damaged by enzymes of psychrotophic bacteria
II	ADDITION OF STARTER CULTURE (*Str. lactis*)	
III	INCUBATE AT ca. 30°C UNTIL ACIDITY (as lactic acid) ca. 0.2%. ADD RENNET (and colour if applicable)	(H) Starter failure
	6–8 hours at ca. 35°C Curd forms during this period	
IV	CUT CURD INTO SMALL CUBES, STIR GENTLY TO SEPARATE THE WHEY	(H) Improper separation of whey
V	HEAT TO < = 42°C	
VI	MATTED CURD CUT INTO SLABS. SLABS TURNED TO REMOVE FURTHER WHEY (cheddaring)	(H) Incorrect texture formation
VII	SALT ADDED AND PRESSED, PLACED IN MOULDS AND PRESSED OVERNIGHT	(H) Off-flavour production
		(H) Gas formation
	RIPENING	(H) Overt spoilage

(H) = Specific quality hazard

79. A flow diagram for the manufacture of cheddar cheese. The basic technology is common to the production of most types of hard cheese.

Following pressing, hard cheese is ripened for varying periods of time from a few months to over a year. During ripening a variety of chemical reactions occur as a result of which the cheese acquires its characteristic flavour (and texture). The nature of the changes is complex and details should be obtained from more specialist discussions such as that of Law and Sharpe (1973).

The ripening of hard cheese is a major factor in determining quality and the factors involved are summarised here. In the past four factors, milk derived enzymes, starter activity, non-starter microbial activity and non-enzyme mediated reactions, have been considered to contribute towards flavour development. Of these factors milk derived enzymes are destroyed by pasteurisation while non-starter micro-organisms, with the exception of specific examples such as *Propionibacterium spp* in **Emmenthal** cheese and *Penicillium spp* in the various mould ripened cheeses, play only very minor roles in the development of desirable cheese flavour (Law and Sharpe, 1978).

In respect of the involvement of the starter organisms the same authors state that their role is restricted to that of producing the correct environment of pH value and (negative) redox potential and to supplying flavour precursors which are transformed by non-enzymic reactions to flavour compounds. A typical example is free methionine which is converted to **methanethiol**, a compound whose concentration in Cheddar cheese correlates with flavour intensity.

Starter organisms may however produce well defined defects in cheese. Cheddar cheese for example with high starter populations in the finished curd (i.e. after cooking) tends to develop bitter flavours possibly as a result of bitter tasting **hydrophilic oligopeptides** being produced from casein in concentrations exceeding their flavour threshold (Lowrie and Lawrence, 1972). Equally starters (usually strains of *Str. lactis*) which survive in the ripening cheese may produce ethanol from acetaldehyde which, in turn, combines with butyric or hexanoic acid to produce the corresponding fruity flavoured esters **ethyl butyrate** and **ethyl hexanoate** (Bills *et al*, 1965). In addition some strains of *Str. lactis* convert certain amino-acids to aldehydes by means of transaminase and decarboxylase enzymes. A particular example is the conversion of leucine to **3-methylbutanol**, a compound which can impart malty flavours to cheese (Sheldon *et al*, 1971).

Non-starter micro-organisms may also cause off-flavours without visible spoilage. *E. coli* and other members of the *Enterobacteriaceae* for example, can produce a faecal taste (as well as gas production – see below) although it should be noted that *E. coli* when present in numbers of ca. 10^5 colony forming units/g produce a distinctive 'sharp' flavour which some consumers consider desirable. Visible spoilage may also occur during ripening. In the past such spoilage has resulted in total loss of product with obvious economic consequences. Davis (1983) for example refers to the total loss of Cheddar cheese in Somerset during 1930 due to development of a stink so strong (probably due to clostridia) that even cattle would not eat it. The condition was colloquially (and apparently aptly) referred to as **'dead man's finger'**.

There is little doubt that such spoilage and other, less spectacular spoilage such as 'rusty' or red spot caused by a carotinoid producing strain of *L. brevis*, were largely a consequence of poor technological control of the cheesemaking process, poor hygiene and poor storage conditions. The most common problem in mass manufactured cheese today is probably gas formation (a desirable feature, of course, in Swiss and related cheeses). Lactate fermenting species of *Clostridium* whose endospores survive the initial heat treatment of the milk are frequently responsible but gas may also be produced by heterofermentative lactobacilli, propionibacteria and members of the *Enterobacteriaceae*.

Spoilage of ripened cheese is usually by moulds such as *Penicillium spp* although yeast may also be important in softer varieties such as **White Stilton** (non-mould ripened) and **Wensleydale**. Mould spoilage is of particular importance in pre-packed cheese and in many cases it is considered that excessive storage lives are given for the storage conditions, frequently resulting in mould spoilage within life. Vacuum packaging may be used to prevent mould growth but cannot be used for softer varieties which are crushed when the vacuum is applied. Some pre-packing units have installed precautions such as air filtration systems and positive pressure packing rooms but in many cases these have had little effect, the prime source of the mould being the bulk cheese which cannot be effectively removed by cutting or scraping. It is noteworthy that those pre-packing units operating the most rigorous acceptance/rejection standards for incoming cheese are the most successful in restricting mould spoilage of the pre-packed product.

Formation of Toxic Amines in Cheese

Group D Streptococci can be present in Cheddar cheese curds in numbers up to 10^6 colony forming units/g (Law *et al*, 1973) and similar numbers may be found in other varieties of hard cheese. The presence of the amines tyramine, histamine, and tryptamine has been associated with the growth of *Str. faecalis* in cheese. Tyramine in cheese may be involved in the onset of **migraine** attacks in susceptible persons (Hannington, 1967) and also produces a hypersensitive response in persons undergoing treatment with monoamine oxidase depressing drugs (Blackwell and Mabbitt, 1965).

Mycotoxins in Hard Cheese

The position with mould ripened hard cheese is the same as with mould ripened soft cheese and is discussed on page 24. The widespread practice, both commercial and domestic, of trimming or scraping mould from affected hard cheese obviously carries a risk of consuming mycotoxin containing cheese since any mycotoxins produced by surface growing moulds will diffuse into the interior. In the absence of definite evidence of risk from mould affected cheese it is probably unrealistic to expect the practice of trimming to be abandoned. The solution must lie in improving handling conditions to minimise the levels of mould contamination.

A number of good quality hard cheeses illustrating the diversity of the product group.

80. Good quality pre-packed Cheddar cheese. The 'plastic' appearance imparted by the packaging film can cause adverse criticism.

81. Addition of walnuts to a basic Cheddar cheese adds value and produces an up-market product. The quality of walnuts or other such additives should be carefully monitored.

82. Blue Cheshire, a traditional English mould *(Penicillium spp)* ripened hard cheese. The lines produced by the inoculating needles are clearly seen.

83. Swiss Gruyère cheese with correctly developed 'eyes' (see **90** and **91**).

84. Danish Havarti cheese. The gas holes which are typical of this cheese would be considered to be a fault in many other types of hard cheese.

85. Norwegian Goat's milk cheese Gytost. This cheese has a high sugar and caramel content and is often eaten as a dessert.

86 and **87. Extensive mould** *(Penicillium spp)* development on pre-packed Dutch Cheddar cheese **(86)**. The close-up illustration **(87)** shows the restriction of development of aerial mycelium by the overwrapping film.

88 and **89. A further example of mould** *(Penicillium spp)* on pre-packed hard cheese (English Cheddar) **(88)**. Mould growth is common on the bulk blocks of cheese used for pre-packing and is usually scraped off prior to cutting. In practice it is impossible to remove all the spores and hyphae and growth may re-occur on the subsequent pre-packed cheese particularly if storage temperatures are high. Vacuum packing is often used as a means of preventing mould growth on pre-packed cheese. Examination of a section through the cheese **(89)** shows that the mould growth is restricted to the surface and that there is no penetration into the interior. This is the usual pattern although occasionally the moulds are sufficiently proteolytic to produce actual 'rots'.

90 and **91. Gas holes or 'eyes'** are a fault in most hard cheeses but are desirable in such cheeses as Gruyère, Emmenthal and Jarlsberg. In these, *Propionibacterium spp* are incorporated at the curd stage and during growth evolve large quantities of gas as well as contributing to the distinctive flavour of the cheese. The illustrations show a general **(90)** and close-up **(91)** view of poorly developed holes in Gruyère cheese. This may be due to slow development of the *Propionibacterium* or to contamination by gas producing species of *Clostridium*.

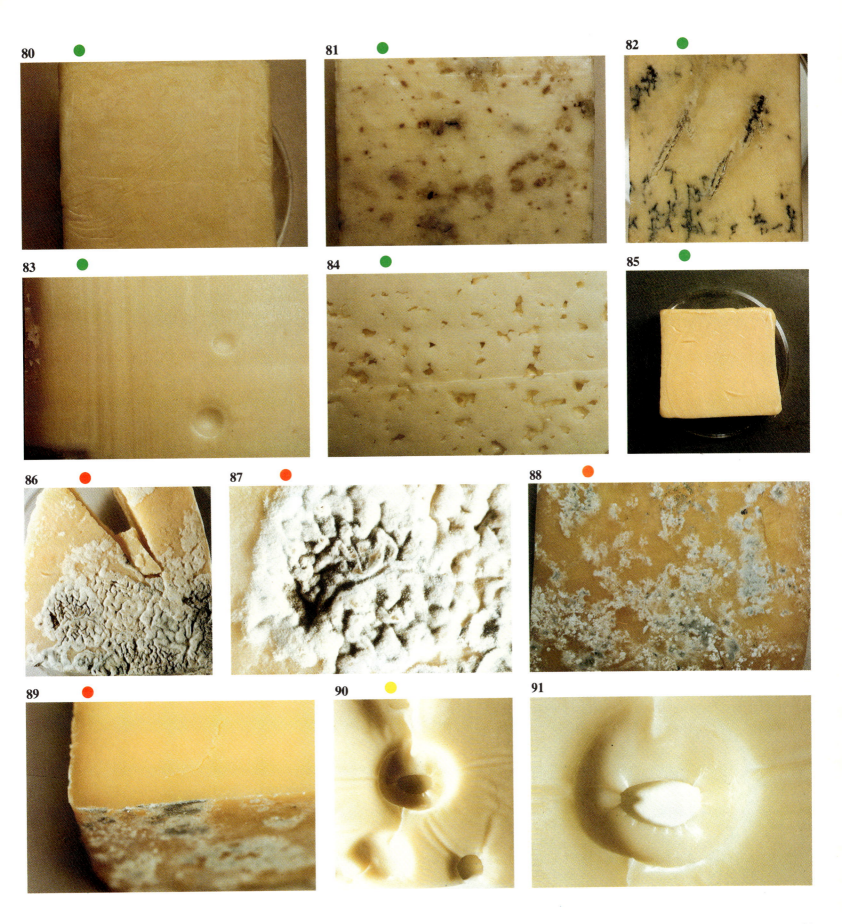

80

81

82

83

84

85

86

87

88

89

90

91

33

92 and **93.** **Two views of gas production** in a hard cheese (Dutch Cheddar). The extent of disruption suggests that clostridia were responsible. The condition is most common in high pH cheese where partial starter failure has occurred or where lactate fermentation by *Cl. tyrobutyricum* has reduced the acidity. Note that the major effects are localised at one end of the cheese.

94, 95 and **96.** **Blowing of Red vein cheese.** Red vein is an 'added value' product made from Cheddar cheese and red wine (note that the red veining is not due to a mythical red mould as has been erroneously stated elsewhere). The first illustration (**94**) shows a general view of the pack. The vacuum pack is not overtly swollen probably due to its having leaked under pressure. A close-up (**95**) of the end of the block shows disintegration and residual material blown onto the interior of the film pack. The effect of gas production in the interior of the cheese is illustrated (**96**). The likely cause of the gas production was *Clostridium spp* although other bacteria including members of the *Enterobacteriaceae* and heterofermentative lactobacilli may also be responsible. The cheese was purchased from an unrefrigerated market stall and is likely to have been subject to considerable temperature abuse. It must however be stated that the basic cheeses used in products such as Red vein are often of poor initial quality.

97 and **98.** **Hybrid cheeses.** A further development in the U.K. market has been the introduction of hybrid cheeses consisting of two or more regional varieties compressed together. These are often described by fanciful names. The cheese illustrated contains Stilton (centre) with red Cheddar (**97**). It was purchased from a market stall and is of obvious inferior quality. Note the 'sweaty' appearance as a whole and the splitting of the Stilton which, in close-up (**98**) is seen to be the site of massive gas evolution. This is bacterial in origin probably by *Clostridium spp*. It is obvious that the cheese has been subject to temperature abuse but as with Red vein (**95** and **96**) such products are often used in attempts to upgrade poor quality cheese.

99 and **100.** **Formation of lactate crystals,** illustrated in a pre-packed piece of Gruyère cheese (**99**) is a common phenomenon in hard cheeses. It has little effect on flavour but may impart a grittiness to the texture. A careful examination is required to distinguish lactate crystals from yeast colonies which they superficially resemble. A close-up view (**100**) shows the crystals to be concentrated around the area where an 'eye' formation has commenced (A) and the edge of the block (B) where a degree of drying out has occurred.

101 and **102.** **A general view of the rind and body** of Parmesan cheese. The white areas are lactate crystals, again readily mistaken for yeast colonies (**101**). The crystalline nature is seen more clearly in the close-up (**102**).

103. **The waxy coating of Edam** cheese is a protective layer against dehydration and mould growth. Before solidification, however, it is a trap for foreign bodies, including insects. This illustration shows a damaged insect embedded in the waxy coating.

Processed Cheese (104–108)

Processed cheese is produced from a blend of hard cheeses together with emulsifying salts, cooked in the molten state and resolidified. Other ingredients such as ham may be present and the final product may be smoked before packing. Cheeses selected for processing are chosen to give a final reproducible flavour and cheeses considered unsuitable for the fresh market due to flavour defects may be used. Cheese containing *E. coli*, for example, which has a characteristic 'sharp' taste making it unsuitable for the fresh market is often selected to impart a 'bite' to otherwise bland blends of cheese.

Although the heating during manufacture of processed cheese is sufficient to kill vegetative bacterial cells endospores survive and gas formation by *Clostridium spp* may cause problems. For this reason the permitted antibiotic nisin is usually added. **Examples of laboratory tests are shown in Table 4, p. 37.**

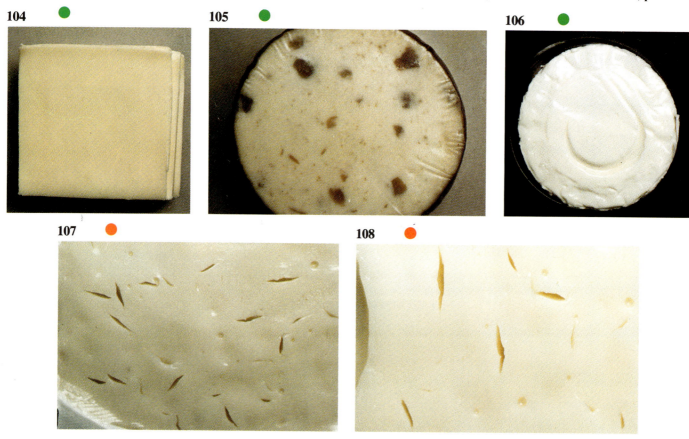

Three examples of good quality processed cheese.

104. Processed cheese slices. Probably the most common form of processed cheese in the U.K. and U.S.A., such slices are bland in taste and plastic in texture. Their convenience means that they are widely used for sandwich making in both domestic and commercial catering.

105. Bavarian smoked processed cheese with ham. European processed cheeses are usually less bland than their American counterparts.

106. Bel Paese, an Italian processed cheese spread. Cheese spreads have a softer texture than processed cheese itself and contain gums or gelatin to promote smoothness and act as binder. The moisture content is as high as 60% and spreads are considerably more perishable than processed cheeses.

107 and **108. Fissures in processed cheese.** A general (**107**) and close-up view (**108**). These are due to gas formation by the anaerobic bacterium *Clostridium*. This organism produces heat-resistant endospores which survive the cooking process and which may then germinate, the vegetative cells growing in the finished product and producing gas. Where sealed outer packaging is used sufficient gas may be produced to swell and ultimately 'explode' the pack. Control is usually by incorporating into the cheese the permitted (U.K.) antibiotic nisin which blocks endospore germination.

Table 4 Laboratory Tests Dairy Processed Cheese

I. Microbiological
Total aerobes
Anaerobes

II. Chemical/physical
Fat
Moisture

Foreign body/infestation examination not usually required.

Notes:
(1) These tests are examples only and are not inclusive.
(2) Examples are for finished products only.
(3) Techniques are detailed in specialist publications such as ICMSF (1978) – Microbiology: Egan *et al* (1981) – Chemistry.
(4) Instrumental techniques such as Impediometry may be substituted for some classical microbiological methods.

Concentrated Milk Products (109–115)

In the present context the term concentrated milk products embraces **evaporated**, **sweetened condensed** and **dried milk**. The basic technology of the concentration process is common to each product: in the case of dried milk concentration is required before the final drying stage.

Evaporated milk is usually manufactured by boiling under a vacuum to remove some 20% of water. Either skimmed or full cream milk may be used, the latter being more common with evaporated milk for direct consumption by the consumer. Various types of plant are used for the process which may be batch or continuous but details of plant are beyond the scope of this book. Evaporated milk may be sold refrigerated in bulk for further manufacturing such as toffee confectionary or ice-cream manufacture, or canned for retail sale. Neither the pre-heat treatment which is equivalent to pasteurisation nor the temperature at the evaporation stage is sufficient to kill endospore-forming bacteria and thus bulk evaporated milk has only a short shelf life while canned evaporated milk requires a full heat sterilisation process. Intrinsic problems with evaporated milk are few although excessive caramelisation may occur due to overheating.

It should be noted that when a small quantity of evaporated milk is made by **ultra-filtration** rather than by vacuum boiling, such milk is of more natural taste than conventional evaporated milk but potential problems of microbial spoilage are greater.

Sweetened condensed milk is manufactured by the same plant as is used for the manufacture of evaporated milk but sugar, usually as sucrose, is dissolved in the milk before evaporation to give a total sugar content of 55%–60% (w/v). Highly refined granulated sugar is normally used but other types including brown and treacle may be added to condensed milk destined for toffee confectionery manufacture. Sweetened condensed milk may be produced for sale as a bulk commodity or packed into cans for retail sale. In either case the low Aw level of the milk ensures a long storage life at ambient temperatures and no heat treatment is applied to cans of sweetened condensed milk. It should be noted that the economic importance of the latter has declined very appreciably since the second world war, most consumers preferring other forms of milk and cream to sweetened condensed.

As with evaporated milk intrinsic quality defects are few although the presence of added sugar leads to greater problems of caramelisation. Three types of microbial spoilage have been reported:

(a) **Gas formation** by sucrose fermenting osmotolerant yeast such as *Torula spp.*
(b) **Thickening** due to proteolytic *Micrococcus spp.*
(c) **'Buttons'** due to restricted growth of mould *(Aspergillus spp, Penicillium spp)* on the milk surface.

In addition physico-chemical age thickening may also occur.

Dried milk is manufactured from evaporated milk. Either skim milk, skim milk with added vegetable fat or whole milk may be used but the majority is made from skim milk. The final drying process reduces the water content to ca. 4%. Drying may either be by the **roller** process where a film of concentrated milk is sprayed onto steam heated contra-rotating rollers, the resulting sheet of dried milk being removed by a knife blade fixed across the width of the roller, or by the **spray** process. In this, the more modern and more widely used process, concentrated milk is sprayed through an atomiser into the top of a large chamber with air at 120°–205°C blown in from the base. The particles dry rapidly and are collected at the bottom of the chamber.

Roller and spray dried milk differ in several ways. The former consists of relatively large irregular particles or flakes, is relatively ark in colour and has a strong heated taste. (The latter may be reduced by enclosing the drier in a vacuum hood.) In contrast spray dried powder consists of small, regular spherical particles, is light in colour and has very little burnt taste. Spray dried powder is generally considered to be of better organoleptic quality than roller dried. The effective degree of heating of roller drying is greater than that of spray drying and roller dried milk is usually of better microbial quality than spray dried. For this reason roller drying is often used for baby foods and for some specialist manufacturing purposes.

Spray dried milk has poor dissolving properties and where sold for domestic use is subjected to a further process, **instantisation**, which agglomerates granules and leads to a faster dissolution time.

The moisture content of dried milk is too low to permit microbial growth during storage, but the product is highly deliquescent and unless suitably stored may pick up sufficient moisture to permit mould growth. In addition full fat dried milk is subject to oxidative rancidity on storage. Where storage is likely to be prolonged in hot climates gas packing in an inert atmosphere usually of nitrogen may be used to reduce rancidity but in temperate climates anti-oxidants are usually sufficient to control rancidity.

Table 5 Laboratory Tests Dairy Concentrated Milk

I. Microbiological
Osmotolerant yeast (sweetened condensed only)
Sterility (canned only)
Total aerobes (dried only)
Staph. aureus (dried only)

II. Chemical/physical
Viscosity (sweetened condensed and evaporated only)
Fat
Sugar (sweetened condensed only)
Moisture (dried only)
Filter test (see p. 39, dried only)

Foreign body/infestation examination not usually required.

Notes:
(1) These tests are examples only and are not inclusive.
(2) Examples are for finished products only.
(3) Techniques are detailed in specialist publications such as ICMSF (1978) – Microbiology: Egan *et al* (1981) – Chemistry.
(4) Instrumental techniques such as Impediometry may be substituted for some classical microbiological methods.

109 and **110.** **Sweetened condensed skim milk** and evaporated skim milk. The former (**109**) is preserved by virtue of its high sugar concentration and is stable at room temperature (see p. 37). Evaporated milk (**110**) has a limited refrigerated life and requires full thermal processing for room temperature storage. (See also Chapter 15, Canned Foods.)

111 to **114.** **Spray dried milk powder** is a fine, pale coloured powder (**111**) which, while suitable for manufacturing and possibly catering use is less than satisfactory for domestic use by virtue of its low 'wettability' and consequent difficulty in dissolving. **Instantising,** a process which agglomerates the powder (**112**), overcomes this problem. Dried milk may be affected by defects the most common of which is burnt particles. Simple grading systems based on this defect are often employed and only the highest grade powder is used for instantising and hence direct domestic consumption.

Burnt particles may be readily detected by filtering a fixed volume of a 10% ($^w/_v$) solution of milk through a medium grade filter paper, counting the burnt particles and then comparing the counts with predetermined standards. **113** shows manufacturing grade milk powder tested in this manner and may be compared with domestic grade (**114**). The burnt particles seen in the manufacturing grade are probably derived from the concentrated milk burnt onto the sides of the spray chamber. This is the most common cause of such problems. Extraneous material may also be derived from the air entering the chamber if filtration is not efficient. Such material includes combustion products from the plant's own boiler and environmental pollutants such as pollen.

115. **Dried milk powder is highly hygroscopic** (water attracting) and if stored in a humid atmosphere in permeable packaging rapidly becomes lumped together. Silica gel may be added to retard this process which can also be associated with flavour changes.

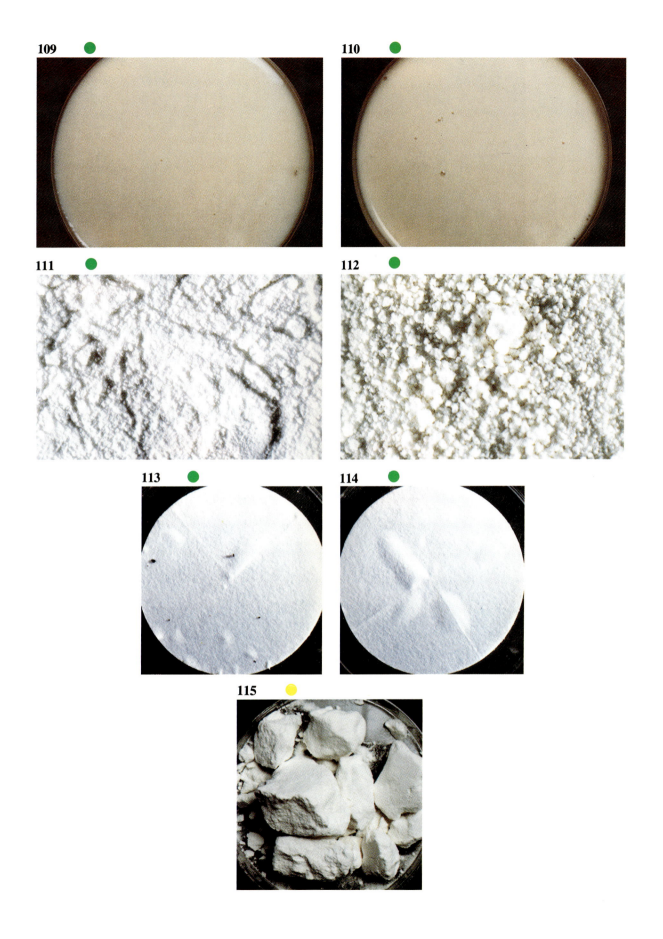

109 🟢

110 🟢

111 🟢

112 🟢

113 🟢

114 🟢

115 🟡

39

Cream (116–127)

Although hand skimming of cream from milk is still practised in small scale operations the vast majority of cream production involves separation of heated milk by centrifugal separators. The separated cream passes to vats and is standardised to the required fat content, the minimum contents in the U.K. being, half cream 12%, single 18%, whipping 35%, double 48% and clotted 55%. Sterilised (retorted) cream has a fat content of 23%. Following standardisation cream is subject to various other processes before packing dependent on the desired physical properties of the cream. These are not necessarily dependent on the fat content; double cream for example may be further thickened, while half cream intended for coffee is processed to avoid 'feathering' when added to the hot liquid. Changes in the physical properties of cream are usually brought about by various homogenisation treatments coupled with, in some cases, high temperature pre-heating and controlled cooling.

Although most cream for the retail market is sold as a fresh product requiring refrigeration and having a relatively short shelf life, increasing quantities of the lower fat content creams are aseptically packed after ultra heat treatment while smaller quantities are canned and frozen.

A special mention must be made of fresh cream cakes since the major potential hazards of these lie in the cream rather than the cake base. Fresh cream cakes have been implicated in food poisoning on a number of occasions (e.g. Hobbs, 1974) and examinations of the microbial status at retail level indicates that microbial numbers are often unacceptably high (authors' unpublished observation). While it is recognised that the filling of cream cakes is necessarily by hand with associated high risk of contamination there is no doubt that many of the problems lie in the inability or unwillingness of many bakers to accept that fresh cream is highly perishable and requires adequate refrigeration and a high level of hygiene. Some commercial equipment used for whipping cream is virtually impossible to sanitise effectively and it is still all too common to see fresh cream cakes on sale in inadequately refrigerated or unrefrigerated cabinets.

The practice of adding the preservative sorbate (a permitted preservative in the pastry base) to cream is not only likely to be ineffective but is illegal and those that select this solution rather than observing basic food hygiene principles are irresponsible at best.

Examples of laboratory tests are shown in Table 6.

Table 6   Laboratory Tests Dairy Cream

I. Microbiological
As for milk with yeast counts in addition for clotted cream.

II. Chemical/physical
Fat
Moisture

Foreign body/infestation examination not usually required.

Notes:
(1) These tests are examples only and are not inclusive.
(2) Examples are for finished products only.
(3) Techniques are detailed in specialist publications such as ICMSF (1978) – Microbiology: Egan *et al* (1981) – Chemistry.
(4) Instrumental techniques such as Impediometry may be substituted for some classical microbiological methods.

---

116 and **117. Physical properties of cream can be changed** by varying the processing conditions. Thus while **double cream (116)** and **extra thick double cream (117)** have the same fat content (minimum 48%) the processing of the latter gives a thicker body. This is commonly associated by consumers with higher fat content and higher quality. The additional processing includes a long post-filling cooling stage and thus extra-thick double cream may be more prone to microbial spoilage than the conventional product.

118 and **119. Whipping cream (118)** has a higher fat content than single cream (**119**) (35% *vs* 18%) but also derives its whipping properties from differences in processing.

120. Thickening of fresh single cream. The cause is the proteolytic enzymes of *Pseudomonas spp* which entered the cream as post-pasteurisation contaminants and were subsequently able to develop during refrigerated storage (see **125** and **126**). The thickening was accompanied by a slight putrefactive odour but since the organisms do not produce acid from carbohydrates curdling was absent.

121 to **124. UHT dairy products are normally aseptically packed** into foil lined cardboard Tetra-brik or similar packing systems. UHT cream however may be found in a number of types of pack in addition to the conventional Tetra-brik. **121** shows UHT cream in a plastic carton with a heat sealed metal foil lid. This type of packing is normally used for fresh (pasteurised) cream and is **not** aseptic. The lid carries the information that the cream is UHT and a 'use by' date (**122**). A more unusual packaging for cream is the aerosol pack (**123**). While aerosols have been in regular use for some years for products such as fly spray, hair spray, polishes etc. their use in the food industry is relatively uncommon. Thus the aerosol container illustrated represents a revolutionary concept in packaging cream.

UHT cream aseptically packed in Tetra-brik or similar containers has a long room temperature shelf life in the sealed pack. However, after opening the advantages of the UHT process are lost and the cream is as perishable as the normal pasteurised product. In a sealed aerosol container the possibility of post-opening contamination of the bulk of the cream is much reduced and a longer post-opening life may be expected, although the nozzle (**124**) appears difficult to clean and may contaminate the cream on removal from the aerosol can.

116

117

118

119

120

121

122

123

124

125 and 126. **Age-thickening of UHT cream** (and more commonly milk) is caused by proteolytic enzymes derived from psychrotrophic bacteria. These bacteria, which are able to multiply in refrigerated farm bulk tanks, are killed by the UHT processing but their proteolytic enzymes remain active. The phenomenon tends to be seasonal in that microbial numbers in bulk milk are higher in summer while the higher ambient temperatures increase the rate of post-processing enzyme activity. Age-thickening results in a solid, unsoured mass (**125**) from which liquid is expelled by syneresis (**126**). The example shown is of UHT single cream, 2 months within shelf life. Age-thickening of UHT cream is less common than that of UHT milk as fat appears to exert a protective effect.

127. **Frozen cream sticks.** Life is limited by a fairly rapid onset of oxidative rancidity of the fat.

Fresh Cream Desserts (128–134)

Fresh cream desserts are a growth product within the dairy industry. They consist typically of an individual pot containing a flavoured base thickened by starch and possibly other thickeners together with pieces of fruit as appropriate. Gelatin may also be used as a structural ingredient. The flavoured base is then topped with a layer of fresh, usually whipped cream. The filling process entails a considerable amount of manual operation with associated risk of contamination while manufacturing expediency may require the base to be produced in a plant distant from that in which the final product is assembled and filled. These factors together with the highly perishable combination of ingredients mean that fresh cream desserts are **highly perishable** with a high spoilage potential.

They also present an unusual food poisoning hazard. Although *B. cereus* is a common cause of spoilage of dairy products (see p. 18) and may be present in large numbers without spoilage, it is rarely associated with food poisoning in such circumstances. Davies and Wilkinson (1973) however suggest that a combination of cream as a source of *B. cereus* and a high starch base has a high food poisoning potential.

In addition to fresh cream desserts frozen desserts are also available. In these products the cream is stabilised by addition of stabilisers such as alginates, carrageen, waxy maize starch and general starches.

Examples of laboratory tests are shown in Table 7.

Table 7   Laboratory Tests Dairy Desserts

I. Microbiological
Total aerobes
Yeast
E. coli
Staph. aureus

II. Chemical/physical
None normally required

Foreign body/infestation examination not usually required.

Notes:
(1) These tests are examples only and are not inclusive.
(2) Examples are for finished products only.
(3) Techniques are detailed in specialist publications such as ICMSF (1978) – Microbiology: Egan *et al* (1981) – Chemistry.
(4) Instrumental techniques such as Impediometry may be substituted for some classical microbiological methods.

128, 129 and 130. **Spoilage of fresh cream desserts. 128** shows a top view of a dessert in good condition which has been stored under correct (≯5°C) temperature conditions. **129** shows spoilage of the cream by a pseudomonad showing clearly the cream structure broken into large fragmented clots. Some acid formation may have taken place but proteolysis due to powerful exo-enzymes is likely to have played a major role. Contamination by pseudomonads is probably derived via poorly cleaned utensils or plant. The organism may also be associated with the water supply.

A second type of spoilage is shown (**130**). This is due to *Serratia* (a member of the *Enterobacteriaceae*) which has been able to develop as a consequence of a high (>15°C) storage temperature. The cream is clotted due to acid formation and gas bubbles are also seen. In this example acid production has been relatively weak and other members of the *Enterobacteriaceae* may bring about a greater degree of clotting.

131. **The cream of a fresh dairy cream dessert** collapsing after storage at −5°C. Stabilisers such as alginates are used in the equivalent frozen product to maintain the desired structure after freezing and subsequent thawing.

132 and 133. **The soft, skim-milk cheese Quark** (Quarg) is often used as a basis for fruit flavoured desserts and such products are popular as part of slimming diets. Good quality Quark dessert flavoured with red cherries is illustrated in **132** compared with the same product spoiled by coagulation of the cheese and gas production (**133**). This is an unusual form of spoilage and was due to growth of a *Bacillus sp* possibly as a consequence of partial starter failure.

134. **Development of a film yeast** on the surface of a mandarin yogurt dessert (arrowed). The growth is not easy to see and the product could easily be consumed in this state.

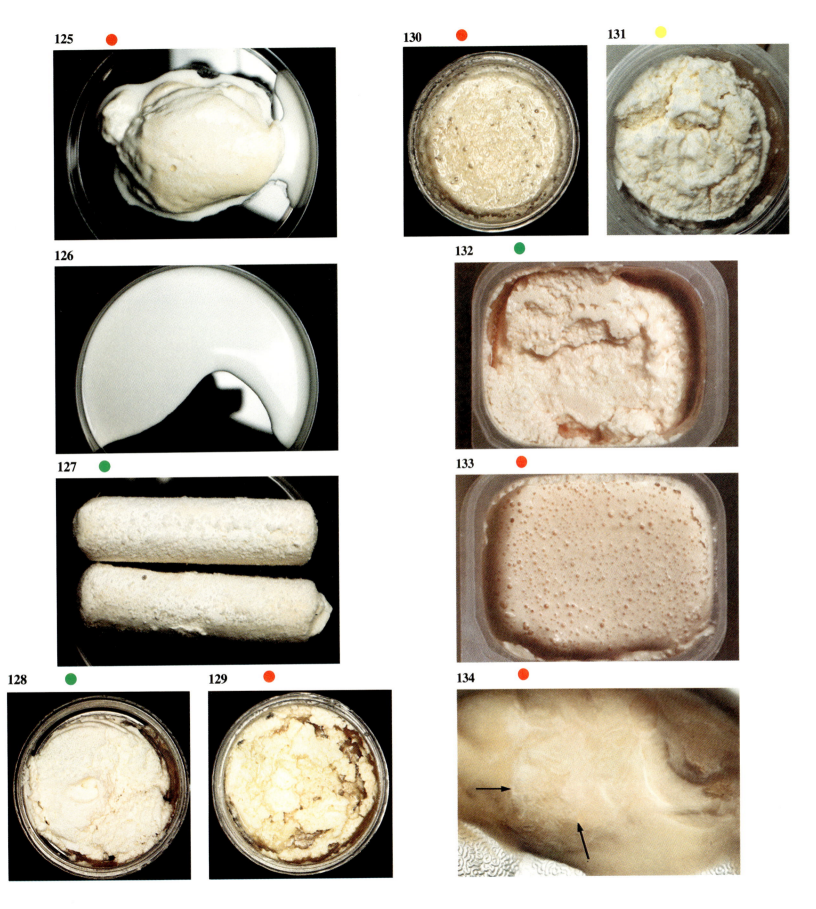

43

The basic operation in **butter manufacture** is the conversion of an oil in water emulsion (cream) to a water in oil emulsion (butter) with the removal of excess water (buttermilk) which also contains small quantities of milk solids (cultured buttermilk is a fermented milk product in its own right, see p. 16). Traditionally the conversion is brought about by 'churning', a batch process carried out in large rotating drums fitted with tumbling bars. After the butter has separated, the buttermilk is drained off and the butter washed. It is then tumbled or kneaded to reduce the water content, and salt blended in. The quantity of salt affects the keeping quality, and high salt butters tend to be produced where refrigeration is not available. The trend in the U.K. is to low salt butters although there is regional variation in tastes with high salt butters being preferred in coalmining and heavy industrial regions. (This preference extends to other foods such as bacon, and miners in the Nottinghamshire and Derbyshire coalfields traditionally added salt to beer!) Although much butter is still produced in batch churns continuous buttermakers which replicate the churning procedures are in wide use.

While the basic technology is the same in all butters there are two main types. The first, **sweet cream** butter, is produced from unfermented cream, while in the second type cream is fermented or **'ripened'** by a mixed culture of lactic acid bacteria usually *Str. lactis* and *Str. lactis subsp. diacetylactis*. The most important compound produced during fermentation is not lactic acid but **diacetyl** ($CH_3COCOCH_3$) produced by *Str. lactis subsp. diacetylactis* from citric acid via the intermediate acetyl methyl carbinol ($CH_3COCH(OH)CH_3$) which is also present in ripened butter. Diacetyl is responsible for imparting to butter much of its characteristic flavour (note that diacetyl production in beers and wines is usually a fault see p. 232).

Although butter may be subject to a number of faults as a result of microbial growth before churning or on the final product (see Frazier, 1967) these are largely of historic concern and with the advent of improved manufacturing hygiene and wide scale refrigeration spoilage is rare. Butter is subject to oxidative rancidity producing tallowy flavours but in practice the most common fault tends to be undesirable flavours 'picked up' from adjacent goods during storage.

Margarine is dealt with here as, in its orginal conception at least, it is a butter substitute, the fat constituent being an animal fat or vegetable oil (see pp. 127–129) rather than milk fat. A wide variety of fats may be used, choice often being dictated by economics. The liquid phase and most of the solid non-fat constituent is fermented skim milk or, increasingly, whey. In manufacture the refined and deodorised fats or oils are hydrogenated to the desired consistency and then emulsified with the aqueous constituent. The ability to vary the consistency by varying the degree of hydrogenation is an important factor in margarine manufacture and can produce advantages over butter. For instance margarine may be produced which spreads directly from a refrigerator or which retains its consistency at high temperature.

At the emulsification stage emulsifying agents, usually lecithin, are added to stabilise the emulsion, and other compounds such as the sodium sulphoacetate derivative of mono and diglycerides are sometimes added to prevent spattering during frying. After formation of the emulsion the margarine is tumbled or kneaded to improve texture, and salt and colouring (if used) is worked in. In many countries margarine is supplemented with Vitamins A and D and in some sodium benzoate is permitted as a preservative. Margarine is rarely subject to microbial spoilage. Oxidative rancidity may occur and it can 'pick up' taints if improperly stored.

Its position in the marketplace is of interest since it reflects changing attitudes to quality. Margarine was in concept a cheap substitute for butter and to be accepted had to resemble butter as closely as possible. Many older people see it as an intrinsically inferior product and the more status conscious would undoubtedly select low quality butter rather than high quality margarine. The situation is rather different with the younger generations. Although the price differential between butter and margarine has been reduced margarine remains significantly cheaper than butter but is increasingly seen as a product in its own right rather than as a butter substitute. This is due partially to improvements in quality and presentation but a significant factor has been public awareness over the role of dietary fats in heart disease. It is not pertinent to comment here on the counter claims of interested parties but it is undoubtedly true that it is easy to produce margarine with a high proportion of polyunsaturated fats and a low cholesterol content which is likely to be preferred by many diet conscious consumers.

A further 'hybrid' product which should be mentioned at this stage is exemplified by a recently developed product based on 40% dairy fat with added vegetable oil. The latter component is of low melting point and increases the plasticity of the dairy fats at refrigerator temperature. The product resembles a butter with the low temperature spreading properties of soft margarine.

Examples of laboratory tests are shown in Table 8.

Table 8   Laboratory Tests Dairy Butter and Margarine

I. Microbiological
Total aerobes
Enterobacteriaceae including *E. coli*

N.B. Counts of lipolytic micro-organisms may sometimes be required.

II. Chemical/physical
Moisture
Rancidity

Foreign body/infestation examination not usually required.

Notes:
(1) These tests are examples only and are not inclusive.
(2) Examples are for finished products only.
(3) Techniques are detailed in specialist publications such as ICMSF (1978) – Microbiology: Egan *et al* (1981) – Chemistry.
(4) Instrumental techniques such as Impediometry may be substituted for some classical microbiological methods.

135

135. Poor quality butter. The block illustrated had been subjected to temperature abuse. In addition to the greasy unattractive appearance it had a rancid odour and flavour. The cause is likely to have been chemical rancidity rather than microbial.

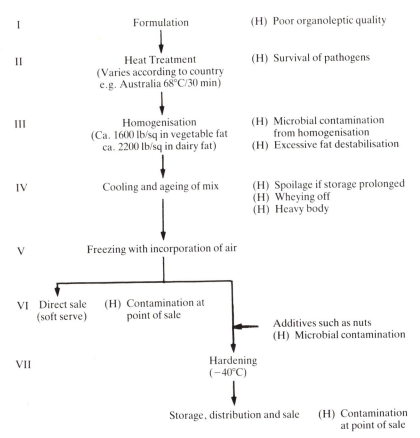

I	Formulation	(H) Poor organoleptic quality
II	Heat Treatment (Varies according to country e.g. Australia 68°C/30 min)	(H) Survival of pathogens
III	Homogenisation (Ca. 1600 lb/sq in vegetable fat ca. 2200 lb/sq in dairy fat)	(H) Microbial contamination from homogenisation (H) Excessive fat destabilisation
IV	Cooling and ageing of mix	(H) Spoilage if storage prolonged (H) Wheying off (H) Heavy body
V	Freezing with incorporation of air	
VI	Direct sale (soft serve)	(H) Contamination at point of sale
		Additives such as nuts (H) Microbial contamination
VII	Hardening (−40°C)	
	Storage, distribution and sale	(H) Contamination at point of sale

(H) = specific quality hazard

136. A flow diagram for the manufacture of ice cream. Details of formulation and processing vary according to the type of ice-cream being produced.

Ice Cream (136–138)

In view of the dairy origins of ice cream it is discussed here rather than in Chapter 14 concerned with Frozen Foods.

In the present context ice cream refers to a number of products including **cream ices**, **ice cream proper**, **milk ices**, **custards**, **sherbet** and **water ices**. It should be noted that the latter category which includes, in the U.K., 'ice lollipops', contain no dairy ingredient whatsoever. The dairy component of other products varies according to product designation and national laws, but most ice cream today is manufactured from non-dairy fat.

Ice-cream manufacture is a relatively complex procedure and fuller details should be obtained from publications such as Hyde and Rothwell (1973). A basic flow diagram is given (**136**). Ice cream produced on a large commercial scale (continuous freezer) contains as major ingredients 14% sugar or sugar/corn syrup mixture, 11% milk solids-non-fat (SNF) and 10% fat. Minor ingredients are stabiliser and emulsifier (total 0.5%–0.75%), colour and flavouring (Rothwell, 1981). Liquid whole milk is used mainly by small scale manufacturers, large scale producers using concentrated skim milk at an SNF content of ca. 30% or spray dried skim milk powder. Fat may also be of dairy origin and the highest quality (and most expensive) ice cream is that made using cream. Most fat however is based on a mixture of partially hydrogenated vegetable fats. Ideally the fat mixture has a well defined major melting peak at ca. 30°C with all fats being liquid below 37°C to avoid a fatty after sensation.

The sugar is normally sucrose but glucose syrups of high dextrose content may be used to improve texture. Sugar substitutes such as sorbitol (a sugar alcohol) may be used in special ice creams for diabetics giving, however, a poor texture.

The first stage in ice cream manufacture after formulation, weighing and mixing ingredients is heat treatment. The severity of the required heat treatment varies according to the country of manufacture but is typically not less than 65°C for 30 minutes or the equivalent at higher temperature:shorter time treatments. Heat treatment is required to destroy vegetative pathogens such as *Salmonella*. (A major outbreak of typhoid which resulted in 210 cases and 4 deaths occurred in Aberystwyth, U.K., during the summer of 1947. The ice cream maker, who had previously suffered from typhoid but had been declared free of the organism, was found to be an active urinary excretor of *S. typhi* of the same bacteriophage type as the persons affected.)

Homogenisation is usually carried out at the same time and as homogenisers are notoriously difficult to sanitise it is recommended that homogenisation should be completed before the final stages of heat treatment. Emulsifying agents such as glyceryl monostearate, polyoxethylene glycol (Tween) and sorbitol esters (Spans) are used to stabilise the fat globules although some degree of destabilisation is now considered desirable to produce ice cream which does not melt too rapidly. Excessive fat destabilisation leads to large buttery lumps being formed.

Following homogenisation the ice cream mix is cooled to ca. 4°C and held for up to 48 hours. Excessive storage may lead to spoilage by psychrotrophic bacteria. During this 'ageing' period milk proteins hydrate, fats begin to crystallise while hydrocolloids, added as stabilisers, absorb free moisture. A number of substances may be used as stabilisers, the purpose of which is to give body to the ice cream, impart attractive melting characteristics and minimise the effects of temperature fluctuation during storage. Excessive quantities of stabiliser produce a gummy or heavy body. Typical substances are gelatin (rare today), alginate, gums, carboxymethyl cellulose, carrageen and other seaweed derivatives. Locust bean and guar gum may cause precipitation of casein and 'wheying off'. This phenomenon can be reduced by addition of carrageen.

Table 9   Laboratory Tests Dairy Ice Cream

I. Microbiological
Total aerobes
Enterobacteriaceae including *E. coli*

II. Chemical/physical
Fat
Over-run

Foreign body/infestation examination not usually required.

Notes:
(1) These tests are examples only and are not inclusive.
(2) Examples are for finished products only.
(3) Techniques are detailed in specialist publications such as ICMSF (1978) – Microbiology: Egan *et al* (1981) – Chemistry.
(4) Instrumental techniques such as Impediometry may be substituted for some classical microbiological methods.

Freezing of the 'aged' mix may be carried out in batch freezers or, more commonly, continuous freezers. The ice cream is vigorously agitated during freezing and air is incorporated to give an aerated product. The incorporation of air increases the volume (the over-run) by up to 120% in large continuous freezers. The degree of over-run may be controlled to produce a particular type of ice cream but is usually as high as possible. This is due to the simple economic fact that ice cream is sold by volume not by weight. The greater the volume increase due to incorporation of air, the greater the profit.

After freezing most ice cream is packed into bulk or individual retail packs. Coatings such as chocolate or chopped nuts may be added at this stage and care must be taken to ensure that additives do not introduce microbial contaminants. Particular care is necessary with chocolate (see p. 142) and desiccated coconut (see p. 142) both of which have been associated with outbreaks of salmonellosis.

The packaged ice cream is then hard frozen at −40°C in either wind tunnels or hardening rooms and stored at ca. −30°C before and during distribution.

A proportion is sold direct from the freezer without hardening. Such 'soft-serve' ice cream is typically produced in small freezers in restaurants and cafés or in mobile vans. Pre-prepared powdered mixes, heat treated before drying, are used, the powder being added to cold potable water and frozen within 1 hour.

Although ice cream is of a complex physico-chemical nature it is

137 and **138.**   **Thawed and refrozen ice cream** results in formation of large ice crystals imparting an unpleasant gritty texture. Refrozen ice cream (**137**) is seen to be of a visibly rougher texture after serving than the equivalent ice cream (**138**) which has been correctly stored. The effects, however, are lessened by the incorporation of stabilising agents such as carrageen.

possible to manufacture a highly reproducible product, and once a suitable formulation and manufacturing protocol have been established specific technical quality problems at a manufacturing level are few. Ice cream is, however, subject to abuse during storage which may affect the quality while taints may be picked up from adjacent foods.

Ice Cream and Public Health

The introduction in the U.K. of heat treatment regulations together with the efforts of local health authorities in improving manufacturing hygiene have vastly improved the public health status of ice cream and it has not been directly implicated in food poisoning since 1955. While outbreaks due to ice cream do occur in other countries, Rothwell (1981) points out that in most cases there are minimum heat treatment standards and standards for post-heat treatment handling.

The major public health risk undoubtedly exists at point of sale. While high standards of hygiene are necessary in all food retailing operations particularly where unwrapped foods are involved ice cream is both difficult and 'messy' to handle and serves as a substrate for rapid microbial growth. The problem is exemplified by two areas of particular concern; the 'soft-serve' freezer which requires a rigorous cleaning schedule, and mobile ice cream sales vans. Even when conscientiously operated, mobile vans present particular hygiene problems and it is not uncommon to see units where there is a total lack of the most basic hygienic precautions with respect to utensils and equipment and a total disregard for personal hygiene. The problem is exacerbated by the large number of unlicensed 'cowboy' operators over whom local health authorities have no effective control and whose premises may, for example, lack the means of effectively cleaning 'soft-serve' freezers even if the will to do so exists.

Examples of laboratory tests are shown in Table 9.

Chapter 3

Fresh Meat

Consumption of fresh meat on a *per capita* basis varies widely. As a general rule meat consumption is highest in the economically advanced countries such as Western Europe and the U.S.A. or in major producing countries such as Argentina. Patterns of consumption are affected by suitability of the climate and terrain for rearing the various meat animals as well as by religious and social factors. Members of the Hindu religion, for example, are vegetarian while Muslims and Jews are forbidden pork. In the U.K. there is a marked abhorrence to horse meat as a human food despite the lack of any religious or economic considerations.

Meat Inspection

Meat for human consumption must come from healthy and, as far as possible, disease-free animals. Meat animals carry a number of human diseases and Wilson (1973) noted that concern about meat as a cause of disease is increasing rather than diminishing. Animals exhibiting overt clinical symptoms should be removed from the lairage before slaughter but not all disease conditions can be detected ante-mortem, therefore post-mortem inspection is of major importance. Meat inspection is a specialised field and reference should be made to specialist publications such as Wiggins and Wilson (1976).

It should be noted that emergency slaughtered animals give greatest cause for concern. The attitude towards emergency slaughtering varies from country to country. Some do not permit such animals to be utilised for human consumption while others rely on veterinary guidance or hold carcasses pending bacteriological or histological examination.

Despite the precautions taken there is no doubt that condemned meat finds its way into the human market. Considerable publicity has been given in the U.K. to condemned (knacker) meat in burgers and meat products (see p. 55) and undoubtedly some knacker meat is also present in the fresh meat market. The underlying cause is economic in view of the difference in price between the fit and unfit meat and scope for malpractice occurs throughout the industry from farm to ultimate consumer.

Animal Parasites

A number of animal parasites may be present in meat. Some of these, if not destroyed by processing, may cause disease in man.

Liver Flukes (Fasciola hepatica)

Liver flukes are shared between the primary host, sheep, occasionally cattle or rarely pigs, and the secondary host, the water snail (*Limnaea trunculata*). The flukes, which cause serious disease often resulting in death of the host, may be detected in the liver and possibly lungs and peritoneum of infected animals at the meat inspection stage (see Wiggins and Wilson, 1976). The adult fluke, which is readily destroyed by heating, does not infect man. Human infections may arise from drinking water containing the **cecarial** stage of the fluke or by eating plants such as **watercress** grown in contaminated water (see p. 191).

(Snails cultivated for human consumption are produced in conditions free from contamination with *Fasciola hepatica*. Uncultivated water snails are occasionally consumed and may be a cause of human fascioliasis. A type of snail found on lock-walls and other canal-side structures of the Kennet and Avon canal in Wiltshire and Berkshire was consumed as a delicacy by boatmen who suffered, as a consequence, a high incidence of fascioliasis.)

Tapeworms (Taenia spp)

Two species of tapeworm are common parasites of man; *Taenia solium*, the pork tapeworm and *Taenia saginata*, the beef tapeworm. **Cystercerci** encyst in the muscle of the infected secondary hosts, cows or pigs, giving the muscle a 'measly' appearance; liver and other organs are occasionally infected (see Wiggins and Wilson, 1976). Ingestion of live cystercerci leads to infection of the primary host, man, producing the pathological condition taeniasis.

Tapeworm infections of food animals are rare in countries with good sewage systems which prevent pigs and cattle being in contact with human faeces. In more backward communities this is rarely possible particularly in subsistence economies where the cow or pig may live in close proximity to its owners.

Nematodes (Trichinella Spiralis)

The nematode worm *Trichinella spiralis* infects man and a number of other hosts including food animals and domestic pets. The most important food animal involved is the pig. **Larvae** encyst in the skeletal muscles of infected animals and if eaten live cause a serious and long lasting illness in man with a mortality rate of up to 30%.

The incidence of *Trichinella* in pigs may be considerably reduced by ensuring that feed is fully cooked and by exterminating rats since pigs will eat dead rats which are frequently infected by the parasite. Such measures are readily implemented in developed areas but not elsewhere. The incidence of *Trichinella* is high in wild pigs in Central Europe, and, in areas where bears are hunted for food, the bear is often a major source of human trichinellosis. Of 11 cases reported in Canada in 1978, for example, 10 were attributed to bear meat (Anon, 1984). Bears in the vicinity of town are notorious scavengers and share the pig's taste for dead rats and other rodents.

Sarcocysts (Sarcocystis bovihominis: Sarcocystis suihominis)

Both cattle and pigs may be infected with sarcocysts. The incidence appears to be higher in cattle, De Kruijf *et al* (1974) finding that in the Netherlands nearly 100% of beef hearts were infected with (unidentified) sarcocysts as opposed to only 3.5% of pig hearts. Sarcocysts are generally thought to have no effect on humans but Heydoorn (1977) noted digestive complaints in persons eating heavily affected beef and pork.

Toxoplasma gondii

Toxoplasma gondii is a protozoan parasite which may infect man through ingestion of infected meat or by taking up **oocysts** from feline faeces. The organism is microscopic and cannot be detected by normal meat inspection procedures, serological techniques being used where necessary. A survey made by serological techniques in the Netherlands showed 22% of cattle, 30% of sheep, 0% of fattened pigs and 11% of sows to be infected (van Sprang, 1983). Equivalent figures for other countries are not known.

In many cases adult infections with *Toxoplasma gondii* proceed sub-clinically but congenital infections are more serious resulting in blindness, mental instability, abortion and in severe cases death of the foetus. At the present time toxoplasmosis is probably the most important food derived parasitic disease in developed countries.

Although the incidence of many parasites in meat may be reduced by adopting practices which break the life cycle of the parasite, parasitic infections cannot be wholly eliminated. In many countries carcasses are examined for the presence of parasites (the action taken varying according to the parasite and from country to country) but meat inspection cannot be considered to be the primary defence against parasitic disease. Apart from *Toxoplasma gondii* which, as already noted, is microscopic and not detectable by normal procedures, inspection of meat for *Trichinella spiralis* is 'laborious and not always successful'. Serological techniques are the most reliable but are not suitable for routine application at meat inspection level. Even with *Taenia spp*, which are generally considered to be easily recognised in muscle, routine slaughterhouse examination may considerably underestimate the incidence of infected animals.

Control of parasitic diseases lies, therefore, in ensuring that cooking or other processing is effective in inactivating the parasites. In domestic situations where most parasitic disease is acquired cooking is the usual means of inactivation. A centre temperature of 60°C is generally considered to be required to kill cysts of *Taenia spp* while *Toxoplasma gondii*, *Sarcocystis spp* and *Trichinella spiralis* require a temperature of 70°C. Such temperatures are readily obtained under normal conditions. Undercooking may however occur in large joints especially if ovens are inadequate or where meat is surface heated only. This may well occur in barbecue or campfire cooking where meat that is heavily charred on the surface may be virtually raw in the interior. It should also be noted that *Trichinella* larvae can survive a temperature of 77°C if the meat is very rapidly heated (Kotula *et al*, 1982). This fact may have implications where microwave or infra-red heating is used.

Similar hazards to those arising from undercooking occur by conscious consumption of raw meat which is often consumed as a delicacy. Steak tartare is a common example in Western Europe and similar delicacies are eaten elsewhere. In Ethiopia and parts of Sudan a combination of a high rate of infection amongst cattle and widespread consumption of a local raw beef delicacy has contributed to a very high level of *Taenia saginata* infection. Meat normally intended for cooking may also be eaten raw especially by children. Thus a Canadian outbreak of trichinellosis was confined to the two children in a family of six who enjoyed eating raw bear meat.

On a commercial scale freezing is often used to inactivate parasites in meats intended to receive only minimal further processing. *Trichinella* larvae are destroyed by holding at −25°C for 10 to 20 days; cysts of *Taenia spp* at −18°C for 7 days; sarcocysts at −20°C for 3 days and *Toxoplasma gondii* at −20°C for 2 to 3 days. (Grossklaus and Baumgarten (1968) found cysts of *Toxoplasma* to survive this treatment in 1 of 35 experiments.)

Pre-slaughter Handling and Quality

Pre-slaughter handling of animals has a direct effect on quality of meat whether it is to be sold in the fresh meat market or processed. Pigs are particularly prone to stress while sheep appear to be relatively immune. Stress may however affect all meat animals. The ideal pre-slaughter state of an animal is for it to be rested and calm with high glycogen reserves. Conditions in the lairage and before slaughter should be designed to produce this state despite the obvious difficulties. The gratuitous violence towards animals in some abattoirs results, quite apart from unnecessary cruelty, in a high percentage of low quality meat. Correct handling of animals pre-slaughter should be seen as part of the overall quality assurance operation. In Denmark for example pigs, which in addition to being generally stress prone have a low tolerance to high temperatures, must be transported during summer months in air-conditioned vehicles.

The direct quality implications of poor pre-slaughter handling are twofold. **Dark Firm Dry (DFD)** meat is meat, usually beef, from a stressed and exhausted animal with minimal glycogen reserves. Typically it has a high ultimate pH value (>5.9) and is dark in appearance with a dry texture. There is usually consumer resistance to DFD meat although in some localised areas of the U.K. such as the North East Midlands it may be preferred to normal meat.

DFD meat spoils rapidly with a characteristic green discoloration accompanied by off-odours. The spoilage flora is also atypical being dominated by *Enterobacter liquefaciens* (*Serratia liquefaciens* see Grimont and Grimont, 1984) and *Alteromonas putrefaciens* (Nottingham, 1982). In contrast **Pale Soft Exudative (PSE)** meat, usually pork, is from a stressed animal retaining significant glycogen reserves. Post-mortem glycolysis and attendant fall in pH value, is very rapid leading to a meat of normal, or sub-normal ultimate pH value, of pale appearance and soft texture along with large amounts of free moisture. In pigs the incidence of PSE meat is genetically linked to the breed.

Slaughterhouse Techniques

Large scale slaughterhouses in industrial nations work on a chain system whereby the slaughtering and dressing operations are broken into a series of unit operations carried out by a number of operators. A typical flow diagram for cattle is shown (**139**). Details for other animals such as pigs and sheep differ but the principles are the same (some animals such as deer are normally shot on the hoof – see below).

In contrast in small scale operations animals may be killed on a one man:one animal solo operation either in a slaughterhouse or, in some countries, on the farm. The superior conditions and facilities of a large mechanised slaughterhouse should in theory lead to improved standards of dressing and a reduction of microbial cross-contamination. In practice these advantages may well be negated by the lesser operator skill required in chain operations and by the increased handling involved. The factors involved in meat carcass contamination are discussed fully elsewhere, e.g. Nottingham (1982).

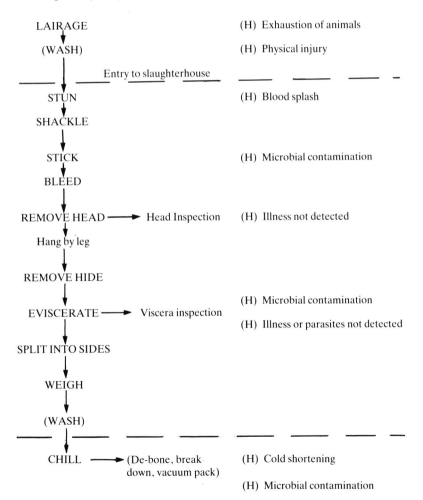

Stages in parentheses are not carried out in all operations

(H) = Specific quality hazard

139. A flow diagram showing the stages in slaughtering and dressing of cattle. The process shown is for conventional butchery and details differ where hot butchery and/or electrical stimulation is used. Large scale slaughter of pigs and sheep differs in detail but a similar series of unit operations are used.

Electrical Stimulation

Electrical stimulation may be applied to cattle and sheep slaughter with the intention of either tenderising the meat or preventing toughening. It has been extensively researched and is applied commercially in New Zealand for lambs and in some abattoirs in Europe and the U.S.A. for beef. The general principle is to apply a pulsed A.C. voltage to the carcass. The voltage may be high (ca. 1000V) or low (ca. 90V) and the means of application vary. Electrical stimulation induces rapid onset of post-mortem glycolysis and thus *rigor mortis* which prevents toughening due to the process of cold shortening during rapid chilling.

This is of considerable importance in the context of European Economic Community regulations which require a deep leg muscle temperature of 7°C within 24 hours of slaughter, a condition leading almost inevitably to cold shortening in beef. Electrical stimulation results in a pH value of <6.2 within 3 hours. Cold shortening does not occur below this pH value and rapid cooling may be applied. There may also be some tenderisation due to stimulation of catheptic enzymes in a low pH value carcass at temperatures close to body heat.

Boning

Traditionally meat is delivered to retail outlets as sides which are then further butchered and broken down into retail cuts. Supermarkets, which in many countries account for increasing sales of meat, prefer primal joints, de-boned where appropriate for breaking into pre-packs at either a central packaging plant or at the individual supermarket. Most plants chill meat for 24 to 48 hours before further butchery. Primal joints of beef, veal, some lamb and occasionally pork are usually vacuum packed prior to despatch. Preparation of primal joints at the slaughterhouse reduces transport and refrigeration costs while primal joints are more suited to unit pre-packing operations using relatively unskilled staff.

Hot Boning

Hot boning is a technique whereby meat (usually beef) is boned, broken down and vacuum packed 1 to 2 hours after slaughter while the meat is still warm. Conventional primal joints may be produced or muscle seaming techniques may be used to separate muscle groups. The advantages of hot boning are saving in refrigeration costs over the cost of chilling whole bone-in carcasses, reduction of evaporative losses, reduced loss through drip and possibly more uniform colour. It should be noted however that some of the economic advantages claimed are difficult to demonstrate in the context of a commercial meat handling operation. The possibility of increased microbial growth rates on hot butchered meat has been investigated but the effects seem unlikely to be significant in practice. However the difficulties involved in handling a pliable, pre-rigor carcass may result in a higher level of initial contamination.

The advantages of electrical stimulation and hot boning may be combined by application of both techniques in tandem.

Conditioning of Meat

Meat requires conditioning to achieve the desired degree of tenderness and this involves holding it for a week or more before butchery for retail sale. Attempts have been made to tenderise meat either by application of proteolytic enzymes to the surface or by immediate pre-slaughter injection of a proteolytic enzyme such as papain. Such meat has not been widely accepted.

Table 10   Laboratory Tests Fresh Meat

I. Microbiological
No meaningful assays for routine quality assurance purposes.

II. Chemical/physical
Fat (mince only)

Meat species determinations may be required for any meat, and presence/absence of soya determinations may be required for mince.

Foreign body/infestation examination not usually required.

Frozen Meat

This has a poor image for quality, often a consequence of poor packaging and appearance. Providing the conditions of freezing are such as to avoid toughening, freezer burn etc. the quality of frozen meat, providing initial quality was satisfactory, will be high. Packaging techniques such as 'skin packing' (Bailey, 1983), which enhances the appearance of frozen meat, may well succeed in changing the consumer image.

Examples of Laboratory tests are shown in Table 10.

Notes:
(1) These tests are examples only and are not inclusive.
(2) Examples are for finished products only.
(3) Techniques are detailed in specialist publications such as ICMSF (1978) – Microbiology: Egan *et al* (1981) – Chemistry.
(4) Instrumental techniques such as Impediometry may be substituted for some classical microbiological methods.

140, 141 and **142.   A typical example of a de-boned vacuum packaged joint** of beef. A significant proportion of beef is now handled in this form throughout the world. Note the purple-red colour of the meat which is due to the de-oxygenation of the normal bright red meat pigment oxymyoglobin to myoglobin. This reaction is reversible on re-exposure to air. The major advantage of vacuum packaging lies in a storage life of up to 3 to 4 weeks, which is made possible by the low oxygen tension significantly limiting the metabolic activities and growth of the major spoilage bacteria *Pseudomonas spp* and related genera.

When required for sale primal joints are opened, drained and butchered into retail cuts, usually packed on polystyrene trays and overwrapped with an air permeable film (**141**). In the presence of air the normal red colour regenerates. In this example the rather dark colour (**142**) is due not to de-oxygenation of the pigments or to a DFD condition but to surface dehydration which serves to concentrate the pigments giving an apparent darkening.

143 and **144.   Minced (ground) beef** may also be centrally prepared and transported to retail outlets in vacuum packs. The shelf life of 7 to 10 days is considerably shorter than that of whole joints. This is due to the micro-organisms being distributed over a greater surface area and to a greater nutrient availability due to disruption of muscle tissue etc. In bulk form it is coarsely minced (**143**) and requires a second mincing before packing for retail sale (**144**). Conversion of myoglobin to oxymyoglobin at this stage is rapid but a patented process is available for injection of oxygen at the point of mincing to improve the percentage conversion of myoglobin to oxymyoglobin and hence improve the meat colour.

145 to **151.   Packaging procedures** adopted by large scale supermarket operations commonly involve placing retail cuts in an expanded polystyrene tray overwrapped in a highly oxygen permeable film such as nylon/polythene although controlled atmosphere (CAP) packing may also be used (see p. 58). Microbial growth on retail cuts of meat is rapid and life at storage temperatures of 2° to 4°C is usually no more than 4–5 days in air or 8–9 days in CAP.

The following sequences illustrate spoilage of retail cuts of beef and pork. **145** illustrates good quality beef sirloin steak held for 1 day after packing at 2°C. Note the absence of staining of the fat detailed (**146**). A further day of storage at 2°C has little or no effect on visual quality (**147**). Two days' storage at an ambient temperature of ca. 10°C, however, results in spoilage. The beef illustrated (**148**) has darkened, lost its attractive bloom, shows fat discoloration and had a detectable odour of proteolysis. Spoilage of the beef was due to *Pseudomonas spp* which raise the pH value contributing towards darkening.

In the samples of pork escalope illustrated the meat lightens between 1 day's (**149**) and 2 days' storage at 2°C (**150**). Two days' storage at ca. 10°C results in a spoiled product (**151**) showing microbial slime formation and having a sour odour. Spoilage of the pork was in this example due to *Brochothrix themosphacta*.

It must be emphasised that on other occasions spoilage of beef may be by *B. thermosphacta* and of pork by pseudomonads. While pseudomonads are favoured by air permeable packaging, the make-up of the dominant micro-flora is determined by several factors. As a generalisation it may be stated that pork and veal tend to spoil more rapidly than beef and mutton and that lactic acid bacteria are more prevalent in the micro-flora of the former than of the latter.

140 ●

141 ● *

142 ●

143 ●

144 ●

145 ●

146 ●

147 ●

148 ●

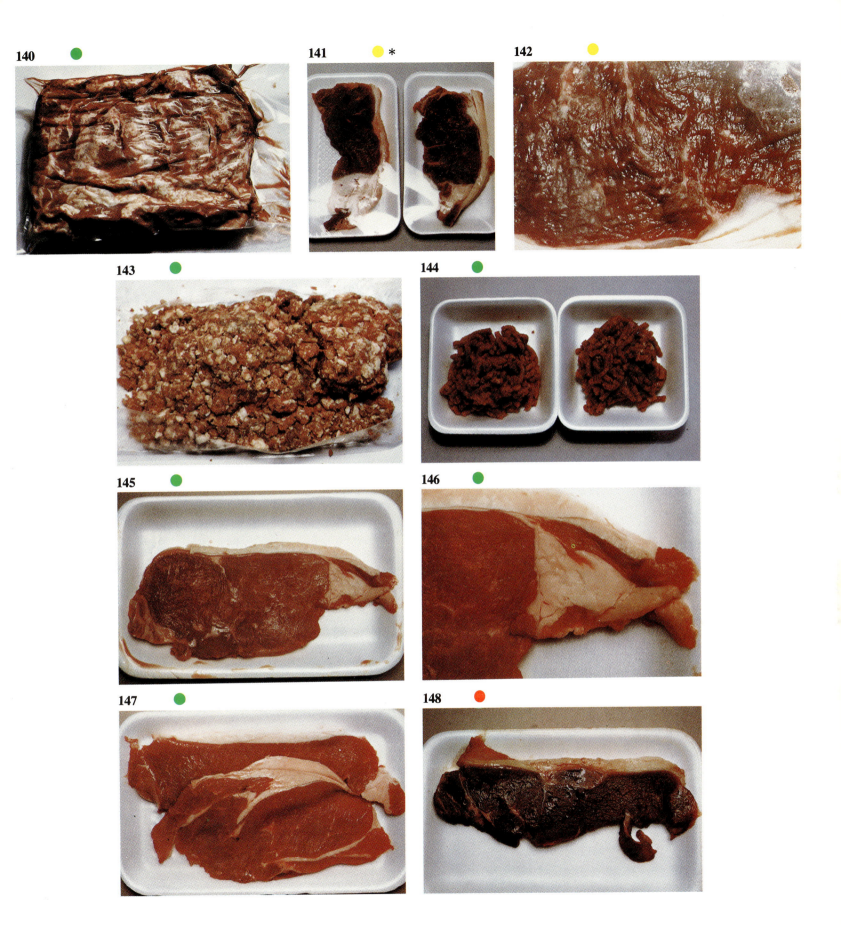

51

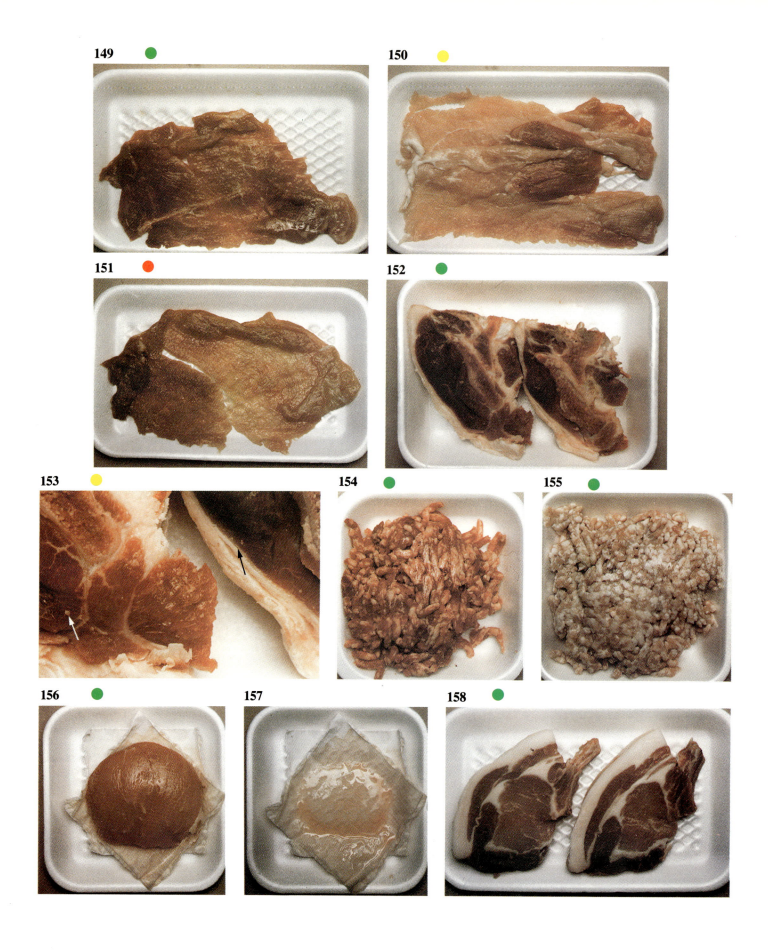

52

152. Lamb is similar to beef in terms of spoilage although its higher fat content and the composition of the fats may increase the likelihood of rancidity during prolonged frozen storage.

153. Bone dust (arrowed) on lamb. Dust is derived from the bone while sawing during butchery. Only small amounts are present in the illustration but larger amounts detract from the visual appearance and impart a gritty texture. The meat in the vicinity of bone dust is alleged to sour more rapidly although documentary evidence for this appears to be lacking.

154 and 155. A comparison of beef mince (154) with veal mince (155). The myoglobin content of beef is considerably higher than that of the pre-ruminant veal which, like pork, tends to fade rapidly in colour under display lights. Pre-prepared mince is often used as a means of selling excessive quantities of fat. This may be detectable by visual inspection but confirmation requires laboratory analysis (see Table 10). Fresh mince may also be extended with a percentage (usually ca. 20%) of **textured soya protein.** Such mince should be clearly labelled.

156. 'Noisette' of veal. Although veal is a popular meat in many parts of continental Europe there is consumer resistance in the U.K., partly as a reaction to publicity concerning production techniques. Therefore veal is often presented as a luxury cut. It is biochemically similar to pork and has a similar pattern of spoilage.

157. Pre-prepared retail cuts of meat, particularly pork and veal, leak considerable quantities of free moisture (drip). This not only detracts from the visual appearance but also serves as a substrate for rapid microbial growth. Absorbent pads are often used to maintain visual appearance, but the rapid microbial growth in the absorbed drip may mean that the pads are the source of initial spoilage odours rather than the meat itself. It should be noted that excessive drip in vacuum packed primal joints (usually beef) is an important source of economic loss.

158 and 159. Butchery techniques vary from country to country and the conformation of a particular cut affects the perceived quality. This is illustrated by comparing the French style pork loin cutlets (**158**) with their English equivalent (**159**). In each case the technical and organoleptic quality of the pork loin is of a high and equal standard.

159 ●

Offal (organ meats and tissues) (160–163)

As little as 50% of liveweight of a meat animal may be converted into carcass meat (ca. 55% steer, 50% lamb, 75% pig; Gerrard and Mallion, 1977). A significant proportion of the remainder is edible and may be used with or without further processing. Offals are of significantly lower value, and in countries such as the U.S.A. have become associated with poor people who cannot afford carcass meat (Forrest *et al* 1975). The lower value of offal means that its handling at the slaughterhouse is of low priority compared with carcass meat. Where profit margins are under pressure successful utilisation of offal and other by-products is often of considerable economic importance.

160 ●   **161** ●

162 ●

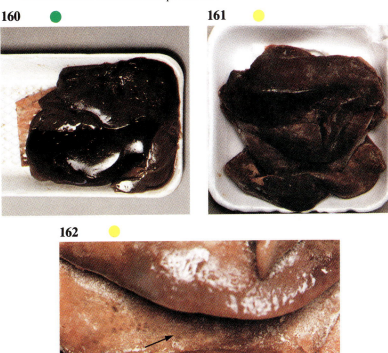

160. Liver, like all offal is generally more highly contaminated than carcass meat and has only a short shelf life under chill conditions. For this reason liver is usually frozen although fresh chill liver may be sold for local distribution or bulk frozen liver may be thawed, pre-packed and sold chilled as in this example. Darkening is a common phenomenon which only rarely leads to rejection. The effect is largely due to the strongly reducing conditions on the liver brought about by a high residual metabolic rate.

161 and 162. The high proportion of haem pigments and their degradation products in liver contribute to a variable appearance (**161**). The brown area shown in detail (**162**) (arrowed) may be due to **freezer burn.**

163. **Tripe** is the cleaned and cooked lining of the rumen and reticulum of the bovine digestive tract. Previously a dietary staple in Northern England tripe consumption has fallen drastically. It is however an important meat industry by-product, the typical bovine stomach yielding ca. 7.5kg and may, therefore, be used as an ingredient in meat products such as sausages although much is processed as pet foods. A similar product may be produced from sheep stomachs while **chitterlings** are manufactured from the porcine stomach and large intestine.

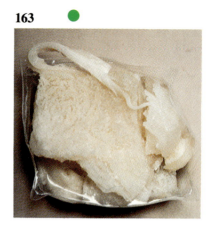

163

Game (164–166)

Venison is classed as game in the U.K. Deer can be shot only within given seasons and must be sold only by licensed dealers. They are ruminant animals and the meat differs biochemically from that of game birds such as pheasant and grouse. The rural and, in some cases, haphazard structure of the game trade can lead to problems of quality assurance. Handling and storage facilities for venison may be primitive and while deer shot in the field are less subject to cross-contamination than, for example, cattle in a slaughterhouse, the use of 'dum-dum' bullets may cause considerable internal damage and spread of gut contents into deep muscle.

Venison requires hanging before eating, a process of tenderisation and flavour development brought about by autolytic enzymes. The extent to which a deer is hung depends largely upon consumer preference. A 'connoisseur', used to eating game for example, will favour a strong 'gamey' flavour. Large scale retailing of venison as an alternative (possibly cheaper) to beef would probably require a relatively bland product. It is thought to be unlikely however that deer will become a major meat animal in the U.K. in the immediate future.

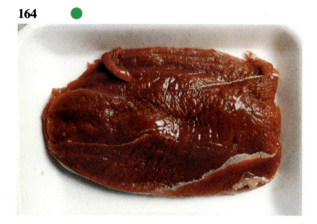

164

165

166

164, 165 and **166.** **Vension for mass retailing** is hung, butchered into primal joints, vacuum packed and transported to point of sale for further butchery into retail joints. The handling chain is analogous to that for beef. **164** illustrates a good quality venison steak. The colour (**165**) is usually darker than that of beef and progressively darkens during storage. This is a biochemical change possibly related to increase in pH value during storage. An example of darkened venison stored as a retail cut for 2 days at 0°–1°C is shown (**166**). This would be unacceptable to the average consumer.

Chapter 4

Fresh Meat Products

The term 'fresh meat products' encompasses a diverse range from fresh comminuted meats such as sausages and burgers through meat pies and puddings to cooked whole meats. Delicatessen products are discussed separately.

It should be noted that some of the products described in the present chapter contain either the curing ingredient sodium nitrite or meat such as bacon which has undergone a curing process. In neither a technical nor a commercial sense, however, are such products considered to be cured since the effect of nitrite is, at most, limited.

For convenience products have been grouped into broad areas of similarity. These are summarised in Table 11 and discussed, individually, below. There are inevitably areas of overlap between categories.

Meat Products, Horse Meat and Knacker Meat (167)

In recent years considerable concern and adverse publicity concerning the meat product industry of the U.K. has been brought about by the finding of horse meat and knacker (condemned) meat in processed products. Attention was focused on burgers, including those from national fast food outlets but there is little doubt that other meat products were involved. There is no health risk with properly slaughtered horse meat (although there may be aesthetic objections in the U.K. as well as illegality if substituted

for other named meats) and the presence of horse meat can be detected by simple serological tests such as the enzyme linked immunosorbent assay (ELISA, Whittaker *et al*, 1983) in raw meat but not necessarily in cooked.

The use of knacker meat is considerably more serious in that a public health risk is presented and the meat is normally undetectable. The only protection afforded to Quality Control personnel is to insist that meat is purchased only from sources who are able to give a direct assurance as to the origin rather than from dealers who purchase on a commodity basis and who have no idea of its history.

Table 11   Fresh Meat Products – Categories

Raw comminuted meat products
e.g. fresh sausages, burgers etc.
Cooked comminuted meat products
e.g. blood puddings, breakfast sausage, meat loaves
Meat pies and puddings
e.g. pork pie, steak and kidney pudding, shepherds pie
Whole cooked, uncured joints and slices of meat

Raw Comminuted Products (168–197)

While in the U.K. the British fresh sausage is the most common product in this category, the burger and burger-type products are of greatest importance world wide. The basic preparation of the meat mix by chopping and comminution is similar in each case although sausages tend to contain more filler and more meat by-products than burgers. Sausages are filled into skins which may be artificial or natural whereas burgers are formed into flat pieces. These are usually round but may be formed to resemble chops, steaks etc.

Other products of a similar nature include roasts which are loaf shaped meat comminutes intended for oven roasting while stuffings are finely comminuted products in some cases containing diced apple, walnut or onions to give a crumbly texture.

Cooked Comminuted Meat Products (198–215)

Cooked comminuted meat products may be retailed whole, e.g. black (blood) pudding links, or sliced either for pre-packing or at point of sale. Vegetative micro-organisms are killed by cooking and while endospores of *Bacillus spp* and *Clostridium spp* survive their numbers are usually small and, if properly cooled and refrigerated after cooking, multiplication is limited. Spoilage of the intact product is usually by yeast or mould growth on the surface. Slicing obviously may involve contamination of the cut inner surface and subsequent bacterial spoilage.

Cooked Comminutes and Public Health

As such products are usually eaten without further cooking they are potential vectors of all types of bacterial food poisoning. Classically the greatest risk lies with *Staphylococcus aureus*. This is discussed in some detail with respect to cooked cured meats (see p. 71) and the same basic considerations apply.

Potential problems are also presented by the survival and possible subsequent germination and outgrowth of endospores of food poisoning bacteria particularly *Clostridium*. Where cooling and subsequent refrigeration has been inadequate outbreaks of *Cl. perfringens* food poisoning and, historically botulism, have resulted particularly with large cooked sausages.

Meat Pies and Puddings (216–225)

Cooked meat pies, puddings etc. are a diverse group of products sharing a basically similar technology. Traditionally such products were manufactured for purely local distribution but in the U.K. today marketing is largely on a national or supra-regional basis. Thus the diversity of products available on a national basis has increased although the overall consumption is static or declining. It should be noted that many of these products such as cold eating pork pies and hot eating Cornish pasties are unique to the U.K. and not found elsewhere.

The meat fillings of pies and puddings consist of meat chopped or comminuted according to the products. **Non-meat** constituents such as onion, egg, starch and even chutney are also present. **Textured soya protein** or **mycoprotein** may be substituted for all or part of the meat particularly in products destined for institutional consumption. It is pertinent to note that pre-cooking spoilage of pastry may occur giving rise to taints attributed to the meat. Such spoilage is largely due to lactic acid bacteria and the problem is a consequence of poor plant management and hygiene.

167. The meat filling of a beef and vegetable pasty. Detection of meats such as horse meat by visual inspection is impossible in products of this nature.

168 to 172. The British fresh sausage differs from most continental sausages in being uncooked and in containing **sulphur dioxide** as a preservative rather than nitrite. (Sulphur dioxide destroys Vitamin B_1 (aneurin) and is itself of debatable toxicological safety. It is not permitted in meat products in most countries other than the U.K. where its use in meat is restricted to comminuted products.) The quality varies widely according to meat content and origin of meats. The organoleptic quality is also affected by degree of chop of the meat. Three examples are illustrated. **168** is of low cost 'economy' sausages. Note the strong pink colour due to the addition of a colouring agent to compensate for a poor appearance. In contrast a colouring agent is not required in a standard quality sausage (**169**) or a high meat content sausage (**170**). A comparison of the fillings shows that of the high meat content sausage to be a coarser chop (**171**). Cheaper sausages contain larger quantities of

Meat Pies and Public Health

The cooking of meat pies is adequate to destroy vegetative micro-organisms but not bacterial endospores. Potential problems of this nature are the same as those affecting cooked comminutes (see this page).

Cold eating pork pies present a particular problem in that gelatin is injected after baking between the pastry lid and meat filling. **Gelatin** provides a good growth medium for bacteria and a number of food poisoning incidents concerning *Staph. aureus* and *Salmonella spp* have occurred in the past (Hobbs, 1974). Such problems may be obviated by holding the gelatin at temperatures above 70°C and the same precaution should be taken with gelatin used for glazing the top surface of pies.

Whole Cooked Uncured Joints and Slices of Meat (226–228)

Whole cooked uncured joints usually of beef or pork are produced for both the catering trade and for retail sale in sliced form. Such products, properly handled, have a long shelf life often terminated by colour changes of a chemical nature.

Whole Cooked Uncured Joints and Public Health

The major areas of public health concern affect whole cooked joints. First, undercooking permitting survival of vegetative pathogens such as *Salmonella* and parasites such as *Trichinella*; second, cross-contamination from raw materials and, third, inadequate cooling permitting multiplication of heat resistant pathogens such as *Cl. perfringens*. All of these hazards are readily avoided by good hygiene and good manufacturing practice. A number of outbreaks of *Cl. perfringens* food poisoning in institutions such as hospitals, (text continued overleaf on page 58).

mechanically recovered meat which imparts a paste like texture (**172**). In extreme cases the filling is effectively a slurry stabilised by a binder such as **carboxymethylcellulose**.

173. Poor quality fresh sausages. The sausages are grey, discoloured and unattractive in appearance. Spoilage is often by souring (*Brochothrix thermosphacta* is usually responsible) which is accompanied by colour deterioration such as shown here.

174 to 178. Specific visual spoilage of the fresh sausages is usually due to production of slime by bacteria or to mould colonies. Occasionally yeast colonies also develop. **174** shows development of a white mould as well as the slimy-looking 'stickiness' (arrowed) which is bacterial slime. Threads of bacterial slime are seen (arrowed) between two sausages in the close-up (**175**). Note in contrast the close-up illustration of good quality sausages (**176**) which have an attractive glossiness but no slime. Other moulds developing on sausages include black mould (in this example *Aspergillus niger* (**177**) and blue mould (*Penicillium sp*) (**178**).

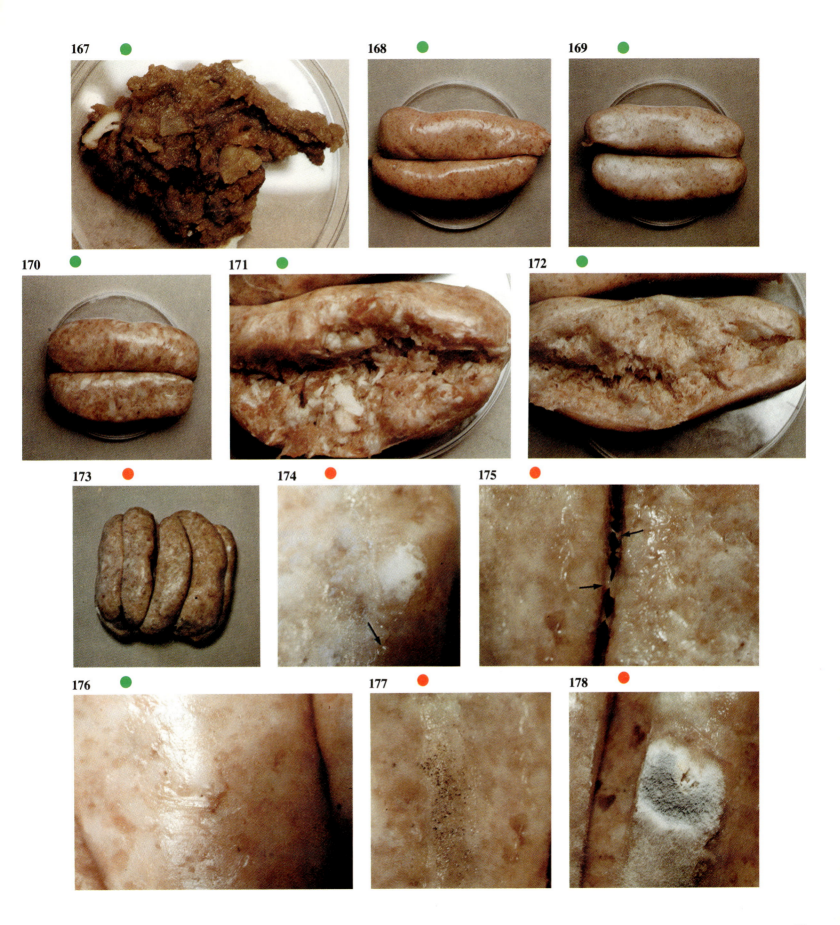

57

old people's homes etc. have occurred due to inadequate cooling after in-house production of such joints (see Pinegar and Suffield, 1982). At the same time concern must be expressed at the increasing number of part-time 'caterers' servicing wedding receptions etc., who operate from domestic kitchens lacking adequate facilities for large scale preparation, cooking and cold storage.

Table 12 Laboratory Tests Fresh Meat Products

A. Uncooked

I. Microbiological
Total viable count

II. Chemical/physical
Meat content

Foreign body/infestation examination not normally required.

B. Cooked

I. Microbiological
Total viable count
Group D Streptococci (whole and sliced meat only)
Staph. aureus (sliced products only)
Lactobacilli (vacuum packed only)

II. Chemical/physical
Nitrite (where added)
Nitrate (where added)
Water content (cooked, uncured joints only)

Foreign body/infestation examination not normally required.

Notes:
1. These tests are examples only and are not inclusive.
2. Examples are for finished products only.
3. Techniques are detailed in specialist publications such as ICMSF (1978) – Microbiology: Egan *et al* (1981) – Chemistry.
4. Instrumental techniques such as Impediometry may be substituted for some classical microbiological methods.

Although fresh sausages are often considered a typically British product their equivalents exist elsewhere in Europe. Three good quality examples are illustrated.

179.   Fresh Italian sausages (Pepata). A coarse chop sausage of high meat content (**180**).

181.   Spanish fresh sausage. The coarse chop filling has a relatively low meat and a high fat content (**182**).

183.   German bratwurst. A coarse chop filling of relatively high fat content (**184**).

Spoilage of the above products is typically the same as the British equivalents – souring by bacteria particularly *B. thermosphacta* and surface growth of yeast and mould (see p. 57).

185 and 186.   Pork sausagemeat stuffing with chestnuts. 185 shows a general view and **186** (arrowed) a close-up of a purple discoloration associated with the chestnuts. Chestnuts contain **anthocyanin** pigments which form metallic salts of grey, blue or purple colour. Sufficient quantities of iron and aluminium, derived from metal handling equipment, are present in sausagemeat to cause such a reaction to occur in the vicinity of the chestnuts.

187.   Quickfries are a burger type product produced from chopped rather than ground meat and shaped to an approximation of a steak. The illustration is a good quality example of lamb quickfries.

188 and 189.   Controlled atmosphere packing (CAP) is a process that has begun to make a significant impact on the fresh meat, meat products and fish trades. The process involves packing the product in a mixture of oxygen, carbon dioxide and, in some cases, nitrogen. The oxygen permits the meat to retain the attractive bright red colour while carbon dioxide inhibits the growth of spoilage micro-organisms (particularly *Pseudomonas spp*) and consequently permits an extended shelf life under refrigerated storage. **188** shows chuck steak burgers in a CAP pack, which are large in volume compared with conventional pre-packs and this may be considered disadvantageous with respect to both display space and visual appeal. Note (arrowed) that the use of CAP packaging does not preclude the inclusion of preservative (sulphur dioxide) as used for conventionally packed sausages and burgers in the U.K. The good colour of the burgers is seen clearly (**189**).

190 and 191. A cross section cut from bacon and pork roll (a comminuted raw pork and bacon product formed into a chub) after cooking showing small pits in the product (arrowed) (**190**). This is caused by gas production by micro-organisms growing in the product when raw. High numbers of heterofermentative lactic acid bacteria (*Lactobacillus brevis*) were present and this organism is likely to have been responsible. The pitting is seen in detail (**191**) (arrowed). It is unusual to be able to recognise fermentative activity in a product of this type after cooking although related organoleptic changes such as souring are readily detectable.

179

180

181

182

183

184

185

186

187

188

189

190

191

59

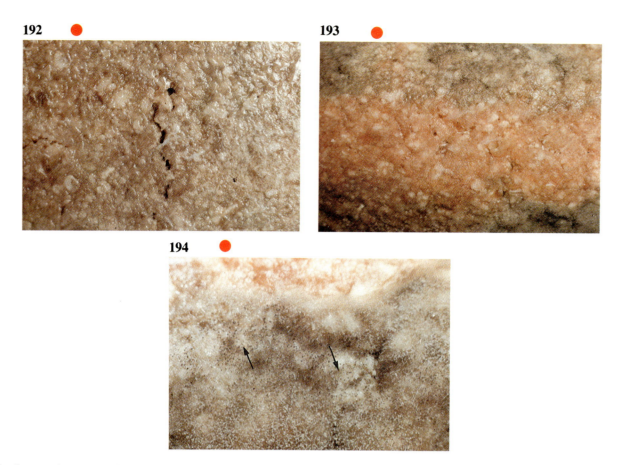

192 to 196. Beef roast is a comminuted meat product normally sulphited, shaped into an oblong block and intended for oven roasting. **192** shows a close view of a poor quality product revealing the grey highly reduced pigment. Note the splitting of the product which may be due to gas production by micro-organisms. Where fully oxygenated (**193**) the normal red colour of the meat is retained, a clearly defined boundary existing between the two pigments. Mould growth is also seen to be present. This is probably *Aspergillus spp* or *Rhizopus spp* and the fruiting bodies can be seen (arrowed) (**194**). Mould growth is not present on all of the discoloured areas and mould is not considered to be responsible for the discoloration (**195**).

A section through the roast (**196**) shows the discoloration to be restricted to the exterior and to a small core through the centre.

197. A damaged insect found inside a beef sausage. The insect is too severely damaged to permit accurate identification but it is probably a fly (Order *Diptera*) which entered the product during preparation of the mix.

Three examples of good quality cooked comminuted meat products.

198. Pork luncheon meat. Note that large quantities of pork luncheon meat are a fully heat processed canned product as opposed to this example which is not fully heat processed and must be stored under refrigeration while in cans as well as after opening.

199. Chopped ham roll.

200. Chopped ham loaf with peppers.

201 and **202. Pork and pepper roll** of generally poor quality. Although the overall colour is satisfactory (**201**) the texture is poor and the pieces of pepper tend to readily fall from the meat matrix. This is shown (arrowed) (**202**).

203. A good example of frankfurters (hot dog sausages) illustrating even size, filling and colour together with an absence of physical damage, discoloration or microbial spoilage.

204 and **205. Spoilage of frankfurters. 204** illustrates spoiled frankfurters. They are of poor quality and show (arrowed) dark red patches of microbial slime. **205** illustrates an area of slime in detail which, in this example, is produced by the Gram negative bacterium *Serratia marcescens*. Under conditions of high ambient temperatures and in the absence of a competing microflora this grows extremely rapidly and, according to biogroup may produce either the red, diffusible pigment prodigiosin, a tri-pyrrole derivative, or the pink, diffusible pigment pyrimine. It should be noted that although pigment production is often considered to be characteristic of *S. marcescens* many strains are non-pigmented. Non-pigmented strains may still, however, be involved in food spoilage (see p. 163). *S. marcescens* is a potentially dangerous opportunistic pathogen and persons handling samples of food contaminated with the organism should observe aseptic precautions.

206 ● **207** ● **208** ● ●

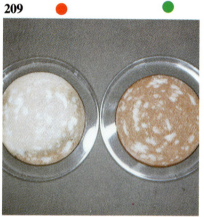

209 ● ●

206 and **207.   English liver sausage showing spoilage by mould.** The white mould colony is seen associated with a zone of greening (**206**). Green pigments of this nature are due to **hydrogen peroxide** formation which oxidises the porphyrin ring of the meat pigment. Hydrogen peroxide is not produced by mould but by some strains of lactobacilli notably *L.viridescens*. It is likely that the mould has developed most rapidly where the pH value has been lowered by lactic acid produced by the lactobacilli, since a close-up (**207**) shows patches of greening where mould has not developed. Spoilage of this nature will normally occur only after extended storage, in this example 30 days at 2°–4°C.

208, 209 and **210.   Discoloration of cooked comminuted meat** products due to production of **hydrogen peroxide.** Sufficient quantities have been produced to bleach the products rather than produce the classic greening (**208** and **209**). The vestiges of a **green ring** may, however, be seen (arrowed) (**210**).

211.   Swelling of a vacuum pack of black (blood) pudding due to microbial gas production. Endospore producing bacteria survive cooking, and growth, together with massive gas production, may be very rapid if post-cooking cooling is delayed or if subsequent storage temperatures are high. In this example the micro-organism responsible was a species of the aerobic endospore forming bacterium *Bacillus*.

212.   Extreme spoilage of a chub pack of a liver and bacon sausage (cooked comminuted meat product). The pack was swollen by gas production and showed large areas of blackening. The exact nature of the blackening is not known but is likely to be related to the production of large quantities of **hydrogen sulphide** probably by **clostridia**. Spoilage of this magnitude is the result of considerable temperature abuse and extended storage, and would not occur under normal conditions.

213, 214 and **215.   Black pudding** showing a white patch on the surface (**213**) which is seen in cross section (**214** and **215**) to be purely superficial. The cause is fat rendering out of the meat mix due to a degree of overcooking. In this case there was also some yeast growth associated with the surface fat.

216.   A good example of a slice of pork pie. Note the good colour and attractive appearance of the pastry. Such slices are cut from a whole 'long' pie and then vacuum packed. Vacuum packing retains

the desired appearance of the meat over an extended period but does not prevent staling and softening of the crust which effectively limits the shelf life. The tight film used is sometimes considered to impart a 'plastic' appearance to the meat when the pack is opened.

217 and **218.   Cured pork pie** contains nitrite and has the characteristic cured meat colour (**217**). In contrast, the natural pork pie appears grey (**218**) although the sample illustrated is of good quality.

219.   A meat pie with lattice pastry. Such pies are traditional, produced now on a mass scale but still involving considerable hand finishing. Visual appearance is an important sales factor in such pies and readily lost by physical damage.

220   An overbaked pork pie. Such a pie would be of poor eating quality as well as being of poor visual appearance and should have been rejected by Quality Assurance systems. Overbaking may occur due to conveyor breakdown in continuous ovens, and in such cases large numbers are potentially affected.

221.   Mycoprotein pie. The filling contains a meat-like analogue produced from the mould *Fusarium graminearum* grown on a glucose substrate in submerged culture (Bull and Solomons, 1983). The flavour and texture of the pie resembled meat but the processing required for the mycoprotein is expensive and the product is intended as a meat replacement rather than an extender (*cf* soya protein page 53).

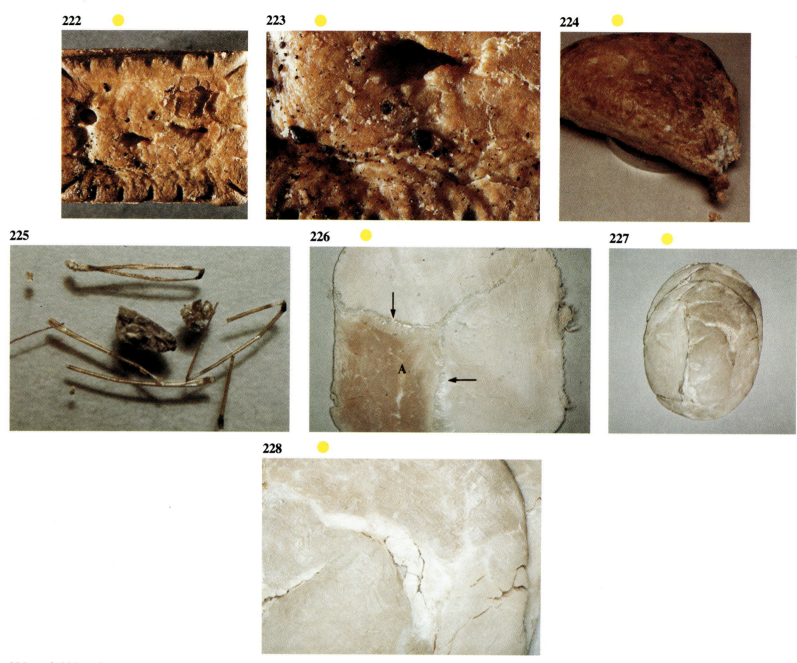

222 and **223.** **Ovenbake on the crust** of a rectangular pork pie (**222**). Ovenbake is carbonised material produced by splashes burning onto oven walls. All baked goods are potentially affected (also see p. 167) and the fault is difficult to eliminate totally, although excessive quantities are indicative of poor housekeeping. The ovenbake is shown in detail (**223**).

224. **Beef and vegetable pasty** (a variant of a Cornish pasty) showing **mechanical damage.** Such damage has no effect on eating quality whatsoever but results in an unattractive display and usually a loss in value.

225. **The remains of a foreign body,** probably a spider, from shepherds pie.

226, 227 and **228.** **Cooked pork shoulder** of generally poor quality (**226**). The colour is patchy and generally unsatisfactory and lines of fibrous material can be seen effectively dividing the slice into three. Such cooked meats as this are often canned in bulk and removed from the can when required for slicing and, in some cases, pre-packing. Note that despite the possibility of a 'cured meat' colour at point A (arrowed) the product was sold as uncured and curing ingredients were not listed on the label. In contrast the cooked pork shown (**227** and **228**), while of poor colour tending to greyness, shows no evidence of the pinkness associated with cured meats.

Chapter 5

Cured Meat

Curing is one of the earliest forms of preserving meat and is believed to have been first practised in the basin of the Tigris and the Euphrates around 1500 BC. **Nitrite** or its precursor nitrate was originally present adventitiously in the salt (NaCl) used but when its role in flavour and colour production was recognised (and later the anti-microbial role) nitrate and/or nitrite were added separately.

The majority of meat for curing is pig meat and cured meats are produced in almost all non-Islamic or Jewish economies (salt beef being produced in the latter). For this reason pig meat curing forms the bulk of this section. An example of the size of the cured meat (bacon) market is that of the U.K. which in 1980 totalled 515 thousand tonnes. The importance of bacon in international commerce is illustrated by the fact that of this figure 211 thousand tonnes originated in Denmark, 43.3 thousand tonnes in Holland, 29 thousand tonnes in Eire, 12.5 thousand tonnes in Poland and 7.5 thousand tonnes in other countries (Gardner, 1982). Overall the U.K. market for bacon is falling, due partly to change in eating habits and reduced consumption of the traditional cooked breakfast.

Cured Meats and Nitrosamines

In recent years considerable publicity has been given to the possibility of reaction between nitrite and secondary and tertiary amines leading to formation of carcinogenic **nitrosamines**. This has led to the possibility of the use of nitrite (and its precursor nitrate) being restricted or barred (Anon, 1972). However the most significant anti-microbial role of nitrite is the suppression of the extreme food poisoning micro-organism *Clostridium botulinum*. Ingram (1976) pointed out that whereas there are still regular outbreaks of botulism caused by home cured meats no botulism arises from the vast volume of commercially cured meats where curing is carefully controlled. Thus the known risk of botulism must be set against the possible risk of carcinogenicity.

Cured Meats and Mycotoxins

The presence of **mycotoxins** (mould toxins) in certain foods has recently been the cause of much concern to the food industry and has attracted a considerable amount of, in some cases ill founded, publicity. Some cured meats such as certain hams are mould ripened while moulds may grow as spoilage organisms on poorly stored bacon etc. and be wiped off prior to retailing or pre-packing.

Moulds from mould ripened hams such as *Aspergillus versicolor, A. ochraceus* and *Penicillium viridicatum* have been shown to produce the mycotoxins sterigmatocystin and octratoxin A and B in laboratory media although studies of specific mycotoxins in the hams do not appear to have been made (see Jarvis, 1976).

A safeguard against mycotoxin production is to inoculate hams with mould strains which have been demonstrated not to produce mycotoxins rather than to rely on natural contamination. Where mycotoxin production may arise from the growth of spoilage moulds the control must stem from good manufacturing practice, particularly the provision of suitable storage facilities and the rejection of mould affected products.

Technology of Cured Meat Products

Dry Curing

Dry curing was the earliest form of curing and consisted of packing sides or joints into boxes and surrounding with curing salts (NaCl, $NaNO_2$, $NaNO_3$). Dry salting is still used for some hams and a small quantity of bacon is produced in this way. The major disadvantages are high cost and poor distribution of the curing salts leading to localised undercuring and spoilage. However, other parts of dry cured products have a very high salt concentration and bacon produced in this way would be unacceptable to most modern consumers.

Brine Curing

The majority of cured meats today are produced using a brine in which is dissolved the curing salts and possibly other curing adjuncts such as **polyphosphates**. There are fundamentally two types of brine cure although each is subject to considerable variation. The first, the **Wiltshire** cure, is traditionally English and now practised in many parts of the world; the second, the **Sweetcure** (Tendersweet, Canadian cure) is a more modern method which has tended in recent years to gain in popularity at the expense of the Wiltshire cure. Each cure will be discussed in relation to quality and spoilage but it must be emphasised that while the generic term curing is applied to both, the products of the Wiltshire and Sweetcure cures are technically very different and present different problems of quality control.

Wiltshire Cure (229–246)

A flow diagram of Wiltshire bacon manufacture together with the hazards associated with each stage is shown (**229**).

Quality defects associated with the incoming pigs are common to all cures. Animals must be subjected to the same meat inspection procedures as those for the retail fresh meat trade. Particular care is necessary to ensure that pigs are not stressed before slaughter, and slaughterers and butchers should meet the required technical standards.

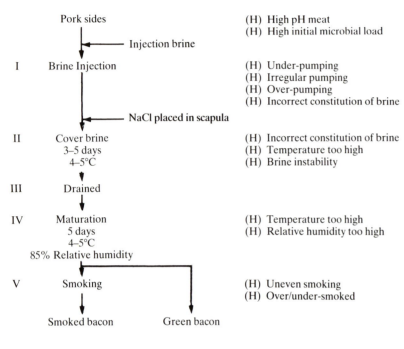

(H) = specific quality hazard

229. A flow diagram for the production of Wiltshire bacon. The process may vary according to the manufacturer particularly with respect to the duration of immersion in the cover brine and maturation.

Brine Injection (Pumping)

Wiltshire bacon is classically produced from sides of pork. Brine is injected into the side to ensure even distribution of curing salts. Control of this stage is of major importance since under-injection (pumping) or uneven injection may lead to various defects while over-pumping can also affect quality. Originally pumping was carried out manually but in recent years multi-needle injections have become almost universal. However, supplementary hand injection is advisable particularly with respect to control of an important quality defect known as **bone taint**, a souring or putre-factive type of spoilage in the deep muscle around the bone or in the bone-marrow itself. Micro-organisms enter the bone marrow at death probably by 'agonal invasion'. Taints may then be produced either by growth of non-halotolerant bacteria before pumping or halotolerant bacteria after pumping. Gardner (1982) noted that micro-organisms can grow in the femur marrow where NaCl levels are low and the pH value high (6.8–7.2) and spread *via* the blood vessels to become established in surrounding tissues.

Bone taints, which may or may not be accompanied by a 'fiery' red ring around the bone (see page 76) were historically attributed to *Clostridium spp* but *Vibrio spp* and *Providencia spp* are now more commonly the causative organisms. (Many bacteria from bacon previously described as *Proteus* are now identified with species of *Providencia;* Penner, 1984.)

Microbial multiplication occurs in injection curing brines particularly with multineedle injections which re-cycle brine. Some of these bacteria such as halophilic *Vibrio spp* can be ultimately involved in the spoilage of bacon but what little evidence is available suggests that the effect of highly contaminated brines on the keeping quality of bacon is minimal (Gardner and Patton, 1978). However, large scale denitrification may occur in injection brines leading to low nitrite levels and instability in the bacon. This problem appears to be of more importance in Sweetcure bacon.

Examples of laboratory tests are shown in Table 13, p. 71.

---

230. A good example of Wiltshire bacon – a gammon joint.

231 and 232. Green discoloration in the fat of Wiltshire bacon (**231**) shown to be a lymph node on close examination (**232**). The node was swollen and such a condition may be found in pigs with sub-acute infections or in animals suffering from lymphosarcoma of viral origin. The dark green pigment should be distinguished from the lime-green pus often associated with *Corynebacterium pyogenes* infections.

233 and 234. A cyst in the fat of a bacon rasher illustrated from above (arrowed) (**233**) and below (arrowed) (**234**). There are a number of possible reasons for the formation of cysts in the flesh of pigs and other animals (congenital cysts may be formed in internal organs). Firstly, they may occur as a result of an injury which, while having healed on the surface, remains as a lesion below the skin. Second, they may form at the site of an injection; third, they may be the result of an infection by, for example, *Corynebacterium*

pyogenes in pigs or *Fusiformis necrophorum* in bovine livers, and fourth, by the cystic stage of tapeworm (in pigs the human tapeworm *Taenia solium*). The latter cause is readily identified after processing but it is less easy to identify the cause in other cases after curing or other processing although in this example a *Corynebacterium* infection is the most probable. Whatever the cause encysted meat should not be sold for human consumption.

235. 'Tiger striping' of Wiltshire bacon rashers. A clear example of uneven distribution of curing ingredients. Tiger striping is usually associated with hand pumping rather than the more modern multi-needle injector.

236 and 237. A Wiltshire gammon rasher showing patchiness of colour (**236**). The close-up (**237**) shows (arrowed) clearly defined demarcation between the pale and dark areas. The cause is normally due to uneven distribution of curing salts possibly as a result of blockage of some of the needles of a multi-needle brine injection.

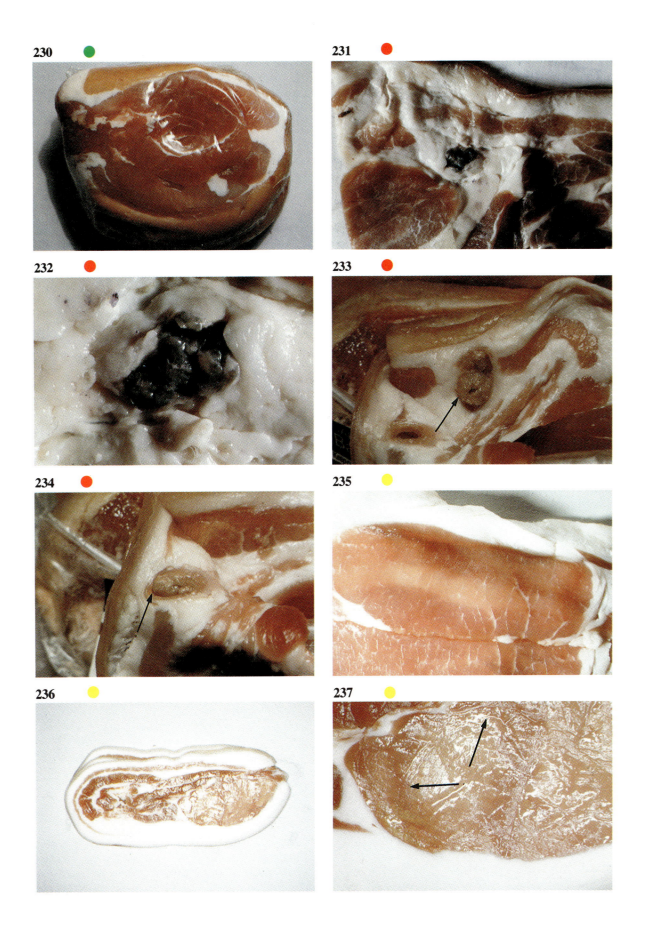

Cover Brine

Immersion in a cover brine is a characteristic feature of the Wiltshire cure although it may be omitted in some modern variants. In general the cover brine is considered to be the heart of the Wiltshire process and cover brines are treated with an almost superstitious regard. The purpose is to cure the shallow unpumped tissues such as rind and belly; to prevent growth of spoilage bacteria during curing and to provide reducing conditions to enhance production of the desired meat pigment **nitrosomyoglobin.**

Cover brines are used repeatedly for many years (allegedly over 100 in some cases) and the maintenance of a microbiologically stable brine is essential for bacon curing. Instability is essentially caused by microbial denitrification exacerbated by poor control of the chemical constitution of the brine and poor temperature control in the curing cellar. Brine microbiology and control is a specialist area and for detailed discussions publications such as Varnam and Grainger, 1973, Gardner, 1973 and Gardner, 1982 should be consulted.

Drainage

Drainage refers to the removal of sides from brine and the draining of excess brine prior to maturation. Failure to drain properly may lead to off-flavours in the bacon.

Maturation

During maturation sides are stacked rind side up on pallets to complete drainage and prevent distortion of the side. Ideally the pallets should be made from anodised aluminium or stainless steel but wooden pallets are still used which may be a major source of contamination. During maturation final equilibration of curing salts occurs and some drying takes place. Temperature and relative humidity should be carefully controlled and failure to do so results in microbial growth and possibly slime formation with adverse effect on the quality of the final product.

Under extreme conditions of total refrigeration failure massive production of purple slime (possibly due to *Chromobacterium*) and red slime (possibly due to yeast or *Serratia marcescens*) has been observed. Illustrations are not available, the slime was hosed off the sides which were then despatched to the retail market!

Smoking

A proportion of Wiltshire bacon sides are smoked. Traditionally smoke is produced from hardwood shavings or sawdust and both **low temperature long time (32°–38°C/15–18 hr)** and **high temperature short time (60°C/2–4 hr)** processes have been used. The low temperature process in particular is difficult to control and quality defects occur due to uneven application of smoke or to under- or over-smoking. In recent years a number of producers have adopted **liquid** smoking where a liquid extract of phenolic compounds from pyrolysed hardwood is sprayed onto the sides with or without heating.

Traditional smoking involved a considerable surface drying with corresponding increase in NaCl concentration and lowering of Aw level and a heavy deposition of anti-microbial phenolic compounds. Such bacon had a high shelf stability and a considerably longer shelf life than its unsmoked (green) counterpart. Many modern smoking processes are less drastic and any anti-microbial effect is offset against the practice of selecting poorer quality bacon for smoking.

---

238 and **239.** **A Wiltshire gammon rasher** showing discoloration in the tail of the rasher particularly the fat (**238**). The close-up (**239**) shows (arrowed) the area of poor colour to be localised in the two long muscles under the rind. The cause is probably a localised area of low curing salt concentration. Some muscles take up curing salts less readily than others (possibly a pH value effect) which leads to a low concentration in the muscles concerned. However, similar effects have been caused by the bacon side not being fully immersed in the cover brine during curing. Note that the above condition is distinct from that due to blockage of injection needles where all muscles in a given area are affected.

240 and **241.** **Holes produced by multi-needle brine injection** seen on a whole Wiltshire bacon rasher (arrowed) (**240**) and in detail (arrowed) (**241**) (N.B. Similar holes may be produced by the retaining spikes on a slicing machine but this rasher was hand sliced and the holes ran deep into the joint). Normally such holes rapidly disappear after pumping. In this example the meat was soft and while of normal colour appeared to be partially in a pale soft exudative (PSE) condition.

242 and **243.** **Break up of Wiltshire bacon** around the rib is shown (arrowed) in general (**242**) and (arrowed) in close-up (**243**). The fault is technological and is probably due to pumping an excessive volume of brine into the bacon prior to curing. This can result in the meat fibres being forced apart and cause separation of the meat from the bones.

244 and **245.** **Wiltshire bacon of very poor quality** showing slime of microbial origin and discoloured patches (arrowed) (**244**). An area of discoloration is shown in detail (**245**). Uneven discoloration may be associated with uneven distribution of curing salts which can be caused by blockage of one or more of the needles on the multi-needle injector. The overall spoilage pattern illustrated is typical for Wiltshire bacon stored at too high temperatures or for excessive periods. The slime formation may or may not be accompanied by off-odours.

246. **Machine dirt** on the rind and sub-rind fat of a rasher of Wiltshire bacon. Machine dirt is normally a mixture of mineral oil or grease (used in maintenance of equipment) and general detritus. With products such as bacon this latter is usually animal fat. In poorly cleaned equipment the machine dirt may be transferred from mechanical parts to surfaces in contact with food.

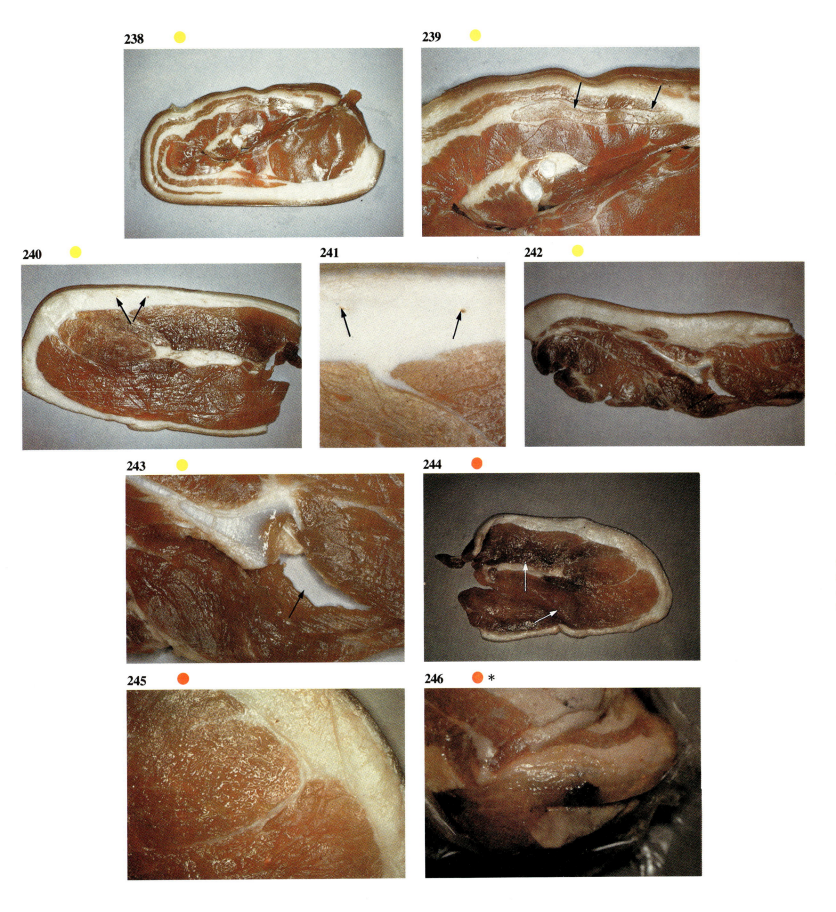

Sweetcure Bacon Manufacture (247–249)

The problems discussed below are specific to Sweetcure bacon. Those common to all cured meat manufacture are discussed earlier (p. 66, Wiltshire cure) although such problems are indicated in the flow diagram relating to Sweetcure bacon manufacture (247). It must be appreciated that by virtue of its nature (low salt, nitrite and containing sucrose or an equivalent carbohydrate) Sweetcure bacon is less shelf stable than Wiltshire and must be treated as a quality control problem in its own right.

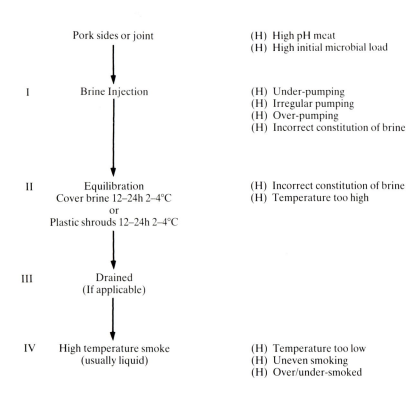

	Pork sides or joint	(H) High pH meat
		(H) High initial microbial load
I	Brine Injection	(H) Under-pumping
		(H) Irregular pumping
		(H) Over-pumping
		(H) Incorrect constitution of brine
II	Equilibration	(H) Incorrect constitution of brine
	Cover brine 12–24h 2–4°C	(H) Temperature too high
	or	
	Plastic shrouds 12–24h 2–4°C	
III	Drained	
	(If applicable)	
IV	High temperature smoke	(H) Temperature too low
	(usually liquid)	(H) Uneven smoking
		(H) Over/under-smoked

(H) = specific quality hazard

247. A flow diagram for the production of Sweetcure bacon. The process is likely to vary according to the manufacturer and details may not be readily available.

Brine Injection

Whole sides are rarely used in production of Sweetcure bacon and cuts for curing may be fully or partially de-boned. The problems of curing salt distribution are considerably less than with Wiltshire cured bacon particularly as tumbling or massaging may be used after pumping to ensure even distribution. Such problems as those discussed for the Wiltshire cure are rare with the Sweetcured product.

The injection brine itself is more prone to microbial problems than the Wiltshire injection brine as a consequence of its having a low concentration of curing salts and a high sugar concentration. Denitrification with attendant loss of curing activity can occur within a working day and the chemical constitution of the brine should be checked at frequent intervals. Halophilic *Vibrio spp* are usually the bacteria responsible.

Equilibration

Equilibration is either by a brief (12–24 hr) immersion in a cover brine or by hanging (usually overnight) in plastic shrouds. Where cover brines are used these are freshly made up although denitrification problems may still occur.

Drainage

Drainage is only required where immersion brines have been employed and is then only for a short period. No specific problems relating to quality are involved.

Smoking

Sweetcure bacon is subject to a high temperature smoking process. Kilns are commonly operated at temperatures above 65°C and thus most vegetative bacteria are killed. Structural proteins are also at least partially de-natured and thus a typical texture is obtained. Liquid smokes are often used.

Correct control of smoking is an important parameter in determining quality.

Cooked Cured Meats (250–258)

Cooked cured meats may be produced from conventionally cured or Sweetcured meat. The latter is widely used in modern processes and it is common to cure the meat in the form of large groups of leg muscle. After injection with brine these are heavily tumbled not only to aid penetration of curing ingredients but also to solubilise the surface proteins of the meat in the brine so that when packed into a mould and heated the pieces of meat are bound together in a protein gel. The ham may also be moulded into any desired shape. Uneven curing of the different pieces of meat leads to patchiness of colour in the product when sliced.

Cooking is carried out in moulds or in sealed cans to an internal heat of 70°–75°C (68°C minimum). This involves heating at ca. 83°C for several hours. It is important to minimise the differential between cooking temperature and product temperature to avoid excessive jelly formation. It is important to note that large hams etc. cooked in cans are not sterile but have a storage life in excess of 6 months at 5°C. **Canned** hams should thus be kept under refrigeration unless the processing is known to be such as to achieve commercial sterility (see Ingram, 1971).

Spoilage of cooked cured meats is usually by contamination at the slicing stage. With vacuum packed sliced products souring by lactic acid bacteria is most common. Other bacteria associated with curing such as halophilic *Vibrio spp* will also grow rapidly on cooked meats and where curing, cooking and pre-packing is carried out on the same premises care must be taken to avoid cross-contamination.

Examples of laboratory tests are shown in Table 13.

Starter Cultures in Ham Curing

Starter cultures of bacteria are sometimes added to ham brines. Initially old brine was added to new (e.g. Buttiaux, 1963) but more recently pure cultures of brine bacteria have been used (e.g. Avagimov *et al*, 1964). Such bacteria are normally *Vibrio spp* which are not only found as part of the normal bacterial flora of ham brines but also as part of the spoilage flora of raw bacon (Gardner, 1982). The heat resistance of these bacteria is low and they are thus readily killed by the cooking of the hams and do not contribute to spoilage.

Cooked Cured Meats and Public Health

The normal saprophytic spoilage microflora of cured meats is absent after cooking and if contamination occurs most pathogenic bacteria can grow providing that the temperature is sufficiently high. *Staphylococcus aureus* is able to develop particularly rapidly as it is selectively favoured by the relatively high salt concentration.

Large scale food poisoning outbreaks have resulted from contaminated slicing machines, one of the more serious being the Aberdeen typhoid epidemic where the slicer was originally contaminated by corned beef. In addition far too many retail outlets ignore the elementary precautions of separate handling procedures for raw and cooked meats.

A number of *Staph. aureus* outbreaks have involved human carriers. Upward of 25% of the healthy population are carriers of *Staph. aureus* and the organism is frequently responsible for infected cuts, boils etc. Typical occasions where food poisoning may occur are buffets at wedding receptions, parties etc. where cooked cured meats are left for long periods at high ambient temperatures.

Table 13   Laboratory Tests Cured Meats

A. Uncooked

I. Microbiological
Total viable count (3%–5% NaCl media)
Halophilic *Vibrio spp* (suspect mis-cure only; see Gardner (1982))

II. Chemical/physical
NaCl
Nitrite
Nitrate (where added)
pH value

Visual examination is useful to determine correct pigment development. Foreign body/infestation examination not normally required.

B. Cooked

I. Microbiological
Total viable count (3%–5% NaCl media)
Group D Streptococci
Staph. aureus
Lactic acid bacteria (vacuum packed only)

II. Chemical/physical
NaCl
Nitrite
Nitrate (where added)
pH value
Water content

Visual examination is useful to determine correct pigment development. Foreign body/infestation examination not normally required.

Notes:
(1) These tests are examples only and are not inclusive.
(2) Examples are for finished products only.
(3) Techniques are detailed in specialist publications such as ICMSF (1978) – Microbiology: Egan *et al* (1981) – Chemistry.
(4) Instrumental techniques such as Impediometry may be substituted for some classical microbiological methods.

248 and 249. **Two good quality examples of Sweetcure (Canadian style) bacon.** 248 illustrates bacon chops and 249 bacon rashers. Note that the colour of Sweetcure bacon is distinct from that of Wiltshire cured (232). This is due to the high temperature smoking of Sweetcure bacon partially de-naturing the protein of the nitrosomyoglobin molecule resulting in partial conversion to nitrosohaemochrome.

250. **A good example of cooked ham sausage** after slicing and pre-packing.

251 and 252 **Additional ingredients** such as black peppercorns are used with cooked cured meats to extend product range and add value to basic products. A good example of ham sausage with black peppercorns is illustrated (251). Despite the consumer having paid for their presence such additives (illustrated in detail, 252) may be mistaken for foreign bodies.

253. **Sliced pre-packed pork lunch tongues.** A good example, where the square shape of the meat is due to the mould into which it is pressed.

254 and 255. **Cured tongue** is particularly prone to discoloration and general quality defects. 254 illustrates a lunch tongue showing darkening round the edge and splitting possibly indicating dehydration, while even lower quality is illustrated (255).

256, 257 and 258. **Metallic foil backings** for cooked cured and delicatessen meats are being increasingly used as a visually attractive alternative to conventional uncoated plastic. The sequence (256, 257 and 258) shows progressive corrosion of such a backing used for cooked ham. The corrosion is due to the salt from the ham attacking the metal foil. The construction of the film is such that the foil is sandwiched between two layers of plastic film. It appears likely that pin holes in the inner layer have permitted contact between the salt of the ham and the metal foil.

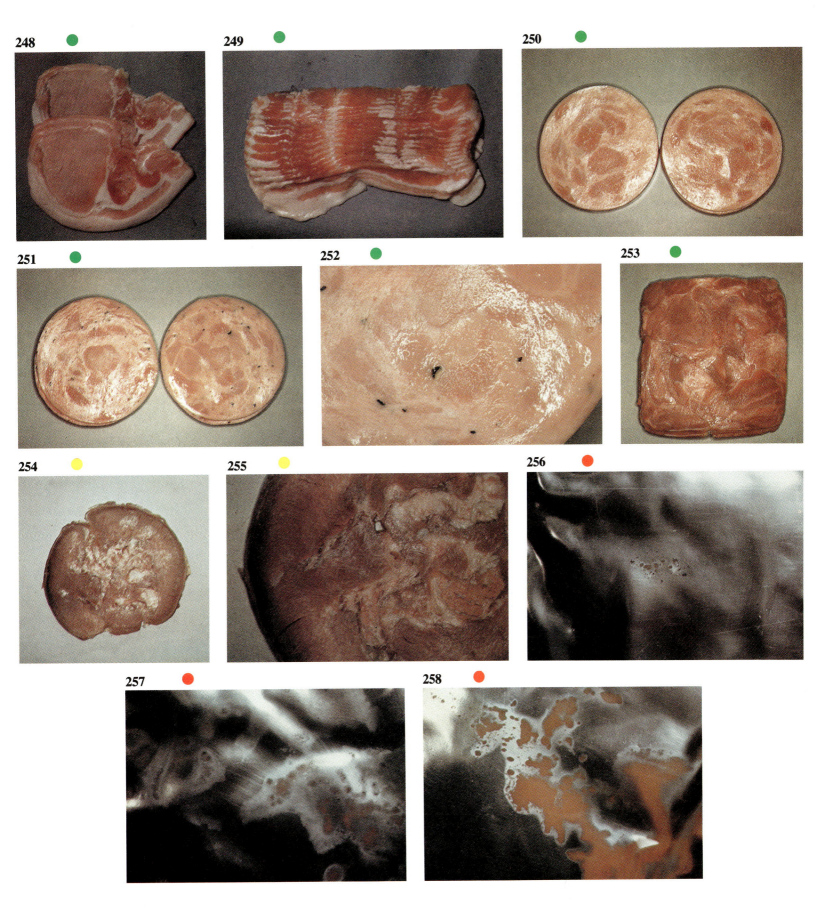

Chapter 6

Delicatessen Meats

Delicatessen meat products comprise an extremely wide variety of commodities, many having a technology in common with cured meats (Chapter 5) and meat products (Chapter 4). They may, however, be grouped into basic product types (Table 14). The delicatessen trade has increased considerably in the U.K. and other countries in recent years. In the U.K., for example, products which were previously restricted to Eastern European immigrants are now on everyday sale.

Delicatessen products present particular problems to persons involved in Food Inspection and Quality Control; not least of these can be knowing what the nature of the product should be. In addition the international nature of much of the trade makes it difficult for a food marketing organisation to exercise control over the transport and storage of products before delivery (see Sutherland and Varnam, 1982) while economic circumstances may force smaller producers to use less than ideal forms of transport, for example, rail parcel and even postal services.

Delicatessen Meats and Mycotoxins

Some delicatessen products such as certain salamis are mould ripened while mould may develop on the surface of other salamis, being wiped off before sale. The possibility of mycotoxicosis has been raised (e.g. Bullerman and Ayres 1968). The considerations are the same as for cured meats as discussed on p. 65.

Delicatessen Meats and Parasitic Infections

As the absence of parasitic infections cannot be guaranteed delicatessen meat products such as salamis which are not heated during manufacture and which are consumed raw should either contain sufficient levels of curing ingredients to ensure safety or should be made from meat which has been held frozen for periods sufficient for the inactivation of parasites (see p. 48 for conditions). The recommendations of the U.S. Department of Agriculture suggest that 3.3% (^{W}/v) NaCl is required to inactivate parasites in raw, unfrozen meat. Most uncooked delicatessen products meet this criterion and have the additional protection of low pH value. Some raw meat products such as the Dutch Ossewurst undergo only a minimal curing process and should be made from frozen meat.

Table 14   Major Types of Delicatessen Meat Products

1. **Cooked fine comminutes** e.g. pâtés and terrines
                                        mortadella sausage
2. **Cooked cured sausages** e.g. Polish sopocka and krajana
                                        German Berliner, Dutch Brabant
3. **Fermented sausages** e.g. salamis
4. **Whole cooked cured products** e.g. katenspek

Cooked Fine Comminutes (259–280)

Cooked fine comminutes are typified by pâtés and sausages like mortadella, and have a technology similar to that of fresh (not canned) luncheon meat, breakfast sausage etc. The cooking process of many pâtés must be particularly carefully balanced between undercooking, causing the survival of vegetative bacteria such as Lancefield Group D Streptococci (Lancefield, 1933), and overcooking which leads to rendering out of the fat and a visually unacceptable product.

Gelatin glazes are often applied to pâtés after cooking and unless correctly prepared and handled (see p. 56) may be a source of both food poisoning and food spoilage micro-organisms. Garnishings such as whole herbs, fruit pieces, raw mushrooms etc. are sometimes added post-cooking to the top of bowl pâtés. These may not be sterilised and are thus an important source of yeast, mould and bacteria.

The vast majority of cooked comminutes contain nitrite, the input level being intended to inhibit bacterial endospore germination (Gould, 1964). Technical control in what is in some cases a 'cottage industry' is often inadequate and the level of added nitrite may vary from the inadequate to the grossly excessive.

Cooked Fine Comminutes and Public Health

Despite the obvious risk of food poisoning from germination and outgrowth of endospores of heat resistant pathogens such as *Clostridium botulinum* and *Cl. perfringens* in poorly cooled and refrigerated pâtés such incidents are rare. Most food poisoning associated with pâté is a consequence of post-processing contamination and the majority of incidents involve either *Staphylococcus aureus* or *Salmonella*.

Of particular concern is the handling of product in service delicatessens and in catering outlets where staff do not adhere to good handling practices. It is fortunate that in many cases contamination involves non-pathogenic lactic acid bacteria which sour the product and overgrow pathogens. In this context it has been stated (Mossel, D.A.A. personal comment to the authors) that the average British person eating pâtés solely in restaurants expects the harsh lactic acid flavour which would be considered to be spoilage elsewhere.

Examples of laboratory tests are shown in Table 15, p. 83.

259 ●

260 ●

261 ●

262 ●

263 ●

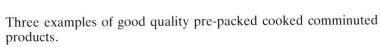

264 ●
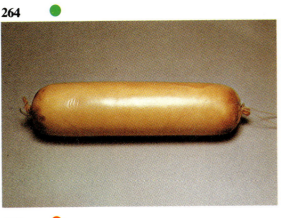

265 ●
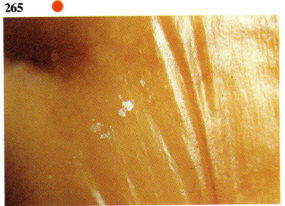

Three examples of good quality pre-packed cooked comminuted products.

259. French pâté d'Ardennes. A fine chopped comminute encased in fat.

260. Italian Mortadella. A fine comminute containing large pieces of fat.

261. French brawn. Coarsely chopped ingredients in a gelatin matrix.

262 and **263. Yeast colonies** developing on the fat of pâté are illustrated (**262**) (arrowed). Unaffected pâté is shown as a contrast (**263**).

264 and **265. Liver sausage.** The centre part of a good example of a liver sausage stick showing the absence of discoloration or yeast and mould growth (**264**). Unlike salami (see p. 80) liver sausage is a highly perishable product of relatively short life. The growth of yeast or mould colonies on the surface (**265**) is indicative of poor temperature control during storage or of excessive storage and is considered as spoilage.

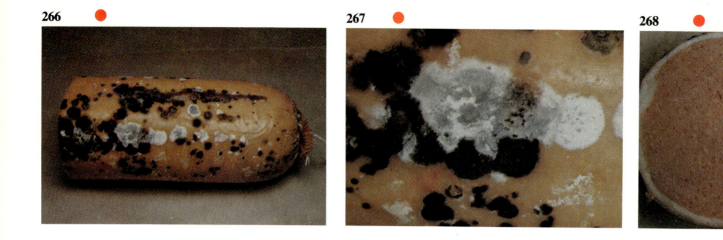

266   **267**   **268**

266 to **269.**   **Dutch liver sausage** showing severe attack by different species of mould (**266**). Mould growth of this magnitude can be attributed only to **temperature abuse** probably coupled with a **high humidity** during storage. (It should be noted that in the U.K. at least liver sausage sticks are often mistaken for, and treated as, salamis.) At least two types of mould colonies are seen (**267**); the blue are *Penicillium spp* and the black *Aspergillus spp*. The degree of penetration by the mould varies. **268** shows penetration to be superficial whereas elsewhere on the sausage the fat is fully penetrated and the meat is becoming affected (**269**).

270, 271 and **272.**   **In this example of liver sausage** the upper end of the stick (loop end) is free from visible yeast colonies (**270**) although there is some discoloration. Yeast colonies have developed at the lower end (**271**) where (arrowed) the weight of the contents have caused restricted leakage onto the impermeable skin of the sausage. These are illustrated in close-up (**272**) and are seen to be distinct from fat seepage.

273 and **274.**   **A cross section of Belgian liver sausage** showing cracking or splitting (**273**). The cause is not known but may be dehydration which appears to work from the ends inward since there is little splitting in the middle of the stick but much at the ends (see **274**).

274.   **Lines of splitting in Belgian liver sausage stick.**

275.   **Liver sausage in cross section** showing a 'fiery' red interior. The coloration is due to bacteria successfully competing with the pigments for oxygen resulting in an excess of the reduced pigment nitrosomyoglobin (Gardner, 1983).

276 and **277.**   **Gelatin glazes** are a common source of microbial contamination both with respect to food poisoning and to food spoilage micro-organisms. Gelatin also serves as a good growth substrate for bacteria derived from other ingredients. **276** shows colonies growing in the gelatin glaze on top of a pâté. The detailed illustration (**277**) shows a very large number of colonies to be present. Gelatin solutions should be held at temperatures above 70°C prior to use to prevent problems of a microbiological nature.

278, 279 and **280.**   **Brawn** consists of discrete pieces of cured meat (and sometimes other ingredients such as peppers) in a gelatin matrix. The example shown (**278**) has been sliced and vacuum packed. The bottom of the pack (arrowed) shows the slices to have coalesced and the gelatin to have collected in the base, illustrated in greater detail (**279**). Gelatin melts at temperatures above 22°–25°C and it is apparent that the pack has been subjected to very high storage temperatures at some stage thus melting the gelatin which re-solidified on subsequent cooling. The presence of gas bubbles (arrowed) (**280**) is indicative of microbial gas production during the period of high temperature storage and the pack had a distinctly sour smell on opening.

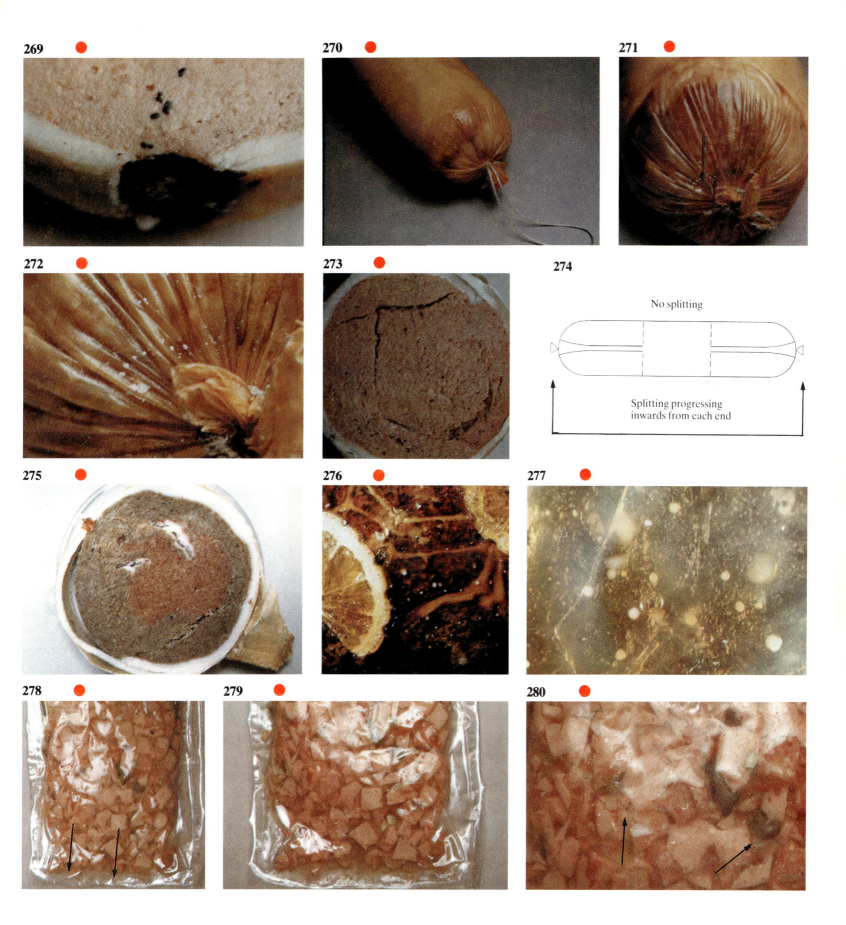

269

270

271

272

273

274

No splitting

Splitting progressing
inwards from each end

275

276

277

278

279

280

Cooked Cured Sausages (281–291)

Examples of cooked cured sausages are Polish **Sopocka** and **Krajana,** German **Berliner** and Dutch **Brabant.** Similar products are native to other parts of the world including China. The appearance of the products varies considerably although the basic technology is similar. The cooking process normally involves a considerable degree of drying with consequent raising of the salt concentration and lowering of the Aw level as well as incorporating a traditional heavy smoke. The products are thus highly stable and spoilage is usually by surface growth of moulds.

It should be noted that before cooking they may undergo a lactic fermentation (see p. 80) either by lactic acid bacteria present adventitiously or by the addition of pure cultures of lactic acid bacteria (e.g. Petäjä *et al*, 1972).

Examples of laboratory tests are shown in Table 15, p. 83.

281, 282 and **283.** **Cooked and coarsely comminuted sausages** are produced in many parts of the world. These are produced in links in an edible casing. **281** illustrates **Chorizos Rosario** sausages, a Spanish product, and (**282**) their Chinese equivalent. The latter is typically a coarse chop with large pieces of fat. Large pieces of fat are also present in Chorizos Rosario (**283**). Note the highly coloured meat due to the addition of a colouring agent.

Such products are highly stable microbiologically, spoilage ultimately being by surface growth of yeast and moulds.

284 and **285.** **Mysliwska** is a Polish cooked cured sausage. **284** shows two good quality examples. On close examination (**285**) white areas (arrowed) are seen which are due to fat being rendered through the permeable casing of the sausage and not to yeast colonies for which they are sometimes mistaken. Excessive surface fat is considered to be a quality defect.

286 and **287.** **Pre-packed cooked continental sausage** showing swelling of the pack and disruption of the meat due to microbial gas production (**286**). The condition of the meat is illustrated in detail (**287**). Note the adherence of meat to the film, the fading of colour due to acid production and fall in pH value and the 'sweaty'

appearance. The causative organism was a heterofermentative *Lactobacillus sp* and contamination probably occurred at the slicing stage. Heterofermentative lactobacilli cause spoilage of other cooked pre-packed meat products including pâtés. In pâtés not containing sodium nitrate a mutualistic relationship between heterofermentative lactobacilli and yeast may occur.

288 to **291.** **Some ethnic and religious communities,** notably the Jewish and Islamic, insist on food prepared with strictly laid down religious guidelines. Until fairly recently such food has been sold only through small shops serving the local community or has been prepared at home. In the case of Jewish foods in particular retailing is now on a much larger scale and modern packaging methods must be used to permit the wider distribution. **288** shows the traditional **Kneidlach** (a suet ball for addition to soups and stews) in a vacuum pack while the labelling (**289**) indicates (arrowed) the supervision of the London Beth Din, the Jewish organisation responsible for ensuring that the provisions of the religious code are met. Such codes extend to ingredients; **Vienna sausage (290),** traditionally a cooked cured pork sausage, has beef substituted for pork (arrowed) in the ingredient list of kosher Viennas (**291**).

281

282

283

285

284

288

KNEIDLACH

286

287

INGREDIENTS: MATZO MEAL, EGG,
EDIBLE FAT, SALT & PEPPER.
227 g 8 oz
TO OBTAIN BEST RESULTS HEAT IN SOUP FOR APPROX. 10 MINUTES.
KEEP UNDER REFRIGERATION

Kneidlach

290

289

INGREDIENTS: MATZO MEAL, EGG,
EDIBLE FAT, SALT & PEPPER.
227 g 8
O OBTAIN BEST RESULTS HEAT IN SOUP FOR APPROX. 10 MIN
KEEP UNDER REFRIGERATION

under the supervision of the London Beth Din

291

INGREDIENTS: BEEF, CEREAL BINDER, SALT, EMULSIFYING SALT, SPICES, MONOSODIUM
GLUTAMATE, PRESERVATIVE: POTA...
SIMMER IN BOILED WATER FOR 5 MINUTES
KEEP UNDER REFRIGERATION
Manufactured under the supervision of the London Beth Din
net weight 8 oz 227 g

Fermented Sausages (292–306)

Fermented sausages are typified by fully dried products such as **salamis** and semi-dried such as **cervelat**. Some care is needed in defining product type in that some salamis, notably American and Canadian, are not made using traditional techniques. Starter cultures are not used, heat treatment is applied and the products have a considerably higher water activity level than do traditional salamis.

Traditional salamis are produced using starter cultures of lactic acid bacteria. These are available commercially and usually consist of lactic (Lancefield Group N) streptococci although pediococci may be used either singly or in conjunction with streptococci (note that the terminology used in describing starter organisms is not necessarily correct in the strict taxonomic sense). The finished products have a pH value of ca. 5.0 and water activity levels as low as 0.6 in some Hungarian salamis, but more typically in the range of 0.7–0.8. They are thus highly stable and only rarely spoilt by micro-organisms other than moulds.

Although starter organisms are not responsible for spoilage *per se* a declining population in a salami stored for an excessive period often produces excessively acid or bitter flavours which lower the organoleptic quality.

Examples of laboratory tests are shown in Table 15, p. 83.

292 and 293. Discoloration of sliced pre-packed salami type sausage due to leaking vacuum packaging illustrated by **cervelat (292)** and **German salami (293)**. In each example the sausage on the right comes from a sound pack whereas that on the left (arrowed) from a pack which has leaked. The greyish colour is due to the oxidation of the cured meat pigment nitrosomyoglobin to metmyoglobin.

Such an appearance would be normal in non-vacuum packed sliced salamis but is considered undesirable where vacuum packing is employed. *Circa* 1% of all vacuum packs leak and this is a cause of considerable economic loss due to reduced shelf life and loss of visual quality. Salami type products may present particular difficulties if fat from the sausage contaminates the film in the area of the heat seal. Under such conditions an improper seal which subsequently admits air may be formed. Other causes include pinholes in the film and physical damage to the pack.

294 and 295. Discoloration of salami due to oxidation of the meat pigment. **294** shows a slice cut from the centre of a stick of salami. A small amount of brown pigment is seen around the exterior of the slice while the interior retains the normal pink colour. In contrast **295** shows a slice cut from the end of the same salami stick where the pigment is uniformly brown. The brown discoloration is due to oxidation of the meat pigments on continuous exposure to air.

296. A slice of Hungarian salami showing browning around the edge due to oxidation of meat pigments.

297, 298 and 299. Darkening and fat accumulation in a slice of Danish salami which had been hung as a cut stick for two days at room temperature (**297**). Darkening is due to oxidation of the meat pigments by exposure to air while fat has migrated down the sausage to the cut end (see **298**). In contrast a slice taken from the interior of the salami has retained the normal pink coloration and does not show excessive fat accumulation (**299**). In practice the end 2cm–3cm of the stick should be cut off and discarded before continuing to sell the salami, but this may not happen at non specialist outlets unused to handling delicatessen products.

300, 301 and 302. Hungarian salami of the type illustrated (**300**) is not intentionally mould ripened. However, as in all products of this type, moulds may develop during the long period during which the salami is hung in cellars to mature. The extent of mould development in this example is not excessive (**301**) and a cross section (**302**) shows no evidence of the mould hyphae penetrating the interior.

Surface mould growth may be reduced by the use of synthetic casings which are relatively impermeable to moisture and thus prevent the diffusion of nutrients to the surface of the sausage. In most cases the more permeable 'natural' coverings are still preferred. The sausage illustrated would be considered fit for sale, a food grade mineral oil being used to remove the mould from the surface. The presence of mould inevitably raises the question of mycotoxin production.

303 and 304. Mould colonies (A) and yeast colonies (B) in the outer synthetic skin of **Italian salami** hung at room temperature in a humid environment (**303**). Removal of the outer skin (**304**) shows the inner 'natural' skin to be free from contamination.

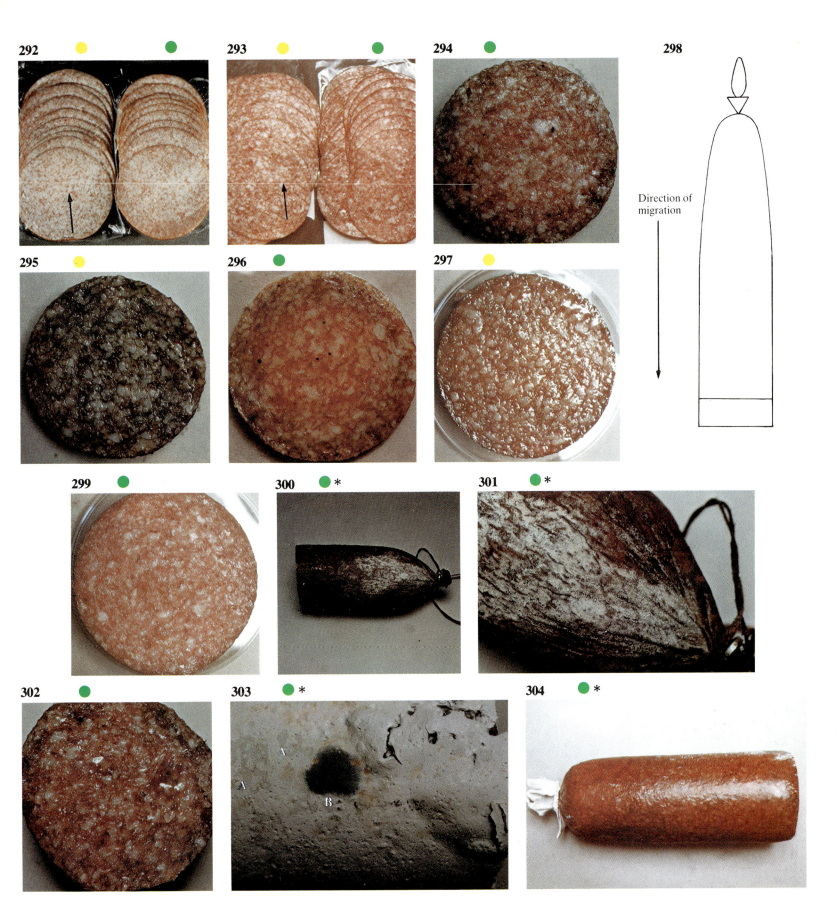

292

293

294

298

Direction of
migration

295

296

297

299

300 *

301 *

302

303 *

A

A

B

304 *

81

305

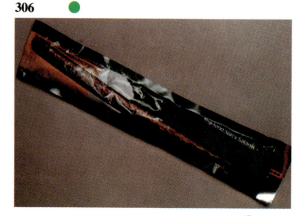

306

305 and **306.** Most snack **products** in Western Europe and the U.S.A. are carbohydrate based. This product is an exception in being a highly spiced salami. Stability at room temperature is achieved by means of a low water activity level (less than 0.7) this being maintained by the highly water impermeable foil pack (**306**).

307

Whole Cooked Cured Products (307–309)

Such delicatessen products are analogous to **cooked ham** and typical examples are German **katenspek** and Polish **bozchek**. Such products often have a very high ratio of fat to lean meat and while highly stable in a microbiological sense may lose organoleptic quality due to oxidative rancidity of the fat.

Examples of laboratory tests, are shown in Table 13B, p. 71.

308

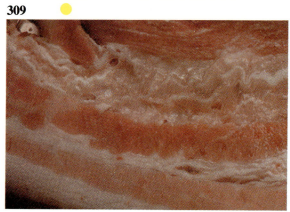

309

307. **Mould ripened cooked Italian hams.** The moulds developing are not cultured but are derived adventitiously in the cellars used for maturation of the hams. Consumption of similar products has been associated with stomach cancer in a nearby region of Yugoslavia possibly as a result of mycotoxin production. (Similar hams such as Parma ham are usually eaten raw.)

308 and **309.** **Bozchek** is a heavily smoked Polish cooked bacon. Similar products are produced in many Central and Eastern European nations (e.g. German **katenspek**). The fat to lean ratio is high, a factor favoured in producing countries but unsuited to modern Western European tastes. It should be noted that much pre-war bacon produced in the U.K. had extremely high fat to lean ratios. The rashers of bozchek illustrated (**308**) are of general poor quality, a consequence of excessive storage at the retail outlet. Microbial growth on the meat is restricted by the high salt concentration, >10% ($^w/v$), but deterioration of the fat is occurring (**309**).

Table 15 Laboratory Tests Delicatessen

A. Cooked Comminutes

I. Microbiological
Total viable count
Group D Streptococci
Clostridium spp
Staph. aureus
Lactic acid bacteria (vacuum packed only)

When sold from bulk, analysis for *Enterobacteriaceae* including *Escherichia coli* may be useful in determining hygiene standards at point of sale.

II. Chemical/physical
Meat content
NaCl
Nitrite
Nitrate (where added)

Foreign body infestation examination not normally required.

B. Fermented Sausages

I. Microbiological
Total viable count. (A large number of starter organisms are usually present. In old product numbers fall and bitter flavours may be detectable. Non-starter contaminants may be distinguished by colonial morphology and are usually catalase positive.)

A visual examination is useful to determine the extent of any mould growth (see pp. 80–81).

II. Chemical/physical
Meat content
NaCl
Nitrite
Nitrate (where added)
pH value
Aw level

Foreign body/infestation examination not normally required.

C. Cooked Cured Products

See Table 13B. Cured meats. Cooked

Notes:
1. These tests are examples only and are not inclusive.
2. Examples are for finished products only.
3. Techniques are detailed in specialist publications such as ICMSF (1978) – Microbiology: Egan *et al* (1981) – Chemistry.
4. Instrumental techniques such as Impediometry may be substituted for some classical microbiological methods.

Chapter 7

Poultry

Since the end of the second world war most of the poultry, particularly chickens and turkeys, in industrialised countries has been produced using large scale agricultural and processing techniques. As a consequence the relative cost of poultry has fallen markedly in relation to other meats and what was once considered to be a luxury item is now a cheap staple food. Consumption of chicken in the U.K. during 1981 was, for example, higher than that of beef or other meats and although turkey and duck are more of a seasonal dish overall consumption has risen markedly (figures were 460 million chicken; 23 million turkeys and 8 million ducks). A considerable business has developed with a degree of vertical integration between feed suppliers, breeders, rearers and processors. In the U.K. the industry has suffered falling profit margins due to over-production, and a range of 'further processed' products such as turkey steaks, flavoured cooked chicken and turkey sausages have been developed.

In consideration of quality it must be stated that mass produced poultry is generally considered inferior to the traditionally reared bird which is hung uneviscerated (New York dressed). There is no doubt that the mass produced bird is bland compared with the traditional type but to a large extent consumer taste is dictated by economics, the New York dressed bird being considerably more expensive.

Poultry and Food Poisoning

For many years poultry, particularly duck, has been recognised as a source of *Salmonella*. The introduction of mass processing techniques exacerbated the situation and the incidence of *Salmonella* in poultry has been of considerable concern to Governments and the industry. More recently poultry and poultry products have been implicated as an important source of the newly recognised enteric pathogen *Campylobacter* (see Grant *et al*, 1980). It is not within the scope of the present work to discuss these problems in detail but it is pertinent to make a number of general observations.

With regard to *Salmonella* it is important to realise that while there is undoubtedly spread of *Salmonella* at the processing plant (see below) and that the spread must be minimised, *Salmonella* often do not originate in the processing plant. Thus *Salmonella* control must be a **total operation** involving **elimination from both the breeding and rearing flocks.** This in turn means eliminating *Salmonella* from feed and from the environment such as bedding, dust etc.

A recent development has been the *Salmonella* exclusion principle where chicks are fed cultures of bacteria derived from the gut of adult chickens in order to establish rapidly an adult gut microflora and thus prevent the establishment of *Salmonella* (Pivnick and Nurmi, 1982). In other countries the use of radurisation (gamma radiation pasteurisation) is recommended to remove *Salmonella* from the processed product. The use of gamma radiation is an evocative topic and its introduction for poultry processing is likely to be controversial from both scientific and socio-economic viewpoints.

Ultimately, prevention of poultry borne salmonellosis must lie in safe handling of poultry after processing. *Salmonella* is not heat resistant and will not survive adequate cooking. The majority of cases therefore involve cross contamination. It is essential in plants producing cooked poultry that there is **total separation** between the raw and cooked production areas. Equally caterers and the home cook must be aware of, and avoid, the possibility of cross contamination.

The situation with *Campylobacter* is somewhat different in that this organism has only recently been associated with food borne disease and that its aetiology is not fully understood. As far as is known only raw milk and undercooked chicken have been definitely associated with campylobacteriosis (Roberts, 1982) and as with *Salmonella* the organism will not survive cooking. Good handling of processed birds will obviate potential *Campylobacter* food poisoning.

Examples of laboratory tests are shown in Table 16, p. 98.

Chicken and Turkeys (310–362)

The processing of chickens and turkeys for sale as oven ready is represented by **310**.

It must be appreciated that the details will vary from plant to plant and that subsequent illustration of equipment is indicative only. Each stage is discussed separately with particular reference to faults which may arise at that particular stage. For a full discussion of microbial cross contamination, particularly with regard to *Salmonella*, reference should be made to a specialist publication such as Mead (1982). Note that 2%–3% of birds are damaged causing total loss or downgrading during processing.

Stages in Processing

Procurement of Birds

Birds are transported in crates from the rearing unit. It is common practice to starve birds for several hours to minimise faecal contamination. There is evidence that stress may cause increased faecal contamination while physical damage such as broken legs and wings can occur either when birds are being crated or in transfer to the processing line. Since birds naturally struggle at this stage airborne microbial contamination may occur.

N.B. Birds that are mechanically damaged during processing are not usually offered for retail sale. However, except where indicated, the meat is suitable for human consumption and damaged birds are used for catering or for further processing.

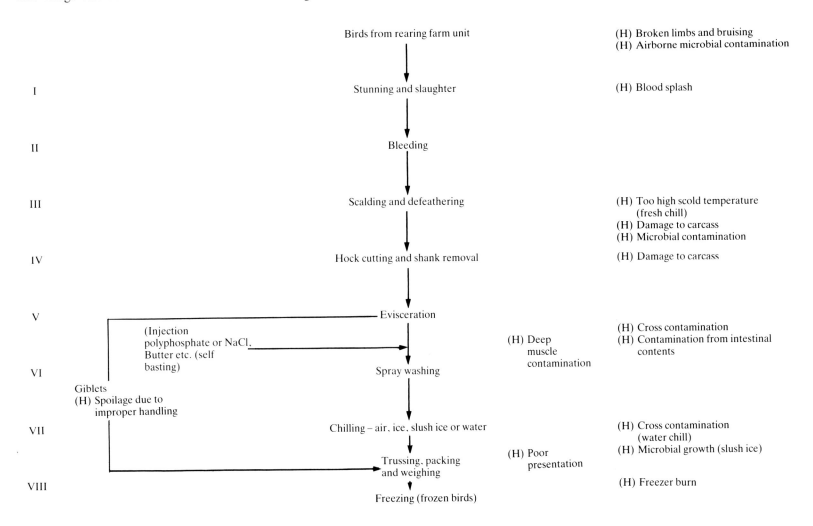

N.B. Microbial cross-contamination refers to spoilage organisms; cross-contamination by *Salmonella* may also occur at these and possibly other points.

(H) = specific quality hazard

310. A flow diagram for the production of oven-ready chickens or turkeys. Large scale processing of ducks is basically similar but treatment with molten wax at *ca* 80–120°C is necessary to ensure complete removal of feathers.

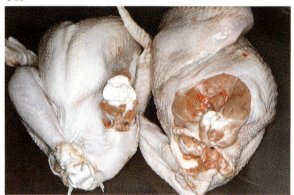

Stunning and Slaughtering

Birds are stunned (usually electrically) prior to slaughter. Stunning occasionally produces 'blood splash' due to rupture of surface blood vessels but this is rare in poultry.

Bleeding

Birds are bled immediately after slaughter. Birds for kosher (Jewish) consumption are killed and bled simultaneously by severing the trachea.

Scalding and Defeathering

Scalding is necessary to loosen the feathers prior to defeathering. Tanks of circulating hot water are most commonly used although

alternative systems using sprays or steam have been investigated (Klose, 1974). A 'hard' scald at ca. 60°C is used for poultry intended for freezing and a 'soft' scald at ca. 50°C for birds to be sold in the 'fresh chill' state. Too high a temperature damages the cuticle and is a quality fault in fresh chill birds. *Salmonella* may accumulate in the scald tank particularly at lower temperatures.

Defeathering is almost invariably mechanical and a series of rubber flails is normally used. There is considerable spread of aerial contamination at this stage and this part of the processing plant should be separated from cleaner areas. The flails themselves may transfer micro-organisms from one carcass to another while mechanical damage to the carcass may also occur.

311. Livestock unloading platform. The birds are attached to the trusses travelling on a continuous loop (arrowed). Physical damage of up to 2% of birds may occur at this stage particularly if they are in an excited state. Note in the background purpose built trailers for carrying crated birds. The crates can be an important source of faecal cross contamination and must be sanitised between journeys.

312. Turkey with a broken leg. The bruising indicates that the bird had not been bled thus damage had occurred prior to slaughter. Normally such birds are detected by inspectors and the limb removed.

313. Two turkeys with leg and wing removed following pre-slaughter damage. Such birds are downgraded (Grade C in U.K.) and sold for catering etc. at a reduced price.

314. An electrical stunner. Birds pass through the stunner on the trusses and after stunning move on to be slaughtered.

315. Scald tank. The circulating impeller may be seen (arrowed) at the end of the tank. The birds remain on trusses and it is worth noting that any *Salmonella* present on the feet survive since the feet do not enter the scald water (Patterson, 1982).

316. Peeling of skin (arrowed) of a turkey due to excessive time in the scald tank. This may have been due to the bird having fallen from the truss or to line stoppage in which case several birds could be similarly affected.

317 and 318. A flail type mechanical defeatherer (plucker) opened for cleaning (**317**). The banks of rubber 'fingers' flail the birds as they pass through the trusses. Organic matter inevitably collects on the flails and in the warm, wet conditions prevailing microbial growth occurs. The flails themselves are serrated and difficult to clean and sanitise (**318**). They can thus be responsible for carry over of *Salmonella* from one processing period to another.

319 and 320. Mechanical damage to birds may occur at the defeathering stage particularly if birds are in the plucker for an excessive period. **319** and **320** illustrate tearing of skin and flesh during defeathering.

321 and 322. Incomplete defeathering shown in general (**321**) and close-up (**322**). This is a quality defect which would lead to the birds being downgraded. Too low a scald temperature or poorly adjusted flails may be responsible. Note that the bird is also severely bruised which in itself would result in downgrading. The cause of bruising is not known in this example.

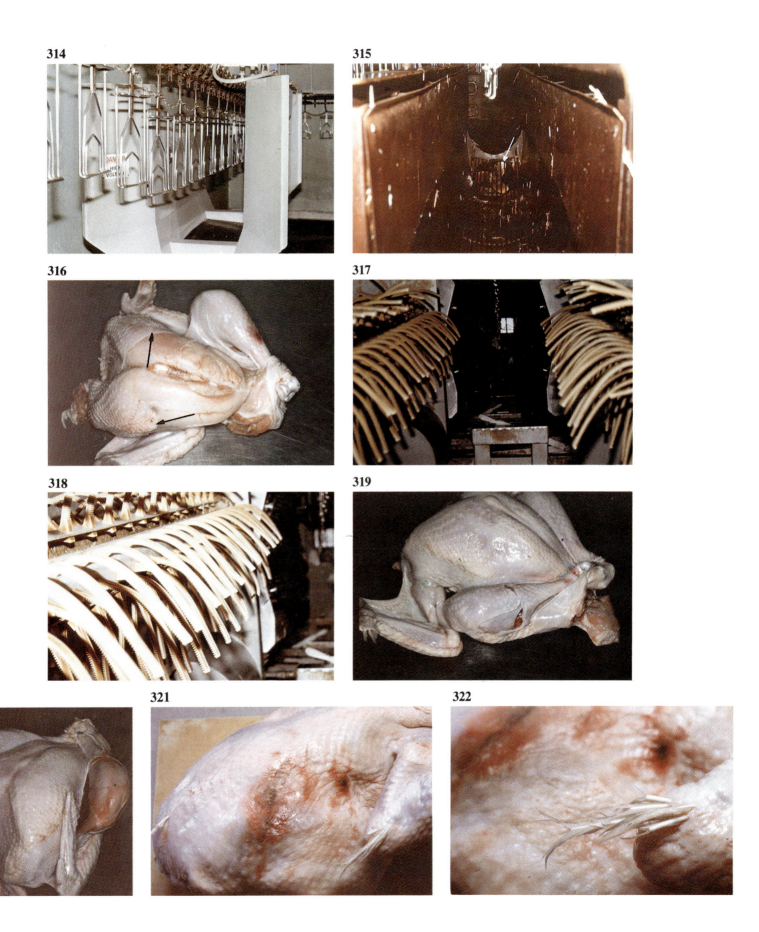

Hock Cutting and Shank Removal

This stage occurs directly before evisceration. The operation is usually manual but mechanical sinew pullers may be used.

Evisceration

Although mechanical evisceration is used at some plants the operation is normally carried out manually in several stages.

Cross contamination by spoilage and by potentially pathogenic bacteria readily occurs at this stage. This affects the outside of the carcass and the body cavity which is less easily cleaned in subsequent stages: Incomplete evisceration can lead both to rapid spoilage and to taints detectable on consumption.

In some cases the freshly eviscerated bird is injected either with a solution of polyphosphates or a brine (sodium chloride) solution. The former practice in particular results in a weight gain (effectively added water) and has been the subject of much critical opinion although processors claim it produces a more succulent bird of higher quality. More recently processes involving injection of butter, oil or a mixture of fats have been developed to produce **self basting** birds which now command a considerable share of the market. Any form of injection into poultry raises the possibility of introducing deep muscle contamination either by transfer of the micro-organisms from the skin or by contamination of the substance being injected. In practice however such problems appear to be minimal.

Spray Washing

Spray washing is an important stage and should be carried out as soon as possible after evisceration to prevent attachment of bacterial contaminants to the skin. Even efficient washing brings about no more than a ten-fold (1 log cycle) reduction in bacterial numbers (Mead, 1982). It should be noted that in the U.S.A. it is a statutory requirement to spray wash so that the carcass 'is wholesome and ready to cook' before chilling (Brant, 1973).

Chilling

Chilling must be carried out promptly after washing; the deep muscle temperature after washing is ca. 30°C and if delayed both spoilage and food poisoning bacteria can develop. In extreme cases of delay spoilage may be initiated before chilling, which is usually carried out either by cold air or by chilled water. The latter system has been the subject of much controversy within the European Economic Community because *Salmonella* and other micro-organisms may build up in improperly operated water chillers. (Report, 1976). The use of water chilling in EEC member countries is controlled by a strict Code of Practice.

In some small plants carcasses are cooled overnight in static tanks of slush ice. This process is often also used as a secondary chilling of turkeys to 'age' the bird and improve eating quality. Psychrotrophic spoilage bacteria may increase ca. 5 fold during such chilling (Barnes and Shrimpton, 1968).

Where packaging does not follow immediately on chilling, birds should be held in a cold store even if the delay is no more than an hour. Some processors prefer to hold overnight at 0°–1°C to dry the bird and to complete the chilling. Longer holding of unwrapped birds is likely to lead to excessive surface dehydration.

Trussing and Packing

To some extent trussing and packing of birds is technically straightforward. It is however an important stage in that it affects the final presentation and hence what the consumer perceives as visual quality, and it is often the final stage at which faults are detected. The type of packaging used depends on the type of bird and whether frozen or fresh chill. Solid carbon dioxide 'snow' may be applied to fresh chill packaged birds as a secondary coolant. This crust freezes the surface and can extend storage life by 1 to 2 days.

---

323 and **324.** **Mechanical sinew puller** in general view and in detail.

325. **A turkey with leg broken** by mispresentation to a sinew puller. Note that there is no bleeding and such damage may therefore be identified as post-slaughter.

326. **The interior of a spray washer** with spray nozzles (arrowed) readily seen. The position of the nozzles is fixed and this makes cleaning of turkeys, which vary widely in size, difficult.

327. **Water chilling (Spin Chiller).** Birds are carried through a tank of water at an entry temperature of <4°C by the rotating vanes which also serve to immerse the birds. In accordance with EEC regulations this chiller is of counterflow type where birds leaving the chiller meet the cleanest (and coldest) water entry.

328. **Mechanical damage to a turkey** caused by the vanes of a spin chiller. Such damage is a consequence of careless loading of the birds into the chiller.

329. **Turkeys in a cold store awaiting packing.**

330. **Where birds are sold with giblets** the placement of these into the body cavity or in the neck of the bird is the first stage of packaging. Giblets usually carry a heavy microbial load and must be handled with as much care as the whole bird. Where excessive numbers of giblets are allowed to stand at room temperature during packaging rapid spoilage may be initiated with development of off-odours during storage.

Giblets are also a common cause of unjustified complaints in that inexperienced cooks neglect to remove them from the bird before cooking with a resultant strong gamey taste on eating.

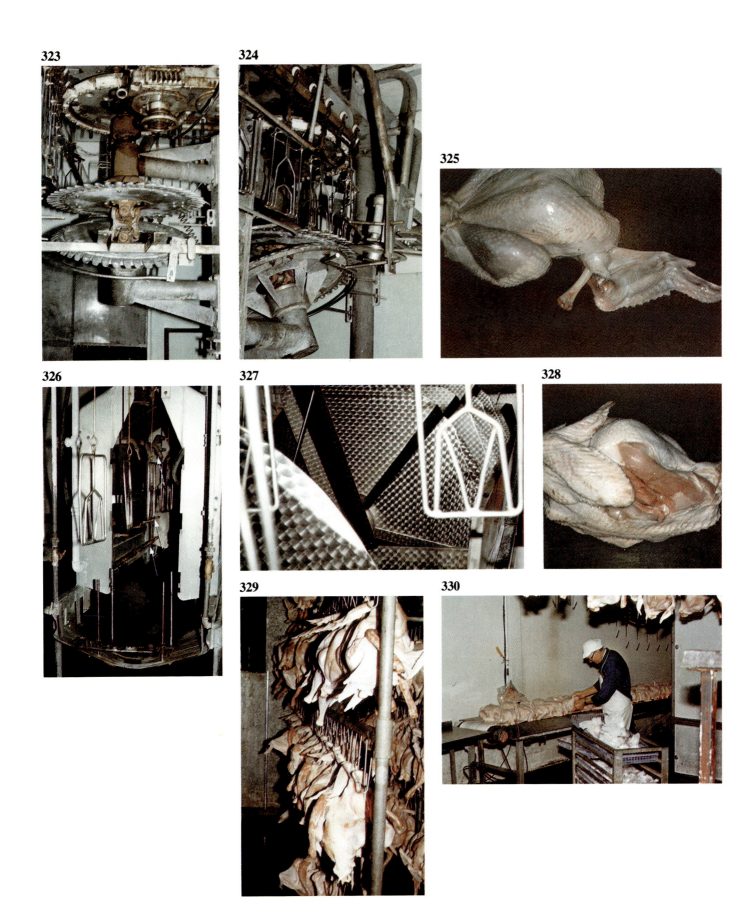

323

324

325

326

327

328

329

330

Freezing

Freezing (where applicable) may be on a continuous basis in a calcium chloride brine or by batch in a blast freezer. The latter is usual with chickens but either method may be used for turkeys.

331. Bagging of turkeys for freezing. Note the operators' correctly worn protective wear and in the background a conveniently placed handwashing station with liquid soap dispenser, pedal operated water tap and paper towel holder. A separate sink for knives etc. is adjacent to the handwashing sink.

332 and 333. Frozen turkeys are usually vacuum packed in a high density film (barrier bag) (**332**). The air is evacuated by a 'snorkel device' (arrowed) and the bag clip sealed. The equipment is seen in use (**333**). During the evacuation the turkey is 'plumped' up to give an attractive configuration.

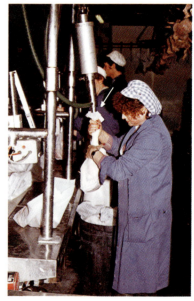

334 to 337. After evacuation and sealing a hot water tank (**334**) is used to shrink the film to the bird (**335**). This gives the classical conformation of a frozen turkey. **336** shows a turkey before·heat shrinking of the film and **337** the same turkey after heat shrinking.

Fresh chill and frozen chickens and fresh chill turkeys are normally packed in non-evacuated nylon polythene bags and heat shrinking is not applicable.

338. Transfer of turkeys to a brine freezer. Care must be taken not to damage the bag which would permit entry of the highly corrosive calcium chloride brine.

339. Turkeys entering an immersion freezer. The birds are carried through the tank of circulating brine by a rotating mesh seen on the left of the illustration.

340. The interior of a blast freezer. In some cases such installations serve both as a secondary freezer and store. Normal holding freezer stores do not have the refrigeration capacity to operate as freezers.

341, 342 and 343. General mechanical damage. In addition to damage which may be attributed directly to given stages in processing general mechanical damage also occurs and can account for up to 0.5% total throughput. Three examples are illustrated.

344 and 345. Freezer burn of chicken drumsticks. The cause is usually excessively long storage, excessively low air temperatures or too high air velocity during freezing. Chicken pieces and other commodities which may be frozen without individual protective packaging are the most vulnerable (**344**). The more seriously affected areas around the feather follicles are illustrated in detail (**345**). Freezer burn does not affect the flavour of the flesh but affected areas are tough and dry.

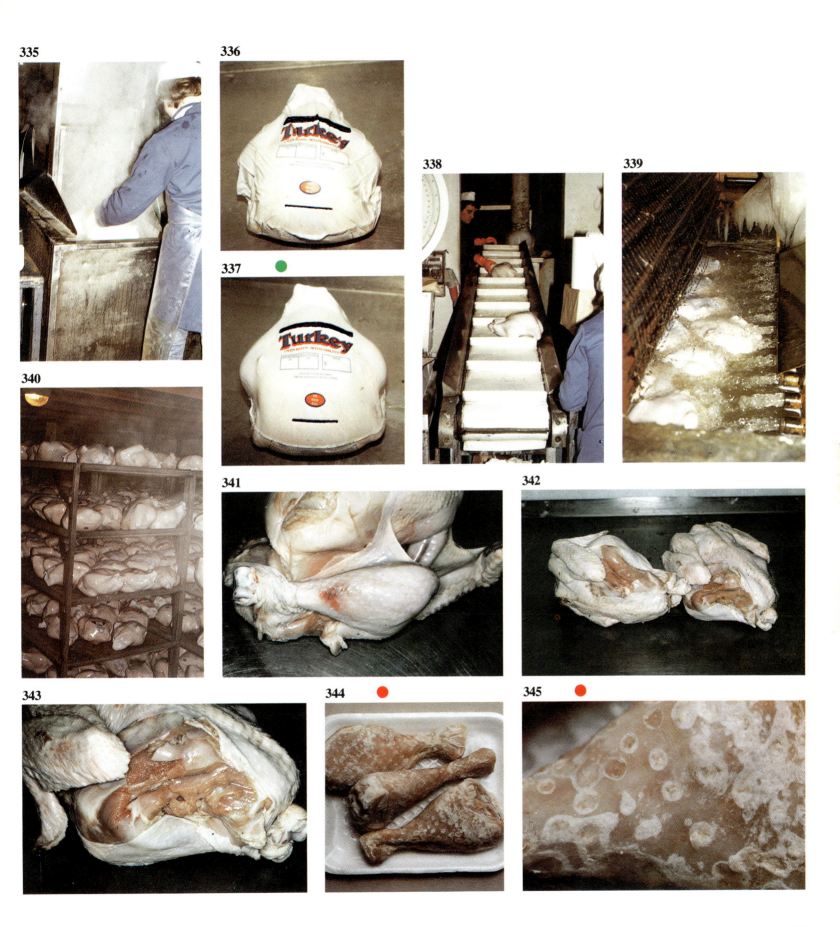

335

336

337 🟢

338

339

340

341

342

343

344 🔴

345 🔴

91

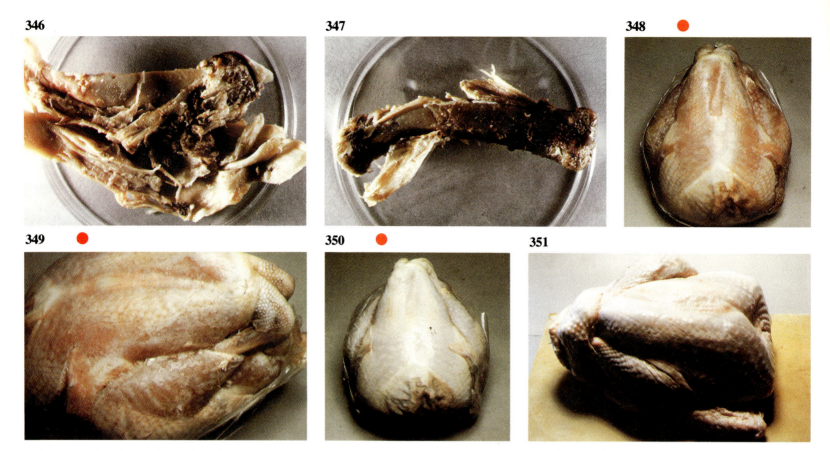

346 and **347. Reddening of flesh around the bone of cooked chicken.** In this example the cause is freezing making the bone more permeable and thereby permitting seepage of bone marrow into the flesh adjacent to the bone. This fault cannot be detected before cooking and care should be taken not to confuse it with growth of *Serratia marcescens* (see p. 96). This phenomenon can lead to the consumer suspecting that the bird is not thoroughly cooked.

348, 349 and **350. Reddish discoloration of a frozen turkey** illustrated on the whole bird (**348**) and in detail on the drumstick where the effect is most marked (**349**). The exact cause is not known but it is associated with poorly controlled freezing. Note also that the bird is covered with a thin layer of ice. The phenomenon is reversible, the discoloration disappearing on thawing (**350**).

351 and **352. Oregon disease (degenerative myopathy)** is an hereditary disease affecting turkeys and to a lesser extent chickens. The incidence is high (up to 13%) particularly in older birds and the economic consequences considerable. The disease affects the fillet breast muscle where lesions due to loss of muscle tissue are formed. The cause is believed to be failure of blood vessels supplying the affected areas.

The meat of unaffected muscle is suitable for human consumption and may be used in catering or for further processing at, of course, a reduced value. The disease is not always easy to detect and some affected birds inevitably pass to the retail market. Typical resultant complaints are of a bird with 'no meat on it' while the greenish discoloration associated with the muscle degeneration (see below) is often mistaken for a cyst.

Advanced cases of Oregon disease may be detected in dressed turkeys by depressions in the breast and, where one side of the turkey is more seriously affected than the other a general 'unbalanced' appearance. **351** shows a whole Oregon turkey. The side nearest the camera is collapsed causing the bird to heel over towards the camera. This may also be seen when the bird is viewed from above (**352**).

353 and **354. A close-up illustration of discoloration** associated with Oregon disease (**353**). **354** shows in contrast healthy breast muscle tissue.

355. A further comparison between healthy and affected tissue made by cutting horizontally through the breast. Degenerated muscle is seen at A (arrowed).

356. A whole cooked turkey affected by Oregon disease with skin removed from the breast showing the lack of muscle on the affected breast (arrowed). The disintegration of the muscle has led to the collapse of the more greatly affected side of the bird.

357, 358 and **359. After cooking** the remaining breast muscle fibres of turkeys affected by Oregon disease readily separate. **357** shows an overall view of this phenomenon which is illustrated in close-up (**358**). A related effect, the complete separation of the breast muscle from the sternum, is shown (**359**).

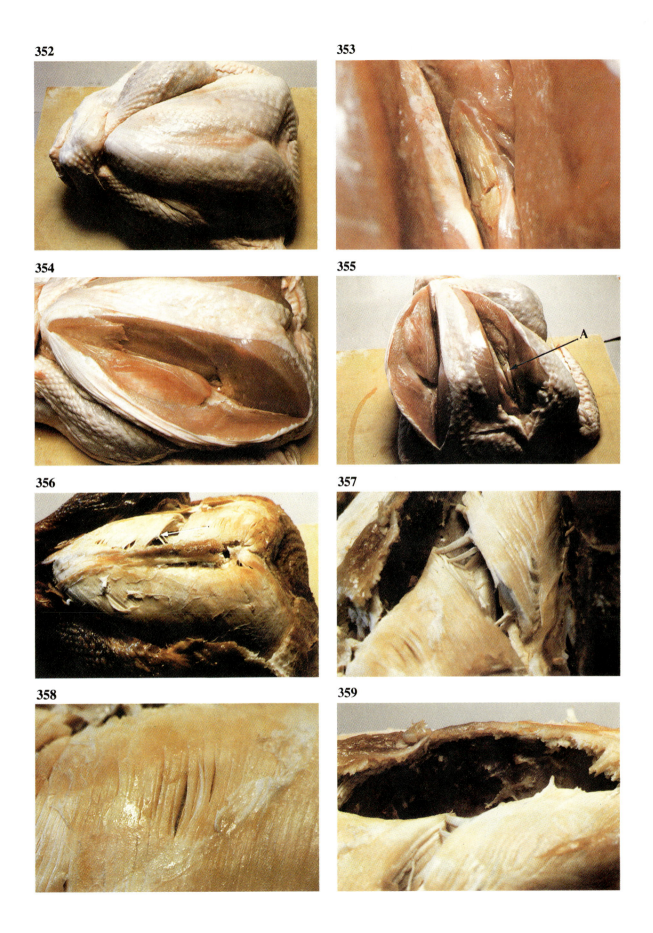

Ducks and Geese

The processing of ducks is similar to that of chickens and turkeys and the quality problems similar. The feathers are difficult to remove and after scalding and mechanical plucking birds are dipped in molten wax at 80°–120°C. After the wax has set it is stripped off removing residual feathers with it.

Ducks have a higher percentage carrier rate of *Salmonella* than other poultry and figures of 50%–100% are not uncommon.

Geese are reared on a large scale in Central Europe but processing is usually by hand and no specific quality problems exist.

Game Birds

Although a number of species of game bird are hunted, in the U.K. at least, pheasant and partridge are economically of greatest importance.

Game birds are hung for 2 to 15 days without being plucked or eviscerated, during which time the flesh becomes more tender and a distinctive flavour develops. Barnes *et al* (1973) found birds hung for 9 days at 10°C to be most acceptable although it should be noted that quality judgement in game is highly subjective.

The nature of game bird procurement means that normal quality control procedures are lacking and where problems of spoilage arise they are usually due to improper handling before hanging or to severe damage to the birds by shot. The presence of lead shot in the tissues of the bird is a hazard to the consumer but is not considered to be a foreign body and in pre-packed game is usually declared on the label.

360 and **361.** **A piece of cooked breast of a turkey affected by Oregon disease** is illustrated (**360**). This may be contrasted with normal turkey breast meat (**361**). Note the green discoloration of the affected muscle which sometimes intensifies on cooking. The texture of the meat is different from that of unaffected flesh, the green areas being very hard.

362. **A cooked leg of a turkey suffering from Oregon disease.** Although Oregon disease normally affects breast muscle, leg may also be affected. In this example of cooked turkey separation of the partly degenerated muscle fibres from the bone may be seen.

363. **Turkey mince.** Poultry meat is now retailed in a number of forms analogous to beef and other red meats. The illustration shows good quality turkey mince similar in appearance to beef.

364. **A good quality example of cured turkey meat.** Pieces of dark meat are cured in a manner similar to that of Sweetcure bacon (see p. 70) and formed into a roll for slicing. In taste and appearance the product resembles ham.

Further Processed Poultry Products (363–373)

In recent years there has been a considerable increase in the range of further processed products available. The degree of further processing and the sophistication of the products varies from simple butchery into joints or stripping of muscles; preparation of comminuted products such as sausages and burgers; curing; preparation of cooked fresh products, often flavoured, and 'complete' products e.g. chicken supreme for the canned or frozen food markets. The technology involved and the quality parameters are largely those associated with the conventional red meat counterparts.

Production of cooked poultry products, particularly whole or portioned birds (usually chickens) involves an enhanced risk of public health hazards from *Salmonella*. The risk of re-contamination of cooked products from raw is always present but it is considerably higher with poultry where *Salmonella* isolation rates from raw material (particularly pieces) may exceed 50%. The only practical way of ensuring safety of the cooked product in a factory environment is to totally separate the raw and cooked areas. This effectively means operating the two areas as separate factories with product transferred between the two *via* double ended cooking ovens, physical separation; separate staff (including supervisors) and separate provision of such staff facilities as toilets and washrooms.

Where staff facilities are shared care must be taken to ensure that cross contamination does not occur in these areas and where people such as maintenance staff have cause to visit both areas extreme precautions must be taken to ensure that they are not the vectors of contamination.

365 and **366.** **Gas flushing** in an atmosphere of carbon dioxide or nitrogen may be used as an alternative to vacuum packing (cf. Controlled Atmosphere Packing pp. 58–59, 104–106). Advantages are claimed with respect to shelf life although leaking packs are less readily detectable than with vacuum packs. The pack illustrated (**365**) contains a smoked chicken product. Note the considerable amount of nutritional information (arrowed) on the pack. The product itself (**366**) would have been considered poor quality in the U.K. market having a texture resembling skin and a pungent aroma of smoke.

At the time of purchase the pack had 8 weeks to expiry of shelf life. Shortly after purchase microbial numbers were in the order of 10^5 colony forming units/g comprising mainly *Brochothrix thermosphacta*. It is considered likely that flavour deterioration would occur as a result of microbial activity even under adequate refrigeration during the subsequent 8 weeks.

367, 368 and **369.** **Cooked whole chicken** and chicken pieces are a common further processed poultry product. The surface of a properly cooked chicken piece is too dry to support extensive bacterial growth. However, moulds are able to develop and are encouraged by film overwrapping the product while still warm as this practice causes condensation on the underside of the overwrapping film. Mould colonies are not easy to detect on casual examination of a whole piece (**367**) (arrowed) but are readily seen on closer examination (**368** and **369**).

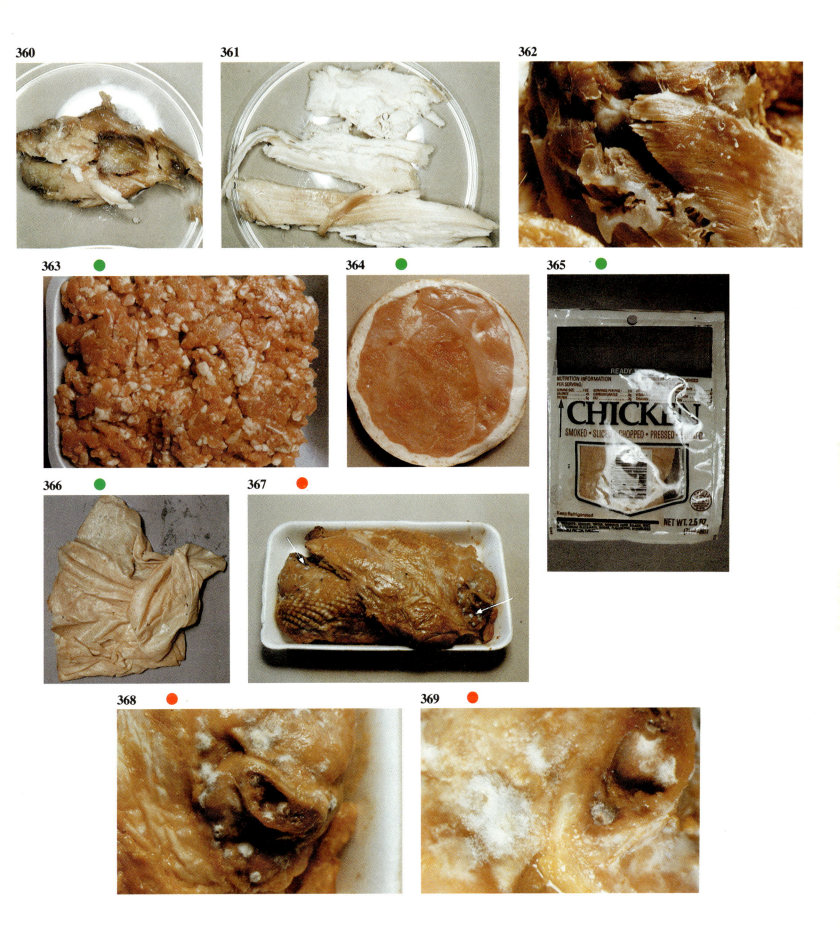

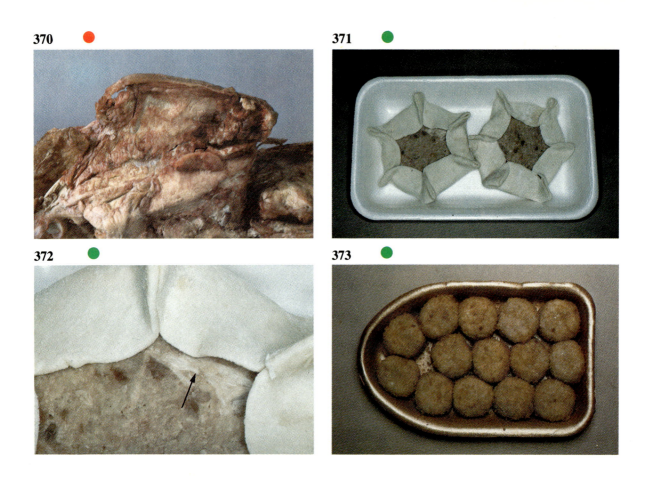

370 ●

371 ●

372 ●

373 ●

370. Growth of *Serratia marcescens* on a cooked turkey. This is a common, environmental organism which grows rapidly at temperatures above 20°C in the absence of competition from indigenous bacteria. (See p. 60 for details of pigments.)

371 and 372. Chicken coronets. A burger type product encased in puff pastry (**371**). The burger market in the U.K., Western Europe and the U.S.A. is large and retailers face considerable competition from fast food outlets such as MacDonalds and Burger King. A close-up illustration (**372**) shows the meat to be of poor quality with a piece of connective tissue clearly visible (arrowed). A high proportion of meat in a product of this type is likely to be

mechanically recovered. It should be noted that the pastry itself may spoil rapidly due to growth of lactic acid bacteria and, possibly, the *Enterobacteriaceae* derived from the flour.

373. Chicken nuggets. These good quality examples of further processed product are cooked chicken meat, rolled, crumbled and flash fried. They are thus effectively a cooked product although the consumer usually reheats them by deep or shallow frying. Such a product has a relatively high potential for growth of *Staphylococcus aureus* between the rolling of the cooked meat and flash frying, and stringent precautions must be taken to preclude any possibility of growth at this stage.

Eggs (374–376)

The majority of eggs consumed are hen eggs although small quantities of duck and turkey eggs are commercially available as are the eggs of some wild fowl such as quail which command a luxury market.

There are two basic ways of rearing hens for egg production, the traditional **free range** system where hens roam freely over a (sometimes) restricted area laying eggs in nests in houses, and the **battery** system whereby hens are restricted to cages in large houses usually with lighting levels and periods artificially adjusted to promote maximum egg production.

The latter system has considerable economic advantages and most egg production in industrialised countries is by this system. In the U.K. at least battery egg production is an emotive issue being taken by its opponents to represent the cruelty inherent in factory farming. It is not relevant to discuss these arguments here but it is necessary to consider the differences in quality which allegedly exist between free range and battery produced eggs.

Eggs are obviously selected for external appearance. In the U.K. and many other countries brown eggs irrespective of means of production are preferred to white. There is, however, no basis for believing that brown eggs are either of superior nutritional status or have better organoleptic quality.

Second to colour is cleanliness and absence of physical defects. Battery produced eggs are invariably cleaner than free range although all types are likely to be washed. **Quaternary ammonium compounds** may be used in an attempt to reduce shell contamination, an important source of micro-organisms causing rots. There is still controversy as to whether washing increases or decreases the likelihood of micro-organisms penetrating the shell. Providing overt spoilage is absent, quality of the egg once broken is judged largely on absence of 'meat spots' (blood clots), general bloodiness, translucent spots in the yolk or evidence of growth of the chick embryo; colour of yolk and general appearance of it and the white and, when cooked, taste. Visual defects should be detected by candling, an important quality control procedure.

Yolk colour is an important parameter, deeper colours usually being preferred and taken to indicate a nutritionally superior egg for high **Vitamin A** content. (In the U.K. some distinct geographical areas prefer pale yolks.) The colour is due to two carotenoids, **lutein** and **zeaxanthine**, which have **no** Vitamin A activity whatsoever. The depth of colour is therefore no measure of Vitamin A content and as carotenoids are feed derived, supplements are available to produce the yolk colour most acceptable to given markets.

Overall appearance of yolk and white is a function of age of egg rather than of means of production. Protein changes (which may be demonstrated by electrophoresis) cause the white to become thinner and the yolk membrane to weaken. When broken the thin white and flattened yolk of an old egg contrasts with the thicker white and upright hemispherical yolk of a fresh laid egg. Supporters of free range egg production contend that as distribution is on a local basis the eggs are fresher than those from battery hens which have been subject to lengthy distribution. If free range eggs are indeed new laid this would probably be correct. There are practical reasons why free range production may not equate with freshness, although there is no doubt that some large scale retailers of battery produced eggs demand excessively long shelf lives.

The flavour of eggs also deteriorates on storage although free range eggs are claimed to have an intrinsically superior flavour as a consequence of the hens having access to food sources such as nettles. Eggs also readily take up off-flavours from such diverse sources as packing materials particularly cardboards or plastics where polymerisation is incomplete, *Streptomyces spp* growing in straw or other bedding material and, where wood shavings are used for bedding, wood preservative chemicals.

In addition to the above quality factors eggs are also subject to microbial spoilage. The factors determining the invasion of the egg by micro-organisms and their subsequent growth is a complex subject and for detailed discussion reference should be made to specialist articles such as Board (1966). The predominant cause of spoilage is bacterial rots which are normally created by a mixed infection of Gram-negative bacteria including both fluorescent and non-fluorescent pseudomonads, *Aeromonas spp*, *Acinetobacter spp*, *Proteus spp* and various other members of the *Enterobacteriaceae*.

Egg Products (377–380)

In addition to the trade in eggs in shell there is a considerable commercial market for liquid egg, dried egg and frozen egg white. Such products are usually intended for ingredient use and are not to be discussed in detail here although dried egg is still used in large scale catering for 'scrambled egg' and similar dishes. The most important aspect of liquid egg is to ensure freedom from *Salmonella* and to this end pasteurisation is mandatory in many countries.

As well as the above, egg products such as 'scotch egg' (boiled egg wrapped in cooked sausage meat) still command a large market, at least in the U.K., while pre-prepared egg dishes such as omelettes are available (see **377**). Eggs may also be preserved by pickling. The traditional English vinegar pickled egg is a retailing curiosity in that a large proportion of sales are through public houses, while the Chinese variety, Pidun, is preserved by a fermentation by indigenous bacteria largely of the *Enterobacteriaceae*.

Table 16 Laboratory Tests Poultry and Poultry Products

I. Microbiological
Total viable count
Salmonella
Staph. aureus (cooked only)
Enterobacteriaceae including *Escherichia coli* (cooked only)

Further tests may be required on more complex further processed products (e.g. *Bacillus cereus* in pre-prepared chicken risotto).

Whole eggs do not normally require microbiological analysis: tests on egg products depend on the nature of the product but should include *Salmonella* and *Staph. aureus*.

Recommended methods for examination of poultry may be found in Parry *et al* (1982).

II. Chemical/Physical
Water content (whole frozen birds only)
NaCl/polyphosphates (where added)

Eggs should be candled to detect internal defects; yolk colour may be measured. Tests on egg products depend on the nature of the product.

Foreign body/infestation examination not normally required.

Notes:
(1) These tests are examples only and are not inclusive.
(2) Examples are for finished products only.
(3) Techniques are detailed in specialist publications such as ICMSF (1978) – Microbiology: Egan *et al* (1981) – Chemistry
(4) Instrumental techniques such as Impediometry may be substituted for some classical microbiological methods.

374. **A boiled egg containing an embryonic chick.** In this example the embryo was at a very early stage of development and positive identification was possible only by histological examination for evidence of cellular differentiations. There is normally no opportunity under commercial egg production conditions for fertilisation to occur. However, all eggs should be examined by 'candling'. In this process the egg is held against a light source (originally a candle). A fertilised egg is partially opaque and may thus be distinguished from the translucent non-fertilised. Fertilised eggs should be discarded.

375 and **376.** **Physico-chemical changes** in the structure of egg white proteins occur during storage and lead to a marked reduction in the viscosity of the white. **375** shows a fresh egg with the extent of the white clearly defined (arrowed). In contrast the white of a stale egg (**376**) has spread across the plate and no boundary is visible.

Although the viscosity of the egg white is often used as an indicator of the freshness of eggs factors other than ageing can lead to reduced viscosity. It is, for example, lower in hot weather when the hen's water consumption increases. Free range hens, or hens housed in poorly ventilated accommodation are particularly affected and eggs produced by such birds in hot weather may be incorrectly judged to be stale.

377. **Omelettes.** Although eggs are widely used as ingredients of foods they rarely form the major component of pre-prepared dishes. The samples illustrated are pre-prepared filled omelettes. Products of this nature require considerable handling during preparation and provide a good substrate for growth of potentially pathogenic bacteria such as *Staph. aureus* and manufacture and subsequent handling must be particularly carefully controlled.

378, 379 and **380.** **Pidun, Chinese preserved duck eggs** sometimes referred to as 1000 year old eggs. The eggs are preserved by coating in a slurry based on soda, salt and slaked lime and either burned straw (**378**) or rice husks (**379**). The eggs are sealed in clay jars and held for ca. 1 month. A natural fermentation occurs classically involving *Bacillus spp* and members of the *Enterobacteriaceae*. The microflora of the egg illustrated (**379**) was, however, dominated by Gram-positive cocci. The interior of the husk coated eggs is shown (**380**). The other egg was entirely liquid in the interior and exploded on cutting! Duck eggs are a recognised source of *Salmonella* and although importation of such eggs to western nations is on a small scale the possibility of introduction of new serotypes cannot be discounted.

374 🔴

375 🟢

376 🟡

377 🟢

378 🟢 *

379 🟢 *

380 🟢 *

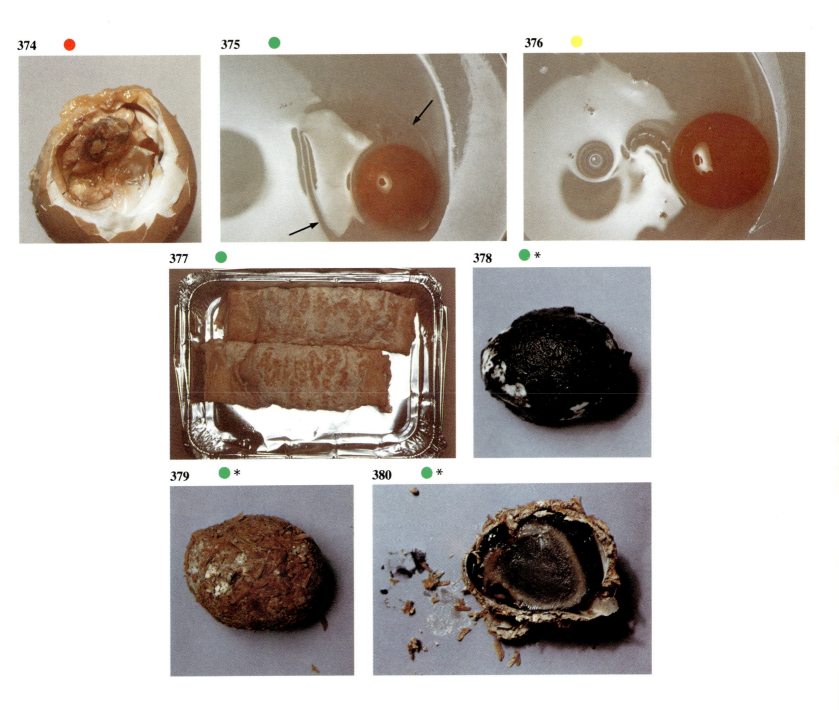

Fish

Fish form an important source of protein and as such are widely utilised. A broad range of fish species are consumed on a large scale by humans, Connell (1983) stating that over 50 species of fish and shellfish are landed regularly and in statistically significant quantities in U.K. ports alone while a much wider variety are utilised world wide. The bulk are salt water fish but freshwater species are a significant source of food in some countries, those of Lake Chad for example being an important source of protein in the African nations of Chad, Niger and, to a lesser extent, Nigeria.

There is also a significant international trade in fish. This is usually in frozen or canned form in developed nations but significant quantities of dried and salted fish are involved in less developed areas. It should be noted that despite generally more cosmopolitan tastes in Western Europe fish consumption has largely been of the traditional species and where hitherto little known fish such as the South American hake have been utilised they have been 'disguised' in the economy ranges of such products as fish fingers.

In recent years the world fishing industry has been significantly affected by two factors. The first and most important has been the reduction of fishing stocks of some staple species. This has frequently been due to over-fishing; the total landing of herrings at British ports of 160,000 tonnes in 1970 declined to 5,000 tonnes in 1980 – a direct consequence of over-fishing about which the industry had been continually warned. Pollution has also resulted in reduction of fish stocks. This originally affected enclosed inland waters but subsequently large scale pollution of tidal waters has occurred. The fishing industry of Northern California for example disappeared as a large scale industry due to a combination of over-fishing and pollution from the fish processing plants, and heavy metal pollution in the Sea of Japan has meant that while actual fish stocks have not diminished the fish caught in some areas are unfit for consumption due to the concentration of heavy metals in the flesh.

The second factor which has partly arisen as a consequence of over-fishing has been an increase in protectionism and a restriction on fishing grounds available to any given nation. Although other factors were also operative the economic effect is exemplified by the massive decline in the U.K. fishing fleet with the numbers of modern long distance freezer-trawlers falling from ca. 50 to ca. 10 over 5 to 6 years.

The combination of economic circumstances has stimulated interest in utilisation of species considered previously to be of no use. The Blue Whiting for example is of potential economic importance in Scottish waters although technical problems of filleting may themselves restrict utilisation.

Fish Farming

Considerable publicity has been given to the development of fish farms. The concept is not new and was practised in the Religious Houses of medieval Europe as well as in China and Japan. Currently trout and salmon are farmed in the U.K. and Scandinavia, flatfish in the U.S.A. and eels in Central and Eastern Europe. The scale of production is small, in the U.K. for example 4,000 tonnes of trout and 500 tonnes of salmon were produced in 1980 to 1984 (Shaw, 1981) and while these figures are likely to increase significantly over the next 5 years the overall contribution from farmed

381

381. Consequences of over-fishing. The most striking example lies in Monterey, California; the setting of John Steinbeck's Monterey Trilogy – *Tortilla Flat, Cannery Row* and *Sweet Thursday*. Monterey was once the third largest fishing port in the world, processing, in peak years, some 254,000 tonnes of Pacific sardines (*Sardinops caeruleus*) in 30 canneries. Between 1945 and 1948 catches fell from 254,000 tonnes for the 6½ month season to ca. 10,160 tonnes. Today the remaining canneries are either derelict or converted to galleries, boutiques and restaurants for affluent tourists. Although factors such as pollution contributed to the decline in the sardine population the major cause was over-fishing: most aptly summed up by Steinbeck's friend, the marine biologist Ed Ricketts 'They disappeared into cans'.

fish is likely to remain small. There are already counterclaims made as to the relative quality of farmed and unfarmed fish. Quality parameters, as far as the consumer is concerned, are based largely on organoleptic quality and are thus subjective. It may be said that farmed fish are likely to be of more consistent (but not necessarily higher) quality and that the nature of fish farming is likely to make application of quality control easier. Farmed trout and salmon have a particular link with *Clostridium botulinum* Type E which is discussed below.

Traditionally farmed fish (carp and goldfish) in China and Japan are the source of human infection by the Chinese Liver Fluke (*Clonorchis sinensis*). Infection arises from consumption of under-cooked fish containing the cercariae of the fluke.

Initial Assessment of Fish Quality and Torry Scores

Fish caught at sea are subject to several days' storage in conditions which are often less than ideal. The spoilage rate varies from fish to fish and a means of assessing suitability which does not require a lengthy laboratory examination is necessary. The most commonly used system is the **Torry Score** developed at the Torry Research Station, Aberdeen, Scotland.

In the Torry Score system fish are scored according to a number of visual and olfactory parameters (see Appendix 1). Although human judgement is the basis of the system scorers are highly trained to give, as far as possible, fully objective judgement, and in this sense are used as 'instruments' rather than as expert tasters as with wine, and no subjective judgement whatsoever should be involved. To ensure maintenance of a universal standard scorers require regular 're-tuning'. Although highly successful, the inherent limitations of human judgement are recognised and attempts have been made to replace Torry scoring with instrumental measurements of fish quality. To date such attempts have been largely unsuccessful.

Fish and Public Health

In recent years considerable publicity in the U.K., Europe and North America has been given to the presence of *Cl. botulinum* Type E in Alaskan and Canadian canned salmon. This problem is seen largely as failures in control of canning technology. Fish do, however, present a number of specific public health problems including the involvement of *Cl. botulinum* Type E. These are itemised and discussed below.

Clostridium Botulinum Type E
Endospores of *Cl. botulinum* Type E are commonly found in the marine environment and may be isolated from fish. Type E isolates differ from other types of *Cl. botulinum* in their ability to grow at 4°C and hence refrigeration is not a protection against the organism. Fish most commonly associated with *Cl. botulinum* Type E are bottom feeding fish such as trout and the incidence of the organism is often high in farmed fish particularly where mud bottomed tanks are used. The organism would not normally be a problem in fresh fish where spoilage due to bacteria such as

Pseudomonas spp would occur, but the risk in smoked fish, which are effectively pasteurised and eaten without cooking after extended storage often in a vacuum pack, present a considerable potential hazard and fatal outbreaks of Type E botulism occurred in the U.S.A. after consumption of Great Lakes smoked fish. Persons involved in the manufacture and retailing of smoked fish must be aware of the potential hazards and ensure that controlling factors such as high salt concentration are not removed for purposes of technological expediency.

Vibrio Parahaemolyticus
V. parahaemolyticus is an halophilic organism of marine origin which causes mild to severe diarrhoea or enteritis (see Barrow and Miller, 1976). Although early reports were restricted to Japan and the Far East the organism is now recognised as being of world wide importance. *V. parahaemolyticus* is not heat resistant and is of importance primarily in seafoods eaten raw. This is common practice in Japan and the Far East but in Europe is largely restricted to oysters and mussels. Oysters and mussels are filter feeders which concentrate bacteria and other food particles from the large volume of water passing through the gills and are thus particularly prone to contamination. Control is effected by growth in sewage-free waters and the use of purification tanks.

Enteric Viruses
Shell fish, particularly filter feeders, have been implicated in transmission of viral disease in several countries including the U.K. and the U.S.A. The cause is invariably sewage pollution of growing beds and the failure to use post-harvest purification. Commercial producers of shellfish are normally aware of the potential problems but a hazard still exists from the activity of small scale and part-time suppliers. Very few laboratories possess the facility to detect viruses and thus control is a matter of ensuring that correct practices are followed.

Fish Products and Anisakiasis
Anisakiasis is a disease of humans induced by larval ascaridoid nematodes such as *Anisakis simplex*. Such nematodes are common in the viscera of many species of fish and under certain conditions migrate into the surrounding muscle.

The extent of migration is increased if the freshly caught fish are held in refrigerated holds. Deardorff *et al* (1984), for example, conducted a survey of Hawaiian fish and fish products and concluded that *Terranova* Type A in locally caught Hawaiian snappers and *Anisakis simplex* Type I from rockfish (*Sebastes spp*) caught in the western Pacific Ocean presented a significant risk of anisakiasis. The extent of migration of nematodes in these fish is further increased by the common practice of brining and cold-smoking; the processed fish being eaten without cooking. Freezing is recommended to inactivate the nematodes, *Terranova spp* being inactivated by 1 day's storage at $-20°C$ and *Anisakis spp* by 5 days' storage at the same temperature.

In addition to the risk of anisakiasis presented by the consumption of raw or minimally processed fish flesh, uncooked viscera are eaten in some parts of the world. In both Hawaii and the Philippines, for example, raw viscera together with fish heads and other parts are allowed to undergo a natural fermentation and then form the basis of local delicacies. Such dishes have been implicated as the source of anisakiasis not only amongst the indigenous population but also amongst gastronomically adventurous tourists.

Although most cases of anisakiasis have involved fish caught in warm oceans a major outbreak of the infection in Holland attributed to consumption of minimally preserved raw herring provided evidence that *Anisakis spp* present in North Sea fish are also potential human parasites. It is also noteworthy that at least one case of human parasitic disease due to the common cod worm (*Phocanema spp*) has also been confirmed although the nematode is usually of low infectivity to man.

Intrinsic Food Poisoning

The most common cause of intrinsic food poisoning from fish is scombrotoxic poisoning. Outbreaks have been known to occur regularly over the past 40 years in Japan, the Pacific Islands and the U.S.A. but the first 50 incidents reported in Britain occurred between 1976 and 1979 (Gilbert *et al*, 1980). **Scombrotoxic poisoning** results from eating partially spoiled fish of the families *Scombersocidae* and *Scombridae* which include **tuna, bonito** and **mackerel**. Such fish contain high quantities of the amino acid histidine which is decarboxylated by bacterial enzymes to **histamine**. The symptoms suffered (most frequently diarrhoea, hot flushes and sweating, rashes, nausea and headache) are essentially those of histamine poisoning although there is evidence that histamine itself is not involved (Arnold and Brown, 1978). Control is by inspection of fish and rejection of any showing incipient spoilage as well as correct post-inspection handling.

The sudden occurrence of a significant number of cases in the U.K. provides an example of the manner in which commercial forces apparently unrelated to food safety and quality may in fact be involved. The decline in herring and the plentiful supply of mackerel led to increased production of hot smoked mackerel as an alternative to kippers. The production was enthusiastically accepted by consumers and demand increased rapidly. This in turn placed pressure on producers, some of whom were effectively cottage industries with little technical knowledge and limited facilities. The pressures on demand were intensified by the large supermarket chains requiring product for widescale distribution and there is no doubt that the outbreaks of scombrotoxic poisoning were due to the processing of partially spoiled fish by persons lacking the knowledge of the implications.

It should be noted that since 1980 there have been a number of incidents of scombrotoxic-like food poisoning from canned salmon originating in several countries including Japan, U.S.A. and the U.S.S.R. Salmon is a non-scombroid fish and the cause of such outbreaks is at present unknown.

In addition allergic responses to shellfish by sensitive persons may produce symptoms akin to scombrotoxicosis.

Other intrinsic poisoning from fish lies in the poisonous nature of parts of some food fish such as the pufferfish in Japan and some types of octopus and squid in the West Indies. Handling and processing of such fish is a highly specialised practice beyond the scope of discussion here.

Dinoflagellate poisoning

Dinoflagellate poisoning results from the consumption of shellfish which have themselves ingested the toxin producing alga *Gonyaulax*. Under certain conditions of high temperature and usually a degree of pollution algal blooms form and shellfish produced in affected areas are not fit for consumption. Tropical and subtropical regions are most commonly affected but in suitable conditions blooms may occur in temperate waters such as the North Sea.

Examples of laboratory tests are shown in Table 17, p. 110.

---

382 and **383. Most shellfish is cooked prior to consumption** usually by boiling in simple coppers (**382**). This process is sufficient to kill vegetative bacteria and viruses. In many cases the conditions under which the cooked fish are handled are primitive and cross contamination between raw and cooked produce may readily occur (**383**).

384. Signs warning of dinoflagellate poisoning are posted on beaches and sea shores in affected areas. The warnings are frequently ignored and sporadic outbreaks occur.

385. Appearance of scales is an important criterion in assessing fish quality. Here, the scales on a good quality fresh herring show a bright 'healthy' appearance.

386 to **389. Deterioration of cod fillets** due to autolytic enzyme activity. **386** illustrates cod fillets on day of purchase from a retail outlet; the arrows show 'gaping' due to separation of muscle fibres, illustrated in detail (**387**). After 4 days' storage at 4°C the gaping has worsened (**388** and **389**) due to autolytic enzyme activity. There is also evidence of slime formation (arrowed) (**389**) due to bacterial growth, and a fruity ester-like odour. Herbert *et al* (1971) ascribed such spoilage to *Pseudomonas spp* of the *fragi* type. (N.B. gaping or softening of some species of fish may be caused by protozoan parasites such as *Kudoa (Chloromyxum) thyrisites* which affects South Atlantic hake ('milky hake disease') and *Henneguya spp* which affects Canadian salmon and which can cause quality deterioration in the smoked product.)

382

383

384

385 🟢

386 🟡

387 🟡

388 🔴

389 🔴

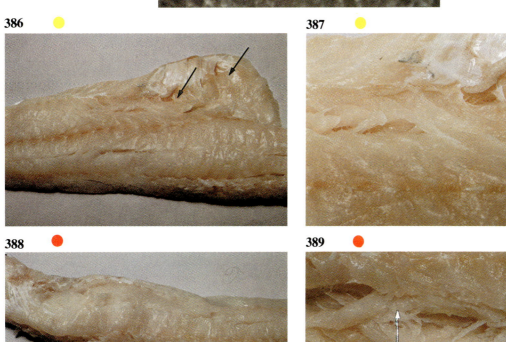

390, 391 and **392. Spoilage of pre-packed (air pack) golden haddock fillets. 390** shows an overall view of the fillets. Although 1 day within 'sell by' date a general quality deterioration is apparent notably discoloration and dryness of the tail of the upper fillet. Discoloration and colour loss is illustrated in detail (**391**), while bacterial slime (arrowed) may be seen between the fillets (**392**). An off-odour of a sulphide nature was present at a low level. Such off-odours are attributed by Herbert *et al* (1971) to *Pseudomonas putida* (*Pseudomonas* Group I of the Shewan *et al* (1960) classification scheme) or to *Pseudomonas*-like bacteria similar to (Shewan *et al*) Group III/IV *Pseudomonas*.

393 to 398. Mackerel is traditionally known for its rapid spoilage rate. This is due to a high flesh pH value and a high level of autolytic enzyme activity. This series illustrates deterioration of a mackerel over 4 days' storage at 4°C. **393** illustrates a whole fresh mackerel on day of delivery from the dock. Comparison with the fish after 4 days' storage (**394**) shows little overall difference although the older fish is losing sheen. Closer examination shows loss of freshness and incipient spoilage. The eye (**395**) of the older fish is glazed and sunken compared with the younger (**396**) and there is incipient microbial slime formation on the skin. This is shown as dullness in the arrowed area (**397**) compared with the skin of the fresher fish (**398**).

No off-odours were detected from the spoiling fish and it is unlikely that taints would be detected on eating. Consumption of a fish in this condition would carry a high risk of scombrotoxic poisoning. (See p. 102.)

399 and **400. Damage to a CAP fish pack** by a sharp object is illustrated (arrowed). Such damage negates the advantages of CAP since the pack atmosphere will rapidly equilibrate with that of the environment. The object has also damaged the fish (**400**) (arrowed) and damage of this nature is a potential source of contamination.

390

391

392

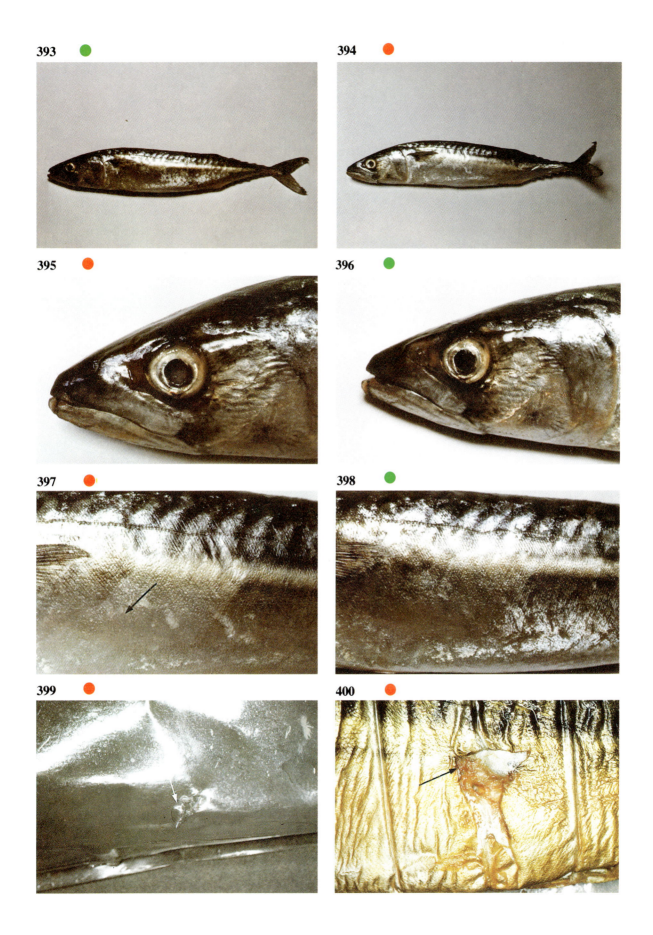

401 and **402. Hot smoked mackerel** has attained considerable popularity in the U.K. in recent years and demands for increased production have frequently strained manufacturers' quality assurance procedures (see p. 102). **401** illustrates a hot smoked mackerel in a CAP pack (note that the advantages of CAP for smoked fish are less than for raw). A blemish (arrowed) may be seen on the side of the fish. A more detailed examination (**402**) showed this to be a lesion in the flesh. This may have been due to a wound or to an infection such as myxobacteriosis (*Flexibacter spp*) and the fish should have been rejected by quality control staff.

403 and **404.   Some degree of 'gaping'** due to separation of muscle is common in fish. **403** illustrates a smoked mackerel fillet (CAP packed) showing excessive gape probably due to the use of poor quality frozen mackerel. A close-up (**404**) shows fat exuding into the gaps.

405.   Underprocessing of a hot smoked mackerel. The flesh in the arrowed red areas along the backbone is virtually raw. Such underprocessing may lead not only to poor quality and rapid spoilage but also to public health problems.

406 and **407. Bloaters (Yarmouth kippers)** are produced by lightly smoking whole herring. Bloaters were popular in the U.K. up to the second world war and were frequently used for further processing into products such as bloater paste. They are highly perishable and the examples illustrated (**406**) are showing signs of deterioration although not actually spoilt. The lifting of scales (**407**) (arrowed) often precedes deterioration.

408 and **409.   Fish fingers** are almost invariably retailed as a frozen product but their quality tends to reflect the fish used rather than the freezing process. **408** illustrates good quality fish fingers (A) and poorer quality (B). The two types differ in external appearance and a section through the two fingers (**409**) shows that the lower quality finger (B) is darker in colour. This is a consequence of the use of lower quality fish such as coley in conjunction with cod whereas the higher quality finger (A) is composed entirely of cod.

410 and **411.   Marinated herrings,** particularly when packed in tomato sauce, are prone to spoilage by lactic acid bacteria. The organisms responsible are usually heterofermentative species of *Lactobacillus* such as *L.buchneri*. Spoilage is readily detected by gas formation (**410**), illustrated in detail (**411**) (arrowed), which may be accompanied by slime. Spoilage usually occurs in the retail pack where the concentrations of salt and ethanoic acid have been reduced. Two routes may be involved in gas formation, carbon dioxide being produced either by **fermentation of carbohydrates** or by the **decarboxylation of amino acids** in the absence of carbohydrates (see Blood, 1975). It should be noted that while marinated herrings and similar products are traditionally regarded as being shelf stable at ambient temperatures consumer preference for blander, sweeter products means they may require refrigeration.

403

404

405

406

407

408

A

B

409

A

B

410

411

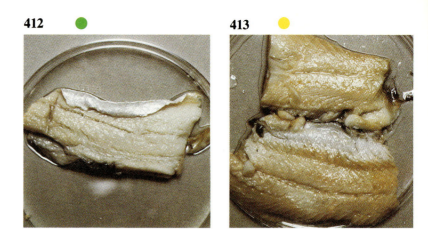

412   **413**

412 and **413.**   **Good quality pickled herring pieces** with typical creamy off-white colour (**412**) compared with poor quality pickled herring which is of a distinct yellow ochre colour (**413**). This fault is attributed to the initial quality of the fish probably in this example rancidity of the fat. The length of time for which the fish has been in the pickling brine has no effect on colour. Problems of this nature have increased since the drastic fall in herring catches in the North Sea necessitated the use of imported frozen herring for pickling.

414.   **Cooked mussels marinated in vinegar.** Although acceptable in appearance the mussels had a strong ammoniacal odour and were unfit for sale.

415 and **416.**   **Dry salted fish such as saith** (**415**) are produced primarily in countries which lack more sophisticated means of preservation and adequate cold storage facilities. Such products are thus of considerable economic importance in underdeveloped countries. The salt concentration is extremely high, free salt crystals being seen on the surface of the fish (arrowed) (**416**) and the product is highly shelf stable. Under damp storage conditions halotolerant moulds may develop and spoilage can also be caused by development of red pigmented colonies of extremely halophilic bacteria such as *Halobacterium salinarium* derived from the salt.

Note that a high consumption of salt fish has been linked with cancer of the stomach and nasopharyngeal region.

417 to **421.**   **Dried, unsalted fish** are common products in Asia. Such fish are sun dried and typically carry a large mixed microflora. Consequently they should be cooked shortly after reconstitution because subsequent spoilage is rapid. Two examples, **cuttlefish** (**417**) and **silverfish** (Bombay duck) (**418**) are illustrated. After storage of rehydrated fish for 13 hours at cool ambient temperature (ca. 10°C) a bacterial slime was present on the water surrounding the cuttlefish (**419**) and red pigmented yeast colonies were visible (arrowed). Massive slime formation occurred on silverfish over the same storage period (**420**) and visible colonies of Gram-negative rod shaped bacteria were present (**421**) (arrowed).

422.   **Anchovies pickled in soya sauce and chilli.** Products of this type, particularly common in the Hunan region of China, are highly shelf stable although spoilage by moulds may occur. The product is pasteurised for the export market.

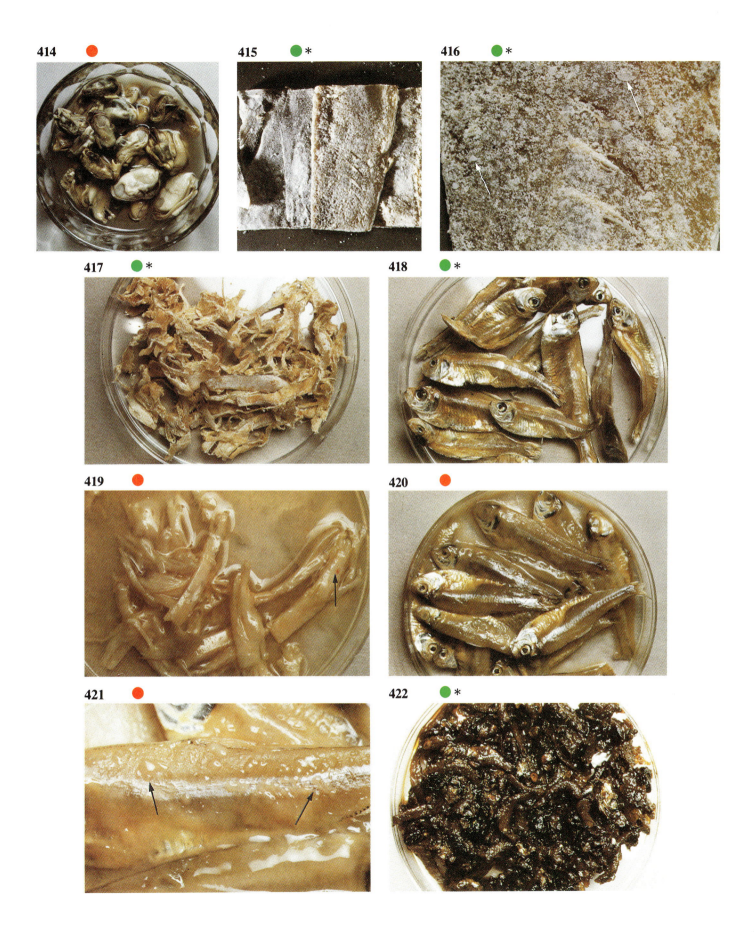

414 🔴

415 🟢 *

416 🟢 *

417 🟢 *

418 🟢 *

419 🔴

420 🔴

421 🔴

422 🟢 *

Table 17 Laboratory Tests Fish

A. Raw (Fresh and Frozen)

I. Microbiological
Total viable count
Escherichia coli
V. parahaemolyticus (where fish to be eaten raw)
Salmonella (freshwater fish caught in warm climates and to be eaten raw)

II. Chemical/physical
Polyphosphates (where added)

Although electronic instruments for determination of fish quality are available the Torry system for grading fish remains the most valuable means of assessing freshness. It should be noted that although assay methods for histamine exist they are not suitable for application on a routine basis and protection against scombroid poisoning lies in selection of good quality fish and in maintaining good handling and manufacturing practices.

Filleted fish should be examined for the presence of parasites.

B. Smoked

I. Microbiological
Total viable count
E. coli
Staphylococcus aureus
Clostridium (Hot smoked fish to be eaten without further cooking)

N.B. The microflora of lightly smoked fish such as kippered herring is similar to that of raw fish.

II. Chemical/physical
NaCl (where a safety determinant)
Aw (where a safety determinant)
Filleted fish should be examined for the presence of parasites.

C. Fermented

I. Microbiological
Choice of analyses is dependent on the nature of the product but safety lies primarily in ensuring that the process is sufficient to either kill pathogens or prevent their multiplication to dangerous levels.

II. Chemical/physical
Choice of analyses is dependent on the nature of the product but analysis should always be made of safety determinants such as NaCl.

Filleted fish should be examined for the presence of parasites.

D. Shellfish

I. Microbiological
Total viable count
V. parahaemolyticus
E. coli
Staph. aureus (cooked only)

N.B. Although shellfish have been associated with viral disease routine examination for viruses is not possible in most laboratories.

II. Chemical/physical
Analysis is not normally required on products of this type.

Shelled (shucked) shellfish should be examined to ensure the complete removal of shells (and other parts where appropriate).

Notes:
(1) These tests are examples only and are not inclusive.
(2) Examples are for finished products only.
(3) Techniques are detailed in specialist publications such as ICMSF (1978) – Microbiology: Egan *et al* (1981) – Chemistry.
(4) Instrumental techniques such as Impediometry may be substituted for some classical microbiological methods.

The Torry Score system is based on the use of separate descriptive tests for sight, smell, touch and taste. For the more discriminating senses of smell and taste a scale of 0 to 10 is used with 10 being an absolutely fresh fish and 0 being completely putrid. Fish having a score of less than 4 to 5 would be unacceptable to most consumers. The senses of sight and touch are less able to detect small changes in freshness and a scale of 0 to 5 is used. This may be related to the 0 to 10 scales by doubling the score.

The example of a score sheet shown below is for round, white fish. Other fish such as flatfish show different spoilage patterns and the descriptive criteria vary. Fuller details of sensory and non-sensory methods of assessing fish freshness may be obtained from Howgate (1982).

Raw Fish

	Score
General appearance	
Eyes perfectly fresh, convex black pupil, translucent cornea; bright red gills (colour depending on species); no bacterial slime, outer slime water white or transparent; bright opalescent sheen, no bleaching	5
Eyes flat, very slight greyness of pupil; slight loss of colour of gills	4
Eyes slightly sunken, grey pupil, slight opalescence of cornea; some discoloration of gills and some mucus; outer slime opaque and somewhat milky; loss of bright opalescence and some bleaching	3
Eyes sunken, milky white pupil, opaque cornea; thick knotted outer slime with some bacterial discoloration	2
Eyes completely sunken; shrunken head covered with thick, yellow bacterial slime; gills showing bleaching or dark brown discoloration and covered with thick bacterial mucus; outer slime thick yellow-brown; bloom completely gone; marked bleaching and shrinkage	0

Appearance of flesh including belly flaps	
Bluish translucent flesh, no reddening along the backbone and no discoloration of the belly flaps; kidney blood bright red	5
Waxy appearance, no reddening along backbone, loss in original brilliance of kidney blood, some discoloration of the belly flaps	3

Some opacity, some reddening along backbone, brownish kidney blood and some discoloration of the belly flaps	2
Opaque flesh, very marked red or brown discoloration along backbone, very brown to earthy brown kidney blood, some discoloration of the belly flaps	0

Odours	
Fresh seaweedy odours	10
Loss of fresh seaweediness, shellfish odours	9
No odours, neutral odours	8
Slightly musty, mousy, milky or caprylic acid like odours, garlic, peppery odours	7
Bready, malty, beery, yeasty odours	6
Lactic acid, sour milk or oily odours	5
Some lower fatty acid odours (for example acetic or butyric acids), grassy, 'old-boots', slightly sweet, fruity or chloroform-like odours	4
Stale cabbage water, turnipy, sour sink, wet matches, phosphine-like odours	3
Ammoniacal (trimethylamine and other lower amines) with strong byre-like (O-toluidine) odours	2
Hydrogen sulphide and other sulphide odours, strong ammoniacal odours	1
Indole, ammonia, faecal, nauseating, putrid odours	0

Texture	
Firm, elastic to the finger touch	5
Softening of the flesh, some grittiness near tail	3
Softer flesh, definite grittiness, scales easily rubbed off skin	2
Very soft and flabby, retains the finger indentations, grittiness quite marked and flesh easily torn from the backbone	1

Cooked Fish

About 200g middle cut of fish steamed in glass casserole over boiling water for 35 minutes

Odours	*Score*
Strong seaweedy odours	10
Some loss of seaweediness	9
Lack of odours or neutral odours	8
Slight strengthening of the odour but no sour or stale odour; wood shavings, vanillin or terpene-like odours	7
Condensed milk, caramel or toffee-like odours	6
Milk jug odours, boiled potato or boiled clothes-like odours	5
Lactic acid and sour milk, or byre-like odours	4
Lower fatty acids (for example acetic or butyric acids), some grassiness or soapiness, turnipy or tallowy odours	3
Ammoniacal (trimethylamine and lower amines) odours	2
Strong ammoniacal (trimethylamine) and some sulphide odours	1
Strong ammonia and faecal, indole and putrid odours	0

Flavour	
Fresh, sweet flavours characteristic of the species	10
Some loss of sweetness	9
Slight sweetness and loss of the flavours characteristic of the species	8
Neutral flavour, definite loss of flavour but no off-flavours	7
Absolutely no flavour, as if chewing cotton wool	6
Trace of off-flavours, some sourness but no bitterness	5
Some off-flavours, and some bitterness	4
Strong bitter flavours, rubber-like flavour, slight sulphide-like flavours	3
Strong bitterness, but not nauseating	1
Strong off-flavours of sulphides, putrid, tasted with difficulty	0

After Howgate (1982)

References to fish are also made in Chapter 15, Canned Foods, as follows:

P. 213 Anchovies
P. 213 Sardines
Pp. 226–227 Tuna

Examples of foreign bodies in fish may also be found in Chapter 17, Foreign Bodies and Infestations as follows:

Pp. 246–247 Unidentified insect in tuna (canned)
P. 251 Trilobite in prawns (frozen)

Chapter 9

Grocery

In the present context the term grocery encompasses a wide and diverse range of products. It essentially refers to non-perishable products which have not received a full thermal processing as have canned goods although by necessity there are exceptions to this generalisation.

For purposes of clarity products are placed into groups related either by the nature of the product and/or by processing (Table 18). Potential problems specific to each group, or members of that group, are discussed individually but it is necessary to make a general observation with respect to quality control.

As a consequence of their relatively non-perishable nature it is too easy to assume that problems are minimal and restricted to infestations. While infestations etc. may be serious in their own right grocery products may also be implicated in food poisoning of various types; at the same time changes in recipe and technology to suit current tastes may alter their intrinsic stability. Persons involved in quality control and assurance of grocery products should be aware of these possibilities at all times.

Grocery Products and Mycotoxins

By virtue of their nature and conditions of production and storage a number of grocery commodities may be subject to mould spoilage and hence mycotoxin formation. Such potential problems are discussed individually.

Grocery Products and 'Health Foods'

In recent years a general increase in dietary consciousness and interest in physical fitness in Western Europe, the U.S.A., Australia and elsewhere has led to an increased demand for so-called **health, natural** and **wholefoods**. These are normally vegetarian products with a lower degree of processing than normal commercial foods, such as wholemeal flours or which contain unconventional ingredients such as fennel tea. It is not intended to enter the discussion of the relative merits of such foods versus the more conventional. It is, however, necessary to make the following points to persons who may be involved in inspection of 'health foods'.

(1) Terms such as 'health food' are not synonymous with good quality. Such foods are often produced by small units who lack the technical expertise, facilities and in some cases product knowledge to ensure a consistent high quality.
(2) Vegetarians and 'health food addicts' are considered by Jarvis (1976) to be one of two groups potentially at highest risk from mycotoxins. He states 'The "pure food lobby" may ensure that their food comes from sources where neither "chemical" residues of fertilisers, pesticides or fungicides nor "chemical food additives" occur in their food, yet without pesticides and fungicides, food stored under *natural* (but not necessarily *good*) conditions may become rapidly contaminated with toxinogenic moulds'.
(3) The term 'health food' and the connotation that such foods are healthier than those sold elsewhere are the subjects of criticism and, in some countries, to possible future legal restraint. Vickery (1983) points out that all foods legally sold are within the context of a normally balanced diet 'healthy foods'.

Table 18 Grouping of Grocery Products

Group	Commodities
Cereal products	(a) Flours and bakery mixes (b) Whole grain products (c) Pasta and pasta products (d) Breakfast cereals (e) Miscellaneous
Dried pulses	
Dried fruit and vegetables (excluding pulses)	
Fats and oils	
Mayonnaise and sauces	(a) Emulsion based products (b) Condiments
Vinegar	
Pickles and preserves	(a) Vinegar based* (b) Salt based* (c) Soya sauce based* (d) Sugar based
Sugar and molasses	
Jams and honey	
Confectionery	(a) Chocolate (b) Boiled sugar (c) Miscellaneous
Snack products and nuts	(a) Potato based (b) Cereal based (c) Nuts
Dried formulated products	(a) Soups, sauces etc. (b) Whole meals
Soft Drinks	
Hot beverages	(a) Tea (b) Coffee (c) Miscellaneous
Miscellaneous	(a) Herbs and spices (b) Jellies, blancmange etc. (c) Oriental fermented products (d) Fillers (e) Extracts and hydrolysates

*Pickled fish products are discussed in Chapter 8.

Cereal products

Flours and Mixes (423–434)

Most flour for domestic use in the western world is produced from wheat (*Triticum sp*). Smaller quantities of oat (*Avena sp*), barley (*Hordeum spp*), rye (*Secale cereale*) and maize (*Zea mays*) flour are also produced. In addition, 'flours' are also produced from root vegetables such as potatoes (*Solanum tuberosum*) and cassava (*Manihot esculenta*) as well as from soya (*Soya max*).

Rice (*Oryzae spp*) is an important source of flour in tropical countries and is estimated as providing the basic food for about half the world's population, while sorghum (Kafir corn, Milo, *Sorghum vulgare*) is the chief food grain in parts of Africa and Asia. It is not possible to discuss milling technology for all grains but a brief description of the production of wheat flour will be given.

Full details for wheat and other important grains may be obtained from publications such as Technology of Cereals (Kent, 1966).

The first stage of milling wheat on receipt from the farm is cleaning. Wheat at this stage is likely to contain impurities including mud and dust, chaff, weed seeds, insects, rodent hairs, excreta and small machinery parts. Mud, dust and hairs etc. which adhere to the grain may be removed either by washing or by dry scouring. Particulate impurities are separated by a variety of methods which include size and shape separation, aspiration, specific gravity separation, electrostatic separation and removal of metal machinery parts by magnetic separation. Insects may be destroyed by an **entoletor** which uses centrifugal force to physically break up insects and eggs (the insect fragments remain in the grain). Grain may also be fumigated with fumigants such as **ethylene dichloride** or by mild heat treatment.

Following cleaning wheat is conditioned. The primary objectives are to improve the condition of the wheat for milling and, in some cases, to improve the baking quality of the flour. Conditioning is achieved by addition or removal of water to achieve the optimal water content and after, by alternate heating and cooling, to obtain the desired average moisture content in the mass of the grain and the desired distribution of moisture in each individual grain.

The objectives of milling white flour may be summarised as follows (after Kent, 1966).

(1) To make, as completely as possible, a separation of the endosperm from the bran and germ so that the flour shall be free from bran specks and colour, and so that the palatability and digestibility of the product shall be improved and its storage life lengthened.

(2) To reduce the maximum amount of endosperm to flour fineness, thereby obtaining the maximum extraction of white flour from the wheat, and at the same time to ensure that the amount of damage to the starch granules does not exceed the optimum.

Modern large scale milling processes make no attempt to achieve either of the stated objectives in a single stage but break down the grain and its parts by a succession of relatively gentle grinding stages. Roller mills are used to fragment the grain and its parts, these consisting typically of parallel, contra-rotating fluted rollers rotating at different speeds. Following each 'grinding' stage the resulting particles are classified into particle size by sieves while air currents are used in purifiers to separate the bran from endosperm particles.

Where it is desired to produce wholemeal or brown flours the white flour is usually blended with all (for wholemeal) or some (for brown) of the offals. However, some wholemeal and brown flours are still produced using traditional stone mills and it is claimed that such flour is of superior quality.

Following milling flour is usually further treated for infestation by entoletion or by fumigation with **ethylene oxide**. Before sale it is customary to bleach white flour using **chlorine** (improves the quality of cake flour but is not permitted for bread flour in the U.K.), **chlorine dioxide** or **benzoyl peroxide**. Improving agents (see Chapter 10, Bakery, p. 160) which accelerate natural maturation properties may be added. Chlorine and chlorine dioxide both have improving properties but non-bleaching improvers such as **potassium bromate**, **ascorbic acid** and **azodicarbonamide** are also used. It should be noted that permitted bleaches and improvers differ from one country to another.

In the U.K. and some other countries flours other than wholemeal are supplemented with iron, Vitamin B and nicotinic acid while flours other than wholemeal and self-raising contain added calcium carbonate (chalk) as a source of dietary calcium.

The importance of flour as a source of micro-nutrients may be judged by the fact that plans in the U.K. to cease supplementation of white flour were abandoned after nutritionalists had protested that bread baked from white flour is of major dietary importance particularly for many old people.

The type of flour used for any given purpose varies according to that purpose. The governing factors are the same whether the usage is to be commercial or domestic. Thus for breadmaking a strong flour would be selected (see Chapter 10, Bakery, p. 160). Many domestic or household flours are intended to be suitable for a wide range of purposes and are typically produced from weak wheats of low protein content, although up to 20% of strong wheats are often included. Similar flours are self-raising which contain chemical leavening agents (usually **sodium bicarbonate** and **acid calcium phosphate**).

Flour may contain relatively large numbers of micro-organisms including non-faecal members of the *Enterobacteriaceae* such as *Erwinia* derived from the grain, as well as endospore-forming bacteria derived from soil. Special grades of flour containing low numbers of bacteria may therefore be required for infant food formulation and for canning purposes.

Flours may also contain mycotoxins produced by mould growth on the grain during storage or, in the case of rye, pre-harvest growth of the ergot fungus *Claviceps purpurea*. Jarvis (1976) has noted that consumption of flour produced from mould spoiled grain is unlikely in developed nations but that a different situation pertains in underdeveloped nations particularly at times of food shortage. (Ergotism in bread is discussed in Chapter 10, Bakery.)

Mixes for bread, cakes, etc., are based on the appropriate type of flour together with the required additional ingredients including, where appropriate, functional leavening agents such as yeast. In some types of cake mixes it is necessary to add some of the ingredients, e.g. fresh eggs and milk. Weighing and mixing of ingredients is carried out automatically and the mix is usually entoleted to remove insects.

Specific problems with bread and cake mixes are few although fat rancidity may occur in prolonged storage and deterioration of the leavening agent may lead to a poor quality finished product.

423. **A number of instruments are available** to physically determine the suitability of flours for production of different bakery products. These include the Brabender Farinograph, the Chopin Alveograph and the Simon Extensometer which measures the resistance of dough to stretching and the distance the dough stretches before breaking. The illustration shows the curves obtained for Manitoba bread flour which has the required high resistance to stretching and a moderate stretching before breaking. In contrast a good biscuit flour would produce a low flat curve indicating low resistance to stretching but considerable stretching before breaking (see also Amylograph pp. 148–149 and Chapter 10, Bakery).

423

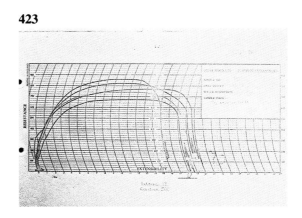

424. A good example of wholemeal flour. The pieces of bran (arrowed) are sometimes mistaken for weevils (**425**).

426 and **427. Grease in wholemeal flour.** A medium (**426**) and close-up (**427**) illustration. The source of the grease is probably the bearings on some part of the mill machinery. An occasional incident of this nature is effectively unavoidable but a large number suggests poor housekeeping (see **428** and **429**).

428 and **429. A medium view of debris** (arrowed) in wholemeal flour (**428**) bearing a superficial resemblance to the grease illustrated (**426**). Close examination shows (**429**) the debris to consist largely of insect webbing. This was detected in the same batch of flour as the grease and while the two incidents have separate causes the standard of housekeeping at the mill must be under suspicion.

430. Semolina is derived from wheat and is an intermediate in the milling of whole wheat grain down to flour. It consists of small chunks of endosperm of up to ca. 1mm^3 in size. The semolina illustrated is of relatively good quality being free from infestation and foreign bodies. A small amount of bran is present (arrowed). In products such as semolina bran is readily mistaken for an insect infestation.

431. Tapioca derived from heated cassava flour and used for puddings etc.

432. A flour moth (*Anagasta kuehniella*) together with webbing found in a Yorkshire pudding mix. Pudding mixes are a compounded product and the moth was probably present in the flour.

433 and **434. Leavening agents,** whether chemical or yeast, present in bread and cake mixes deteriorate with storage leading to a poor quality product after re-constitution and baking. **433** illustrates a dough made with wholemeal bread mix containing yeast of low activity. The mix has failed to rise correctly and is not suitable for baking. In contrast dough made with a mix containing active yeast is illustrated (**434**). (See also Chapter 10, Bakery.)

Examples of foreign bodies in flour and mixes may also be found in Chapter 17, Foreign Bodies and Infestations as follows:

424 ●

425 ●

426 ●

427 ●

428 ●

429 ●

430 ●

431 ●

432 ●

433

434

Whole Grains (435–442)

Although international trade in bulk grains is largely of the whole grain most cereals retailed in Europe and the U.S.A. have been milled into flours. The exception to this is rice and, to a lesser extent, barley. Elsewhere in the world home milling or grinding is more common and grain is more commonly retailed whole.

Whole grain is subject to attack by rodents and insects. In underdeveloped nations a significant proportion of the crop is lost in this way. If storage conditions are not suitable or if drying facilities are not available for damp grain, significant mould growth may occur. A number of the species of mould which infest stored grain are mycotoxigenic, and mycotoxins subsequently found in the derived flour or meal may not be destroyed by cooking.

435. **Barley** is used primarily for animal feed and for malting and brewing (see Chapter 16) in Europe and North America and relatively little is used for direct human consumption in these areas. Although more widely used in this manner in Asia consumption there is also falling. Much of that which is consumed in Europe is in the form of **pearl barley** which is ca. 58% of the whole barley after blocking (shelling) and pearling (rounding) by scouring or abrasive processes involving emery based stones. It should be noted that bleaching of barley is not permitted in the U.K. but is practised in some European countries notably West Germany.

436 and 437. **A commercial sample of long grain rice (436).** Close examination (**437**) shows a number of damaged grains (arrowed) due to **rodent attack** at some stage after hulling.

438. **Rice weevils** (*Sitophilus oryzae*) infesting brown rice. The damage to the grains with the typical powdery appearance is fairly advanced and the extent of damage suggests a long storage period after packing (see **439** and **440**).

439 and 440. **Merchant grain beetle** (*Tribolium mercator*) in pudding rice. The attack on the grain differs from that of weevils (see **438**) and in the absence of the insects may be confused with rodent damage (see **437**). Infestation usually occurs during storage or at hulling but in common with many other infestants the merchant grain beetle is increasingly found in domestic premises.

441 and 442. **A tooth,** identified as a pig deciduous tooth found in rice (**441**). The means of contamination is not known. Such contaminants are difficult to detect since the tooth resembled a discoloured grain of rice before isolation (**442**).

Examples of foreign bodies in whole grain products may also be found in Chapter 17, Foreign Bodies and Infestations as follows:

P. 248 Cockroach in brown rice

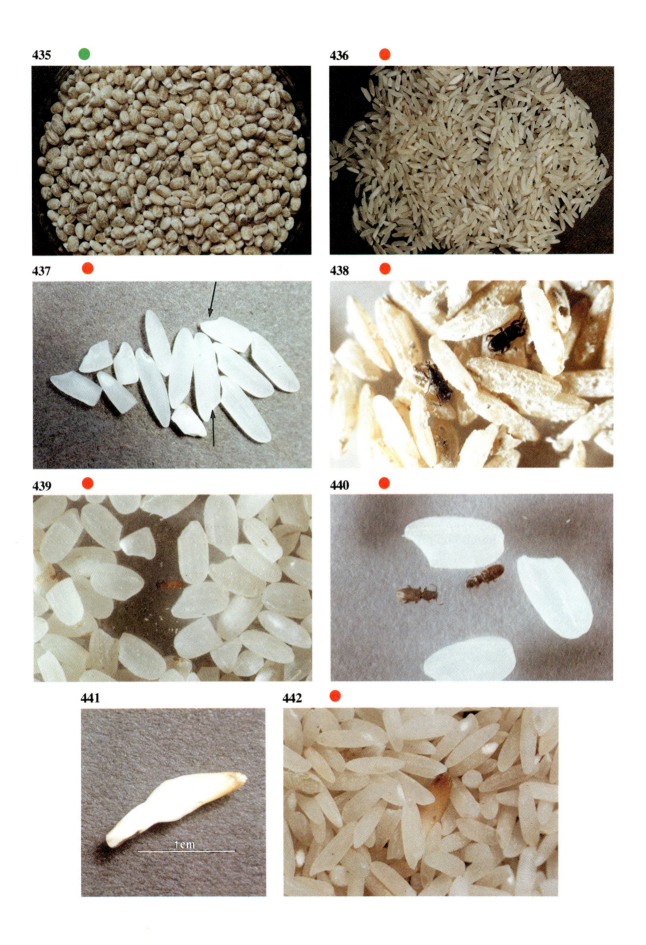

435 🟢

436 🔴

437 🔴

438 🔴

439 🔴

440 🔴

441

442 🔴

Pasta Products (**443–453**)

Pasta products are usually produced from durum wheat although rice and soya flour are also used. Pasta is a generic term covering a wide variety of products ranging from thin sticks of spaghetti, through the flat sheets of lasagne to shells and spirals.

Commercial pasta is produced from an unleavened dough usually based on water and flour which is extruded, moulded and cut to the correct shape and size and then dried.

Fresh chill pasta is also now available on a large scale.

Although dried pasta products are considered to be of low risk with respect to food poisoning hazards a number of cases of **staphylococcal intoxication** due to egg pasta occurred in Western Europe during 1984. The focal point of contamination in the producing factory appears to have been the liquid egg handling although there is no evidence that the egg itself was the original source of the micro-organism. Large numbers of *Staphylococcus aureus* were isolated from the pasta and it seems likely that not only did growth of *Staph. aureus* take place in the raw pasta dough but that the subsequent drying stage was insufficient to kill the organisms. Preformed enterotoxin would, in any case, have survived a more rigorous drying stage and is not destroyed by domestic cooking.

443 and **444.   Good quality pasta shells.** An example of an Italian product increasing in popularity outside Italy (**443**). The small white flecks (arrowed) shown in detail (**444**) are pieces of endosperm and are normal. They may, however, be mistaken for yeast or mould colonies.

445.   Authentic Italian style (as opposed to canned) pasta products have previously been sold as dry shelf stable products (e.g. **443**). A recent trend has been the production of **moist pre-cooked pasta**. Such products represent the more extreme and possibly illogical examples of convenience foods whereby a shelf stable product with effectively no microbiological spoilage problems is converted to a short life product requiring refrigeration to prevent rapid microbial spoilage. The only practical benefit to the consumer is a few minutes reduction in preparation time.

446 and **447.   Straight spaghetti** with some broken sticks (**446**). A close examination (**447**) shows the broken ends to be clean breaks with no indication of tooth marks. The cause is simple mechanical damage and not rodent or insect attack.

448.   A massive infestation of spaghetti by the drugstore beetle, *Stegobium paniceum*. The product was returned to the retailer by a consumer. It is inconceivable that it was purchased in this state and the degree of damage indicates long storage. Under such circumstances it is not possible to determine whether infestation originated before or after purchase. (See also Chapter 17, pp. 244–245.)

449 and **450.   Japanese long life pre-cooked noodles.** The noodles are parboiled, packed and subjected to a heavy pasteurisation (**449**). Bacterial endospores survive but germination and outgrowth is limited despite the lack of preservatives and room temperature storage. This shelf stability may be attributed to the relatively low water activity level of the pasta although the process is seen as being marginal in terms of safety. Note the gelatinous nature of the product (**450**) which is preferred by Japanese but disliked in the West.

451.   **Pasta products of European origin are invariably produced from durum wheat flour. However, rice or soya bean flour (see **452) may be used elsewhere. The illustration shows good quality dehydrated rice flour noodles of Chinese origin.

452 and **453.   Vermicelli produced from soya bean** (**452**). A close examination of the threads show the product, manufactured in mainland China, to consist of extruded soya protein rather than soya flour (**453**).

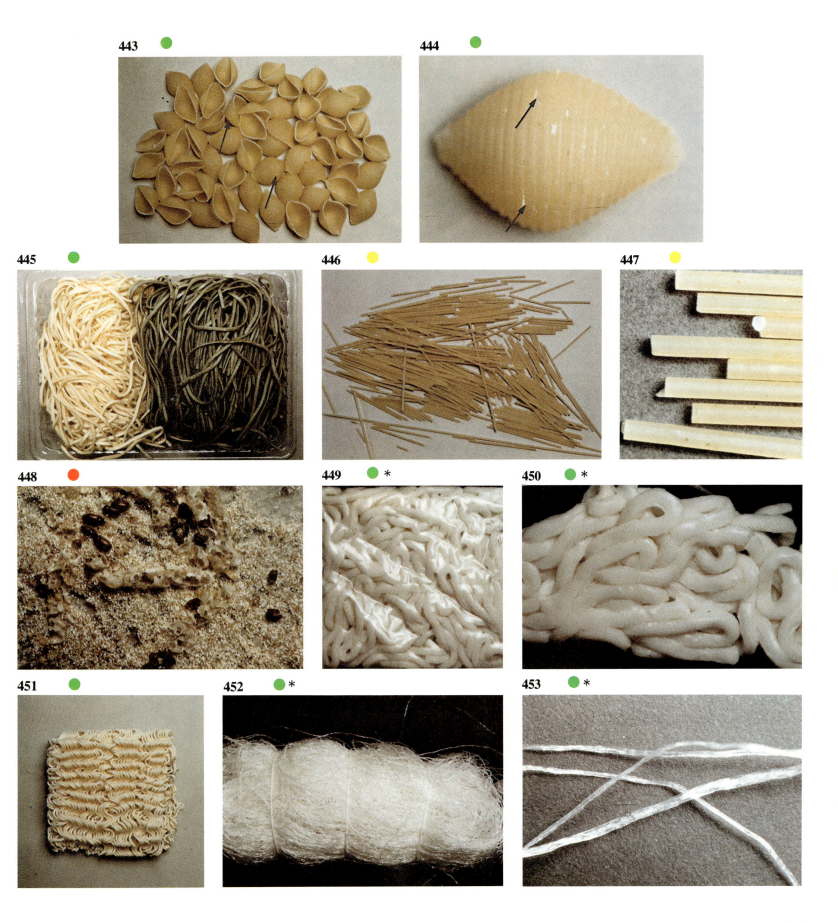

443

444

445

446

447

448

449 *

450 *

451

452 *

453 *

121

Breakfast Cereals (454–463)

Breakfast cereals are diverse in nature but may broadly be classified as **uncooked** and **ready to eat**. Uncooked breakfast cereals are typified by **porridge** (porage), which traditionally is made from oatmeal or rolled oats. Coarse oatmeal is not cooked during manufacture unless heated to inactivate lipases and requires considerable cooking. Rolled oats are partially cooked during manufacture and require less domestic cooking.

Ready cooked or instant porridge is made from rolled oatflakes which have been heat-treated to a higher temperature and subjected to a higher roll pressure. The starch is completely gelatinised. Similar products are a combination of ordinary rolled oats together with roller dried flakes of an oat flour and water batter, and products of this type are often used for infant feeding.

Oat based products are highly prone to deterioration due to fat rancidity. Oats have a fat content in the kernel 2 to 5 times as high as wheat (ca. 6.2% in oat kernels) and the **pericarp** contains an active **lipase** which, unless inactivated, hydrolyses the fat to **glycerol** and **free fatty acids** primarily oleic, linoleic and palmitic which produce a bitter flavour. Heat inactivation is commonly applied to oats for porridge, the process also appearing to stabilise the fat with respect to oxidative rancidity.

Ready to eat breakfast cereals comprise **flaked, puffed, shredded** and **granular** products usually made from wheat, maize or rice although oats and barley are also used. The basic cereals are often enriched with sugar, honey etc., malt extract, flavourings such as chocolate, vitamins and trace elements, while an admixture with dried fruit, nuts or bran may be made. All of the processes used tend to cause dextrination rather than gelatinisation of the starch.

Flaked products are made from whole wheat or rice grains cracked by lightly rolling, or from maize (corn) 'grits'. These together with flavourings are cooked, usually at high pressure, dried to 15%–20% ($^w/_v$) moisture content and conditioned by holding for 24 to 72 hours. The conditioned material is flaked on heavy flaking rollers, toasted in a continuous oven, cooled and packed.

Puffed products are made either from whole grain wheat, rice or oats or from an extrusion pelleted maize or oat dough. Whole prepared grains or pellets are fed into a pressure chamber which is then sealed and heated both externally and by steam injection to produce an internal pressure of ca. 200lb/in^2 which is then released suddenly by opening the chamber or 'puffing gun'. Expansion of water vapour on release of the pressure puffs up the material to several times the original size and disrupts the internal structure in the case of whole grains. (A similar procedure is used to 'instantise' rice for use in snack meals, see p. 149.) After puffing the cereal is dried to a moisture content of ca. 3% ($^w/_v$) by toasting, cooled and packed.

Shredded products are prepared from wheat of high starch content. Whole prepared wheat is cooked to yield a soft rubbery grain with fully gelatinised starch and a moisture content of ca. 45% ($^w/_v$). After cooling and resting for several hours the grain is fed between shredding rolls emerging as long parallel shreds which fall onto a slowly moving belt. A thick mat is built up on the belt by superimposition of several layers of shreds. The mat is cut into tablets and baked for 20 minutes at ca. 200°C, dried to 1% ($^w/_v$) moisture content, cooled and packed.

Three types of good quality porridge.

454.   Oatmeal, the basic product of oat milling. Relatively coarse flakes are required to give the granular structure associated with good quality Scotch porridge but these require prolonged cooking. Such oatmeal may be used in baking although for this purpose finer grades are available.

455.   Flaked oats produced from oatmeal by steaming and passing through flaking rolls. The heating gelatinises the starch and less cooking is required than with oatmeal.

456.   Instant porridge – a combination of oat flakes and roller dried oat flour and water batter. On addition of hot water the roller dried batter mix forms a smooth paste while the oat flakes give a granular structure and provide 'body'.

Good quality examples of ready to eat breakfast cereals.

457.   Puffed and flaked products. Puffed rice grains are shown on the left (A) and maize (corn) flakes on the right (B). (See also p. 149 Instant Snack Meals.)

458.   A typical English wheat-based breakfast cereal.

459.   Bran. This is an important constituent of many high fibre breakfast cereals. High fibre diets have become fashionable in Western Europe and the U.S.A. and bran itself may be consumed. Although having a high vitamin and mineral content it is poorly utilised by humans and is of limited value as a nutrient *per se*.

460.   Hazelnut Granola – a wholefood or healthfood breakfast cereal.

461.   Good quality wheat flakes. A further example of a wholefood or 'natural' product.

462.   Damage to a flaked wheat biscuit such as this is often attributed to insect infestation. Frequently, however, as in this example, the cause is physical damage due to poor handling.

463.   Scotch porridge (porage) oats containing compacted mould contaminated material. Although the mould may have developed after packing as a consequence of damp storage conditions it is more likely that the compacted material was derived at the mill as a consequence of improper cleaning of machinery.

Examples of foreign bodies in breakfast cereals may also be found in Chapter 17, Foreign Bodies and Infestations as follows:

Pp. 246-247   Earwig in cereal

454 ●

455 ●

456 ●

457 ●　　●

458 ●

459 ●

460 ●

461 ●

462 ○

463

1cm

123

Granular products are made from a yeast leavened wheat or malted barley dough baked in large loaves then broken up and ground to a pre-determined degree of fineness. The ground material may be reformed into various shapes.

In addition to the above products some **wholefood** breakfast 'cereals' are based on nuts. In these examples the nuts are usually roasted and chopped or ground. Production is very small compared with conventional breakfast cereals.

The major quality deterioration associated with ready to eat cereals is fat rancidity. Oat based cereals are most affected (see above) and the other cereals which have a lower fat content usually have a longer storage life. (N.B. Whole maize has a fat content of 4.4% (^{w}/$_w$) but most is contained in the germ which is removed during milling. Grits from which cereals are made have a fat content of 1.5%–2% (^{w}/$_w$) similar to wheat, barley and rice.)

The stability of the fat is reduced by heat treatments such as puffing and toasting which can destroy natural antioxidants or induce formation of pro-oxidants. Other treatments such as the roller drying of batters may improve fat stability by production of antioxidants such as **Maillard reaction products**. Synthetic antioxidants are permitted in some but not all countries.

Miscellaneous Cereal Products (464–466)

A variety of part-processed and processed cereal products exist in various parts of the world to meet a variety of requirements. The quality problems associated with such products are largely those of infestation and mould spoilage associated with other cereals.

Dried Pulses (467–474)

Drying as a means of preservation of pulses is an old established process which has to some extent been superseded in technologically developed countries by alternative means of preservation such as canning and freezing. Dried pulses are dietary staples in many less developed countries of Africa and Asia and are also of dietary importance in the poorer areas of developed nations e.g. the U.S.A., an example being black eyed peas (*Vigna unguiculata*) and beans in the southern states. Dried pulses are thus economically important both in the internal and external economies of many nations. There has also been an increased consumption of dried pulses in countries such as the U.K. in recent years partly due to demand from immigrant communities and partly due to increased interest in 'health' and 'whole' foods.

Dried Pulses and Intrinsic Food Poisoning

A number of outbreaks of gastroenteritis were reported in the U.K. in the years between 1976 and 1980 associated with the consumption of **raw red kidney beans** (*Phaseolus vulgaris*) (Noah *et al*, 1980). The factor responsible for gastroenteritis was considered to be a **haemagglutinin** which is destroyed by adequate cooking. Kidney beans, and possibly other pulses, contain other heat labile toxins including **trypsin inhibitors**, a **goitrogen** and **cyanogenetic glycosides** which in extreme cases may induce cretinism. The pericarp and testa of some pulses may also contain **procyanidin polymers** which significantly reduce the biological value of protein ingested (Hulse, 1983).

In addition some types of **lima bean** (*Phaseolus lunatus*) contain significant quantities of **cyanide** but such cultivars are not normally grown as food plants.

Dried Pulses and Mycotoxins

Dried pulses may be subject to contamination by mycotoxin-producing moulds if storage conditions are poor. This problem is not unique to pulses; control lies with provision of adequate storage facilities and rejection of affected commodities.

---

464. **Bulgur** consists of parboiled whole or crushed wheat grains dried to ca. 10% moisture level. It is used in place of rice in Eastern Europe and the Middle East. Bulgur is highly resistant to insect attack, a considerable advantage in areas where storage conditions are poor.

465 and 466. **Husk remaining on flaked rice** is a technological fault which, if excessive, lowers the quality of the product. It is frequently mistaken for an insect infestation, a considerably more serious defect, an example of which is illustrated (**466**). As in many such cases precise identification of the insect is not possible but infestation probably occurred during storage or at milling.

467 and 468. **Poor quality butter beans** (*Phaseolus lunatus*). The beans are dirty and have been damaged (**467**). This may have been a result of careless handling during processing and packing. Nevertheless beans of this quality are frequently retailed in the U.K. They should be contrasted with the satisfactory (rather than good) quality illustrated (**468**). (*N.B.* The hyacinth bean (*Lablab purpureus*) and a white seeded runner bean (*Phaseolus coccineas*) may also be described as butter beans.)

469. **Red lentils** (*Lens culinaris*) of poor to average quality. Some darker types of lentil are present and some are slightly damaged around the edge of the flat surface.

470. **A stone in red kidney beans** (*Phaseolus vulgaris*). Foreign bodies occur at a high level in dried pulses particularly where products are air dried on the ground. A high incidence of stones indicates a potentially serious failure of quality control procedures.

464 ●

465 ●　**466**

467 ●　**468** ●

469 ●　**470** ● *

Dried Fruit and Vegetables (475–478)

Although drying as a means of preservation has been to some extent superseded by other methods such as canning and freezing the method is still widely used. In hot climates sun drying is still in wide use for fruit and to a lesser extent vegetables (spices are also often sun dried; see p. 155). In other climates artificially heated air is used in either batch or continuous processes. Although the quality of dried vegetables is considered to be inferior to frozen (and often to canned), sophisticated drying techniques such as fluidised bed drying (see also dried pulses p. 124) have improved quality markedly. In spite of most vegetables being dried as pieces, dried potato is commonly in the form of a roller or spray dried powder.

471. Dried peas (*Pisum sativum*). Many dried vegetables such as peas bear little relationship to the fresh product when dried by conventional slow batch techniques. More modern fast drying methods utilising continuous processes such as fluidised bed drying result in a product which when re-constituted and cooked resembles the fresh product and is of equivalent quality to frozen peas (see pp. 210–211). The nutritional quality of such products is also considerably higher than that of conventionally processed peas.

472, 473 and 474. A severe infestation of red kidney beans by *Bruchidae* (bean weevils) (**472**). An infestation of this severity is rare and indicates a complete breakdown of infestation control procedures. Note the distinctive circular holes in the beans (arrowed) (**473**). The insects themselves (**474**) are in this example dead.

Examples of foreign bodies in dried pulses may also be found in Chapter 17, Foreign Bodies and Infestations as follows:

Pp. 246–247 Fly parts in lentils

Edible Oils and Fats

Vegetable Oils and Fats (479–486)

With the exception of rape seed oil vegetable oils have traditionally been the products of southern climates. They are of considerable international economic significance, being of considerable importance to the economies of many developing nations while olive (*Olea europeanea*) oil remains an important factor in the agricultural economies of Greece, Spain and Southern Italy.

According to the U.S.A. Department of Agriculture world production of vegetable oils totalled 39,282,000 tonnes in 1980–1981. Of this more than 12,000,000 tonnes was soya (*S. max*) and over 5,000,000 tonnes palm (*Elaeis guineensis*) oil. Future trends indicate increased production of these two oils with lesser increases in rape seed (*Brassica campestris oleifera*), sunflower (*Helianthus spp*) and coconut (*Cocos nucifera*) oils. Refining technology is such that many may be rendered bland enough for European tastes while others such as rape, hitherto unfit for human consumption (see p. 128) may now be fully utilised. It should be noted that oils for local consumption in traditional producing areas often still have strongly rancid flavours.

The technology of vegetable oils is a complex topic and beyond the scope of this book. The main stages are listed (**479**) and two are of particular importance. **Hardening** is a process by which the degree of unsaturation of the fats is decreased by hydrogenation. Originally used to prepare vegetable fats for margarine manufacture, hardening alters the physical properties of fats and permits interchangeability of natural fats. Accompanied by the

475 and **476. Two approaches to dehydrated vegetables. 475** illustrates good quality hot air dried onion flakes while the same commodity fried to a low moisture content is illustrated in **476**.

477. Good quality dried mixed vegetables, a common form of retailing such products in the U.K. **Sulphur dioxide** is usually present primarily to retard browning but also to preserve ascorbic acid, reduce microbial load and to discourage insect infestants.

478. Drying of fruit remains an important means of preservation although in developed countries dried fruit are usually used for their particular characteristics in, for example, cake baking rather than as a means of obtaining long term storage. Sun drying is still used for fruit such as apricot and prunes. Dried fruit is usually treated with sulphur dioxide (see **477**) and where subject to **enzymic browning** may be blanched before drying.

Examples of foreign bodies in dried fruit and vegetables may also be found in Chapter 17, Foreign Bodies and Infestations as follows:

Pp. 246–247 Myriapods in dried fruit
P. 251 Cobweb in dried onion

more recently developed process of **interesterification** it is thus of considerable economic importance. Groundnut (*Arachis hypogaea*) oil, soya bean oil and sunflower oil for example may each be processed to acquire the particular physical properties desired in manufacture of a given food. Coconut oil has specialist uses as a consequence of its containing a high proportion of short chain saturated fatty acids and it is therefore less easily interchanged with other oils.

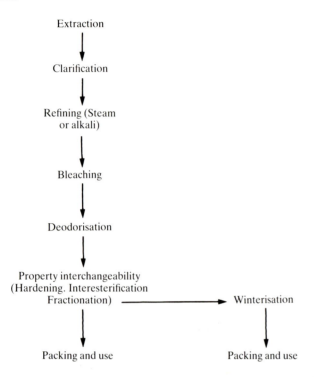

Extraction

↓

Clarification

↓

Refining (Steam or alkali)

↓

Bleaching

↓

Deodorisation

↓

Property interchangeability
(Hardening. Interesterification
Fractionation) ⟶ Winterisation

↓   ↓

Packing and use   Packing and use

479. A flow diagram showing the main stages in refining vegetable oil. The processing will vary according to the source of the oil and the desired properties of the end product. Process related quality defects are rare in vegetable oil production.

Rape is the only oil yielding plant which is grown on a significant commercial scale in northern climates. The oil, however, contains the toxic **erucic acid** and until the recent development of suitable refinery techniques it has been largely used as lamp or industrial oil. It is most widely grown as an oil plant in Germany, Poland, Sweden and Japan (the related rape kale *Br. napus* is widely cultivated elsewhere either as green manure or for animal, and to a lesser extent, human consumption). World production of rape seed oil (including industrial oil) amounted to some 3,860,000 tonnes (9.8% total vegetable oil production) in 1980–1981 an increase from ca. 2,710,000 tonnes in 1975 (8.9% of total production). Further increases in production are expected to be small.

The use of industrial rape oil, denatured with **aniline**, to adulterate olive oil was responsible for an epidemic of chemical food poisoning in Spain in 1981 with more than 15,000 persons affected and 200 deaths (Ross, 1981; Trench, 1981). The epidemic serves as a reminder that although chemical adulteration of food,

rife in Victorian England, is now only a minor problem in countries such as the U.K. and the U.S.A. it may still occur on a large scale elsewhere. The source of the epidemic was traced to oil sold 'on-tap' from tankers by itinerant salesmen. The sale and quality of olive oil in Spain is nominally subject to strict control but the high cost of 'licensed' oil means that many of the poorer people purchase from the unlicensed itinerants. Oil sold by itinerants is commonly adulterated with up to 50% of cheaper edible oils such as soya. Industrial rape oil was introduced purely because of its cheapness without regard for, and presumably in ignorance of, its toxicity. A noteworthy feature of the epidemic is that the economic attraction of cheap oil was such that purchase of the adulterated product continued long after its dangers had been widely publicised.

Spoilage of vegetable oils is by oxidative rancidity and detected only by taste or chemical analysis. Few visual defects occur.

Animal and Marine Oils and Fats (487–489)

Animal and marine oils and fats are still of considerable importance although in each case production is stagnating and market share declining. Production of marine oils in 1980–1981 totalled 1,252,000 tonnes and animal fats (including butter) totalled 14,801,000 tonnes. Up to the 1950s most marine oil was derived from whales. The massive decline in the whale population led to fish oils, suitably refined and processed, being used.

Extraction of animal fats is usually by heating the tissues under pressure to 'render out' the fat. Fish oils are highly unsaturated and require extensive processing including hydrogenation but animal fats less commonly receive sophisticated treatment after extraction.

As with vegetable oils spoilage is by rancidity.

480 and 481. A typical highly refined bland tasting vegetable oil (**480**) compared with high quality first pressing olive oil (**481**). The former will have consistent properties throughout the year although the proportion of the different oils used is likely to vary considerably.

482 and 483. Winterised groundnut oil (**482**) compared with non-winterised (**483**) after storage at ca. 10°C. Winterisation is considered to be of particular importance in preventing oil-in-water emulsions such as salad dressing and mayonnaise 'breaking' (see **490**) at low temperatures as well as maintaining an attractive appearance when used as a dressing without blending. Crystals in oil are often mistaken for spoilage.

484. Cocopalm oil. This sample is relatively unrefined and imported into the U.K. for the immigrant market. The oil is solid at average U.K. ambient temperatures. The free fatty acid content of such products can be as high as 30% due to lipase activity during processing and the taste would be unacceptable to European and American consumers.

480 ●

481 ●

482 ●

483 ●

484 ● *

485 ● *

486 ●

487 ●

488 ●

489

485. Crude sesame seed (*Sesamum indicum*) **oil.** High grade oil is removed after a first pressing, the essentially unrefined oil from the subsequent pressings being sold together with the seed debris. The product is strongly flavoured and widely used in North Africa and the Middle East in the manufacture of hummus, a dietary staple based on chick peas.

486. Vegetable based fats. Although liquid vegetable oils are preferred for many purposes solid fats are required for some applications such as preparation of pastry dough. Hydrogenation of vegetable oils, in this example a mixture of soya and palm oils together with interesterification, produces a solid fat with the baking properties of pork lard (see **487**).

487, 488 and **489. Animal fats** such as pork lard and beef dripping may have their properties modified by the same processes, such as hydrogenation, which are widely used for vegetable oils. With the exception of fats to be blended into shortenings animal fats often receive little treatment after rendering and refining and retain the characteristic textural properties derived from the differing properties of crystalline high melting point triglycerides. **Pork fat** (lard) for example (**487**) has a naturally smooth texture and is suitable for baking pastry, biscuits and cakes, whereas **beef fat** (**488**) has a more granular structure and does not make good cakes or biscuits. The contrasting properties are easily demonstrated by smearing small quantities of each fat onto a dark paper (**489**). Beef dripping (A) retains its granular properties whereas pork lard (B) spreads smoothly.

Mayonnaise and sauces

Emulsion Based Products (490–495)

Emulsion based products comprise shelf stable mayonnaises and oil in water emulsions such as pre-prepared French dressing. The basic technology of forming a stable oil in water emulsion is the same although it should be noted that in some of the higher moisture content dressings emulsion stability is low.

Products of this nature are traditionally shelf stable, microbial growth being controlled by a combination of low water activity level, low pH value and relatively high concentrations of ethanoic and (possibly) other acids, sugar and salt. A formula technique for predicting shelf stability exists (Tuynenberg-Muys 1971) and is of value in new product formulation.

A trend towards less acid formulations means that some products require refrigeration before and after opening (cf. mayonnaise based fresh salads p. 172) while the creation of new products by addition of 'novel' ingredients to a base or emulsion may create problems either by introducing high levels of micro-organisms or by an interaction between the ingredient and an anti-microbial factor destabilising the preservative system.

490. **Total separation of an emulsified salad dressing** into oil (upper) and aqueous phases. A number of factors may be responsible for 'breaking' oil-in-water emulsions. In this case crystal formation in the oil due to prolonged storage at low (<6°C) temperatures is thought to have been responsible (see **483**).

491 and **492.** **Early stages of** *Moniliella acetoabutans* **spoilage** of sandwich spread (**491**) and seafood sauce (**492**). In each example the affected sample is arrowed. The change in appearance is small but is accompanied by a 'fruity' odour due to **ethyl acetate** production.

493. **Spoilage of mayonnaise based seafood sauce** by the mould *M. acetoabutans*. This mould (the vinegar mould) is able to tolerate low pH values and high concentrations of ethanoic acid. The source is usually infected vinegar. In this example the mould colony is just visible (arrowed) on the surface of the sauce and in many cases visible growth is not present. Spoilage is by conversion of ethanoic acid to ethyl acetate with a characteristic odour. Potential spoilage may be detected at a very early stage by assaying for ethyl acetate in the headspace using gas liquid chromatography.

494. **Deterioration of blue cheese dressing** (arrowed) due to **syneresis.** This example of blue cheese dressing is of low acidity and low salt content and unlike most other types requires refrigerated storage.

495. **Mould developing on the surface** of mayonnaise based blue cheese dressing (arrowed). The mould probably gained access after opening the container. This type of dressing was low in acidity and NaCl and required refrigerated storage.

Condiments (496–499)

Condiments comprise chutneys, relishes, sauces and strongly flavoured products eaten as an accompaniment to other foods. They differ from mayonnaise type products in being a mixture rather than an emulsion. The technology typically involves cooking vegetables and fruit together with ethanoic acid, salt and sugar. Starches or gums are usually present as thickening agents. The apparent nature of the products differs, sauces being of thinner consistency than chutneys and relishes, which also contain particulate ingredients. All are basically similar and highly shelf stable as a consequence of low pH value, high levels of inhibitors such as ethanoic acid and the cooking and hot filling process. Spoilage is rare although fermentation by yeast may occur in certain types.

Four good quality examples of chutneys, relishes and pickles.

496. Tomato chutney. Chutneys were originally of Indian origin popularised in the U.K. during the Indian Empire. They are highly spiced although those produced commercially in the U.K. are usually bland in comparison with the original products.

497. Sweet-corn relish. A product of American origin now widely available in Europe as an accompaniment to burgers etc.

498. Piccalilli. An English product containing cauliflower as major consituent in a mustard-containing base.

499. Brown pickle. An English product consisting of a mixture of vegetables in a relatively sweet base.

496

497

498

499

Vinegar (500–504)

Vinegars may be produced from any raw material containing sufficient sugar or alcohol. Examples are fruit juices (apple, orange etc.); starchy vegetables (potato, sweet potato); malted cereals (barley, rye); sugars (molasses, honey) and alcoholic beverages or dilute ethanol. In the U.K. most table vinegar is derived from beer manufacture (technically alegar), in France from wine and in the U.S.A. from cider or apple juices. In the Far East vinegar derived from pure ethanol is more common.

Vinegar manufacture is a two-stage process, the production of alcohol from carbohydrate and the subsequent oxidation of alcohol to ethanoic acid. The first stage may be carried out separately as part of an alcoholic drink manufacturing process (see pp. 230, 236) but if not, a high alcohol yielding strain of *Saccharomyces cerevisiae* var. *ellipsoideus* is likely to be used. The second stage is carried out by a mixed culture of *Acetobacter spp*, and possibly, *Gluconobacter spp*. Commercial production of vinegar involves continuous or semi-continuous processes whereby the alcoholic liquid to be converted is trickled over a film of bacteria growing on a packing material providing a large surface area, traditionally beech wood shavings. The details of the technology involved vary considerably and reference should be made to a specialist publication.

Some spirit vinegar is produced by diluting ethanoic acid and lacks the distinctive tastes of traditionally produced vinegars.

Pickles and Preserves

Vinegar Based Pickles (505–515)

Vinegar is used to preserve a wide range of vegetables. The raw material may undergo a preliminary fermentation in salt or a salt brine as with some types of dill pickle and pickled onions but more commonly is steeped, after cleaning and preparation, in vats in a vinegar pickle before filling into jars. Pasteurisation is usually applied after filling to reduce the potential for spoilage and in some modern processes the pickling takes place in the jar. In addition to vinegar giving an ethanoic acid concentration of 1.5%–3% ($^v/_v$), salt is also present in the brine as well as lactic acid and other preservatives where permitted. Sugar, herbs, spices and colouring may be present according to the product.

Although pickles are relatively stable at room temperatures spoilage may occur due to film yeast and, more rarely, moulds and bacteria such as *A. pasteurianus*. Some pickles made to modern recipes and having low salt and ethanoic acid contents require refrigeration after opening and have a limited post-opening storage life. Pickles have been associated with staphylococcal food poisoning due to growth of the organism and elaboration of toxin on the prepared vegetables prior to pickling. It should be noted that *Staph. aureus* grows well on onions which are strongly bacteriostatic to most genera.

Three examples of vinegars.

500. **French white wine vinegar.** These vinegars are sometimes flavoured by addition of herbs such as tarragon.

501. **English malt vinegar (alegar).** Caramel is added to produce the dark colour.

502. **Chinese 'red' vinegar** produced from fermented rice. The ethanoic acid content is low and the flavour unlike that of western vinegars.

503 and **504.** **Some strains of *Acetobacter* produce cellulose slime** during growth in vinegar and appear as a discrete body rather than as a cloudiness or haze. This growth is often referred to as **Mother of Vinegar.** An example (arrowed) is shown in the bottle (**503**) and in more detail isolated from the vinegar (**504**). When storage is prolonged a bacterial mass may develop which entirely fills the container. The appearance is such that this is often referred to as a **vinegar eel.** Care must be taken not to confuse this with the

nematode worms *Anguillula acetii* which may be found in vinegar. These latter were common when vinegar was retailed from difficult-to-clean wooden casks but are rare today.

505. **Pickled mixed vegetables** are popular in Eastern Europe and are often bottled with a mild heat treatment for export.

506 and **507.** **Vinegar based pickles** enjoy considerable popularity particularly in the U.K. The most popular type is pickled onions (*Allium cepa*) (**506**). Vinegar pickled products of this type do not usually undergo a fermentation. Pickled gherkins (cucumbers *Cucumis sativus*) (**507**) are also a popular type of vinegar pickle. A lactic fermentation is sometimes employed in their manufacture.

508. **Pickled beetroot** (*Beta vulgaris*) containing small white flecks (arrowed). These often appear during refrigerated storage after opening. In the example illustrated the flecks consist of amorphous vegetable material the nature of which has never been fully investigated or elucidated. Early growth of film yeasts such as *Candida* have a superficially similar appearance.

500 ●

501 ●

502 ●

503 ●

504

505 ●

506 ●

507 ●

508 ●

133

509

510

511

512

513

514

515

516

517

518

134

509 and 510. **Good quality pickled dill cucumbers** free from film yeast and other faults (**509**). Dill (*Anethum graveolens*) was originally added as a preservative but now is added purely for flavour. The dill (**510**) (arrowed) is often mistaken for a foreign body.

511 to 514. **Contamination of dill (cucumber) pickles.** Pickles and other delicatessen items sold from bulk containers are subject to both aerial contamination and to cross contamination from improperly cleaned containers. Dill (cucumber) pickles appear to be particularly prone to spoilage by film yeasts both when retailed in bulk or in jars contaminated after opening. **511** illustrates growth of a film yeast (*Candida sp* or *Debaryomyces sp*) on dill pickled cucumbers which were purchased in this condition from bulk in a large London department store. The growth is seen in greater detail (**512**). Film yeast grows well at refrigerator temperatures and **513** shows virtual total coverage by yeast 2 days after purchase. Growth of film yeast is a superficial phenomenon and there is no significant penetration of the interior of the cucumber (**514**).

515. **Film yeast** on the surface of an opened bottle of pickled dill cucumbers. This particular type is a 'sweetcure' with low ethanoic acid content and added sugar. Film yeast growth is particularly rapid once contamination has occurred.

Salt Based Pickles (516–520)

Manufacture of salt based pickles may involve a fermentation of the product or merely steeping it in brine. In the case of fermented products the fermentation is usually uncontrolled, that is, conditions are set up to favour the development of a flora composed of lactic acid bacteria and the fermentation is then allowed to take its course. Members of the *Enterobacteriaceae* including *Enterobacter spp*, *Hafnia spp* and *Klebsiella spp* are of importance in the initial stages.

Vegetables may be dry salted as with sauerkraut where 2.25%–2.5% of dry salt is added or fermented in a brine of ca. 10% salt as with cucumbers. The salt concentration should be sufficiently high to prevent the growth of undesirable micro-organisms including pathogens but low enough to permit growth of lactic acid bacteria and acidification. A wide range of lactic acid bacteria have been isolated from fermenting vegetables but under commercial production conditions *Pediococcus damnosus*, *Lactobacillus brevis* and *L. plantarum* appear to be of greatest importance. Defects may arise during production, including gas production in cucumbers by heterofermentative lactic acid bacteria and red pigmentation of sauerkraut by species of *L. brevis* in addition to harsh or bitter flavours.

Brine fermented vegetables are prone to spoilage by film yeast both during and after fermentation. Fermented vegetables are frequently packed in cans or bottles and pasteurised for long term storage purposes, and considerable quantities of brine cured products are desalted and further processed into condiments such as relishes (see p. 131).

In brine pickled non-fermented products the salt concentration must be sufficiently high to prevent the growth of spoilage micro-organisms as well as that of pathogens including *Clostridium botulinum*. The process must be carefully monitored to ensure that this condition is met. Dry salting without fermentation is occasionally used for preservation of vegetables of low carbohydrate content such as peas and lima beans. It is not possible to ensure even salt distribution in such products and a number of fatal outbreaks of botulism have occurred as a result of their consumption. Such products cannot be considered safe unless fully heat processed.

516. **Brine pickled garnar root.** Produced in India and East Africa the brine contains turmeric (*Curcuma longa*) which may have anti-microbial properties in addition to those of colour and taste. The brine surrounding the garnar supports a large mixed population of yeasts and bacteria without apparently affecting product quality.

517. **Chinese brine pickled radish** (*Raphanus sativus* var *longipinnatus*). Such vegetables are removed from the brine and partially dehydrated before final packing. Although salt is the only preservative such products are highly shelf stable and microbial spoilage is rare.

518. **Quercetin crystals** on the surface of capers (*Capparis spinosa*) in brine. Quercetin, a flavone compound, and its derivatives are present in the tissues of many plants including olives, hops (*Humulus lupulus*) and onions. In high salt concentrations discrete crystals (arrowed) may form. The phenomenon is harmless but may be mistaken for yeast colonies.

Soya Sauce Based Pickles (521–523)

Soya sauce (see p. 156) is often used as a pickling medium in Oriental products. The product to be pickled may undergo a preliminary fermentation (see above) or merely be steeped in soya sauce. Preliminary cooking is sometimes required to soften the tissues. Soya sauce is an effective preservative by virtue of its high salt content, low pH value and alcohol content, and spoilage of soya sauce based products is rare. On prolonged storage surface mould growth can occur and products for export may be pasteurised. Soya sauce based pickles are usually vegetable in origin although fish may also be preserved in this manner (see pp. 108–109).

Sugar Based Preserves (524–528)

In addition to its role in jam manufacture (see p. 140) sugar is also used as a preservative in syrup form. There are basically two types of product. The first is where the preserved material is held in a high sugar concentration syrup, the second where after equilibration in a syrup the preserved material is removed, drained and, in some cases, packed in dried sugar. The first type is usually a luxury product like peaches, which may also have received a mild thermal processing. The second includes cake decorations such as glacé cherries and crystalline fruit, which are normally preserved in syrup, but roots and stems such as ginger (*Zingiber officinale*) and angelica (*Angelica archangelica*) are also preserved in this manner as are some Caribbean sweet beans.

Spoilage, which may involve massive gas production, is usually by osmo-tolerant yeasts and it should be noted that the effects on micro-organisms of any heat treatment are reduced by the protective effect of the sugar. Mould growth may also be a problem and strains of *Aspergillus* and *Penicillium* have been isolated from crystalline fruit with a sugar concentration in excess of 67%.

519 and 520. **Vine leaves** (*Vitis sp*) pickled in a weak brine and pasteurised in large cans are exported from Greece for re-packing. In the example illustrated (**519**) the leaves have been re-packed into plastic pouches which have been stored at ambient temperature and growth of yeast has occurred in the brine (arrowed). Examination of the leaves themselves shows visible colonies of yeast developing (**520**).

521. **Lettuce stem (Chinese lettuce, *Brassica cerua*) pickled in soya sauce.** Such products are pasteurised before export from mainland China and are highly shelf stable.

522 and 523. **Tofu** may be preserved for long term storage by pickling in soya sauce and pasteurising in bottle. **522** shows spoilage of this product by moulds and bacteria which had entered by a faulty cap seal. The spoiled tofu at the top of the bottle is seen (arrowed) in contrast to the relatively unspoiled product at the bottom (**523**). (See p. 156.)

524, 525 and 526. **Sugar preserved fruit** are often used as cake decorations and ingredients. Although generally shelf stable at ambient temperatures **osmotolerant yeasts** may develop particularly if water uptake dilutes the sugar and raises the Aw level. **524** shows good quality glacé cherries (*Prunus sp*) which may be contrasted with spoiled (**525**) showing visible yeast colonies (arrowed A) and a large amount of exudate (arrowed B). The cherries also smelt strongly of alcohol. A close-up of the colonies is shown (**526**) (arrowed). These cherries had been removed from their original container and had probably been stored for a prolonged period in warm, moist conditions. The identity of the yeast is not, in this example, known.

527. **Stem ginger** – a good example of a luxury sugar preserve.

528. **Sugar preserves** are invariably fruit in the Western Hemisphere. In the Far East vegetables may be found as a constituent of mixed sugar preserves. The example illustrated is a Philippine produced sweet bean pickle containing various fruits and two types of bean (*Phaseolus spp*).

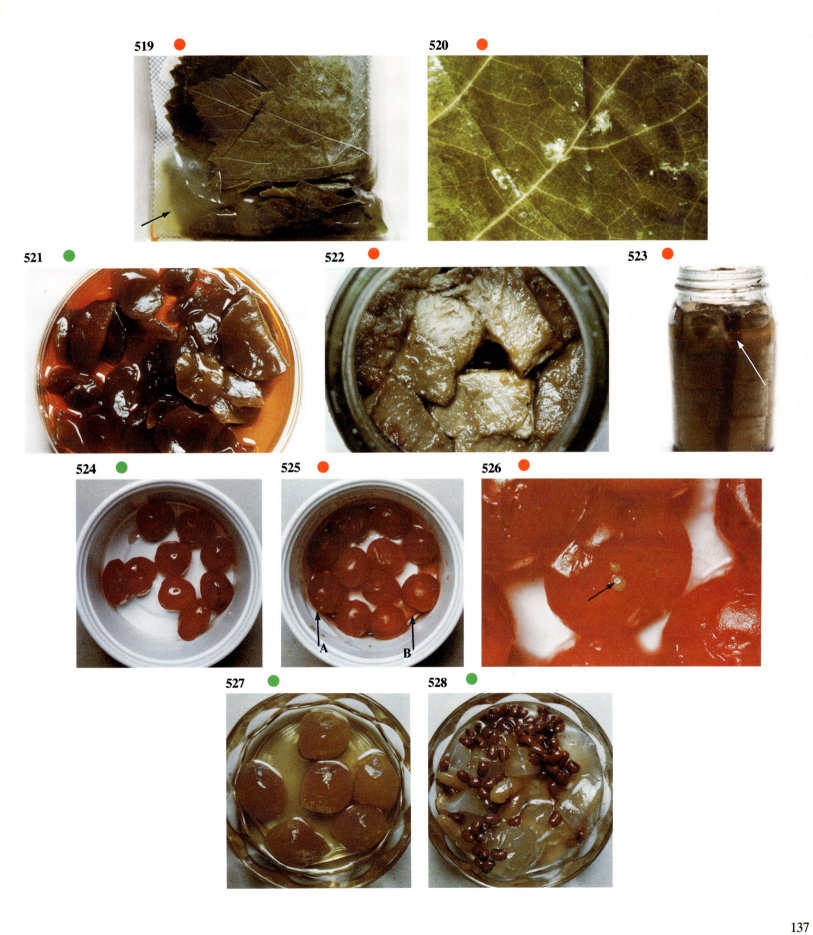

519 ●

520 ●

521 ●

522 ●

523 ●

524 ●

525 ●

A   B

526 ●

527 ●

528 ●

Sugar and Molasses (529–534)

Sucrose

Except for special purposes (see below) the majority of sugar produced for domestic consumption is sucrose. The major sources are sugar cane (*Saccharum officinarum*) and sugar beet (*B. vulgaris*), although significant quantities are derived from palm (*E. guineensis*) in the Far East and from maple trees in the U.S.A. and Canada. It should be stated that cane sugar and beet sugar are not distinguishable; any difference in crystalline structure is due to refining technology. While sugar beet is produced largely in the developed nations of western and central Europe sugar cane is grown in tropical and sub-tropical climates. A number of the major producing countries such as Barbados, Cuba and Mauritius are highly dependent on sugar and its ancillary rum as sources of export currency and may be seriously affected by crop failure due to climate or disease. The sugar industry is thus closely linked to both economic and political stability; a Rastafarian group for example attempted to destabilise the Jamaican economy by burning cane fields.

Highly refined sugars have few microbiological problems and although some may occur with molasses, the major problems lie in post-harvest and processing stages. A detailed discussion of the microbiology of sugar manufacture is beyond the scope of this book and reference should be made to publications such as Tilbury (1973) (cane sugar) and Klaushofer *et al* (1971) (beet sugar). It should be noted that biodeterioration of harvested cane by *Leuconostoc mesenteroides* (sour cane) leads to a mean loss of 4.75% of recoverable sucrose for each day elapsed between cutting and milling. At one Jamaican estate studied by Tilbury (1972) this was equivalent in monetary terms to ca. £340,000 per annum. Further losses of up to 1.5% occur due to microbial activity at the sugar mill while dextran slime in the juice and in mill machinery cause additional processing problems and sucrose loss through blockage. Similar problems occur with beet sugar. Lactic acid bacteria may survive processing or re-enter the refined product by contamination and it is necessary to ensure that sugar supplied to breweries, wineries etc., is free from such organisms. In addition refined sugars may contain large numbers of highly heat resistant endospores of thermophilic bacteria derived from high temperature parts of the process plant. For many canning purposes it is necessary to ensure that the spore count is low and special grades of sugar are produced for this purpose. Some of the metabolites produced by microbial activity during production may impart a particular desirable characteristic to certain sugars. The characteristic flavour of Barbados molasses and some palm sugars has been attributed to fermentation of the raw materials by *Zygosaccharomyces* species of yeast and '*Clostridium saccharolyticum*' growing in a mutualistic relationship (Frazier, 1967).

Microbial spoilage of the final product is confined to relatively moist and unrefined sugars such as palm sugar and to syrups. Osmotolerant yeasts for example are able to grow in sugars with a concentration as high as 62°–72° Brix and some moulds may also be able to develop. Bacteria are rarely involved in spoilage of sugars the exception being maple syrup which is subject to spoilage by several types of bacteria including lactic acid bacteria, pseudomonads and members of the *Enterobacteriaceae*.

Some molasses and syrups are subject to chemical instability and decomposition particularly at high storage temperatures (>30°C), which are rarely encountered in temperate climates but are common in tropical countries. Decomposition results in production of carbon dioxide and considerable pressures may be generated in small domestic containers. Two possible reactions are involved, the first an interaction between amino acids and carbohydrates (Maillard type reactions); the second the decomposition of calcium gluconate. It should also be noted that some types of molasses are sufficiently acidic to attack the steel containers with associated production of hydrogen ('Hydrogen swell').

Glucose

Most glucose is produced by chemical hydrolysis of corn (maize) starch, normally as an ingredient for foods rather than as a product in its own right, and is often handled as a concentrated syrup. Manufacture of glucose provides little opportunity for microbial growth and microbial numbers in glucose syrup are usually low. For this reason, other technological considerations apart, it is often preferred to sucrose for alcoholic drink manufacture and canning. Spoilage by osmotolerant yeasts may however occur. It should be noted that hydrolysis of sucrose (inversion) yields both glucose and fructose.

Lactose

Lactose is derived from milk powder and is widely used as an ingredient in food manufacture. No specific spoilage or quality problems are associated with it.

Other carbohydrates

Small quantities of other carbohydrates are used in foods to meet particular technological requirements. These include sucrose substitutes such as sorbitol in dietetic foods. No specific spoilage or quality problems are associated with them but it should be noted that **sorbitol** has been linked with **allergic symptoms** in some consumers.

Artificial Sweeteners

Artificial sweeteners such as the sulphonamides **saccharin** and **sucaryl** possess considerable sweetness and for various dietetic and technological reasons may be substituted for sugars at both domestic and manufacturing level. Products made with synthetic sweeteners are sometimes considered to be 'harsh' or 'artificial' and may even be thought to have a 'bitter' aftertaste.

In addition to organoleptic shortcomings saccharin has also been shown to cause cancer in laboratory animals and legislation in the United States is to require foods containing saccharin to carry a health-warning.

A more recently developed artificial sweetener, **aspartamane**, avoids the problems of saccharin. Aspartamane, a low calorie nutritive dipeptide (L-aspartyl-L-phenylalanine methyl ester) which is two hundred times more potent than sucrose is further noteworthy in that while its precursors such as aspartic acid can be produced by fermentation the most efficient means of production is by methylation of aspartyl-phenylalanine synthesised using genetic engineering (recombinant DNA) technology (Pellon and Sinskey, 1983).

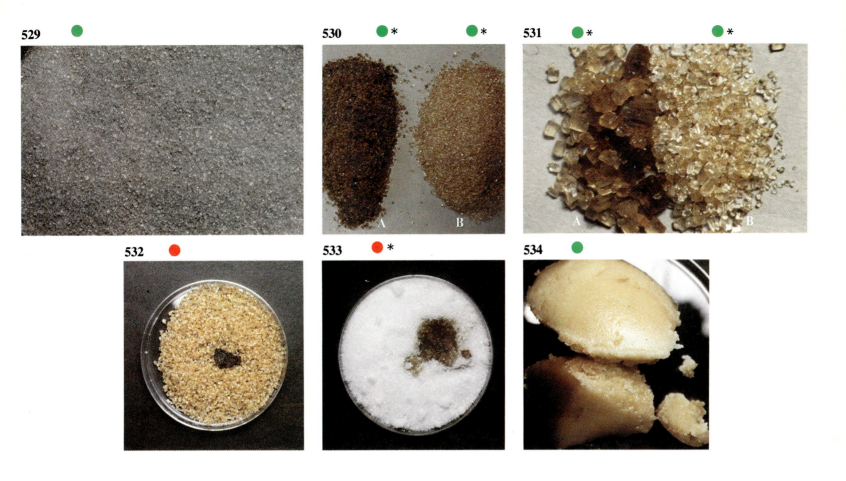

529. ●

530. ●* ●*

531. ●* ●*

A   B   A   B

532. ●

533. ●*

534. ●

529. Highly refined white cane sugar. The common form in the western world. Note that cane and beet sugar are indistinguishable.

530 and 531. Semi-refined cane sugars such as Demerara may vary considerably in appearance while being of perfectly acceptable quality. **530** compares Demerara sugar from Mauritius (A) with that from Barbados (B). The former is seen to be very much darker than the latter. In addition the sugar crystals differ in structure (**531**), those from the Mauritian sugar (A) being larger and coarser than those from the Barbados (B).

532. A foreign body in Demerara sugar. Examination showed it to be caramelised sugar which had probably been 'burnt-on' to the side of the sugar boiler or related equipment.

533. A brown discoloration in sugar. Most of the sugar had been used and the brown area smelt strongly of vanilla essence. It is probable that the sugar was accidentally contaminated with vanilla or a similar essence in the consumer's home.

534. Thai palm sugar as used in a semi-refined state. Although most sugar is produced from cane or beet the palm is an important source in Asian countries such as Thailand. The product is relatively moist and can be subject to spoilage by osmotolerant yeasts.

Examples of foreign bodies in sugar may also be found in Chapter 17, Foreign Bodies and Infestations as follows:

Pp. 244–245 Psocids in sugar

Jam and Honey (535–543)

Jam is produced by boiling fruit in water together with sugar (usually sucrose) and (usually) pectin. The final sugar concentration is normally ca. 65% because higher concentrations may lead to crystallisation, a quality defect. The jam is set by formation of a pectin gel. Sufficient pectin substances may be derived from the fruit itself but in commercial practice pectin is frequently added. The rheological properties of the gel are important with respect to quality because a weak gel is too syrupy whilst an over-strong gel has poor spreading properties. Mould growth on jams can occur when sugar concentrations are as high as 70% and modern low acid, low solid formulations, should be considered perishable once opened.

Jams formulated for diabetics and some weight reducing diets often contain **sorbitol** instead of sucrose. Sorbitol gives rise to allergic symptoms in some people and this fact should be considered both in formulating such products or in investigation of illness related complaints.

Honey is variable in composition being either crystalline or liquid but should have a water content not greater than 25%. The sugar content of honey is 70%–80%, mostly as glucose and levulose and its pH value is typically in the range 3.2–4.2. Spoilage is rare particularly in commercial honey which is pasteurised, but where it occurs osmotolerant yeasts such as *Zygosaccharomyces spp* and *Torula spp* are usually responsible, the typical spoilage pattern being a slow accumulation of carbon dioxide, ethyl alcohol and non-volatile acids which produce off-flavours.

As a natural product honey may contain large numbers of micro-organisms. Commercial honey is usually pasteurised but bacterial endospores survive and honey has been implicated as a causative agent of **botulism.** Unsubstantiated reports of mild food poisoning by honey containing large numbers of *Bacillus spp* have also been made (authors' unpublished observation).

535. Raspberry jam. A good quality example. Unlike lower quality jams this example is of high fruit content and contains no artificial colouring or flavouring.

536. Marmalade. Jams prepared from citrus fruits are commonly referred to as marmalade. The example shown is of good quality and like the jam (535) contains no artificial colour or flavouring.

537. Damage to the lid of a jam jar which probably occurred prior to or during capping. Such a fault, which should be detected by visual quality control procedures, may lead to contamination of the jam before opening (see above).

538. Mould growth (arrowed) around the inside of a jar of jam. The mould is readily distinguishable from machine dirt on close examination (see 539). Contamination probably occurred after opening the container and moulds developed during a long storage period (>2 months). Although jams are considered to be room temperature stable the manufacturers of some low acid (high pH value), low sugar (high Aw level) jams recommend refrigerated storage after opening.

539. Jam containers appear to be particularly prone to contamination with machine dirt (arrowed) caused by improper cleaning of the capping machine. Machine dirt is often mistaken for mould or for 'corrosion'.

540. Air bubbles in jam. This is not spoilage but merely due to incorporation of air when the jam was tipped to one side.

541, 542 and **543. Honey** may be packed as clear (liquid) honey or as set (crystallised). The flavour and to some extent colour is determined by the plant on which the bees have fed. Clear honey is usually from a single source, the example (541) being Hungarian acacia honey. In some cases a piece of honeycomb is included in the pack.

Set honey may be from a single source but is often blended (542). The sugar content of set honey is higher than that of clear and it is less prone to microbial spoilage, which is caused by osmotolerant yeasts and is a slow process lasting for several months. Crystallisation and darkening are visual indications that fermentation is occurring (543) although crystallisation of set honey may be caused by other factors including storage at low temperatures. In the case of a fermentation the crystallisation is accompanied by an odour of alcohol.

Examples of foreign bodies in jam may also be found in Chapter 17, Foreign Bodies and Infestations as follows:

Pp. 244–245 *Curculionidae*
Pp. 244–245 Insect leg

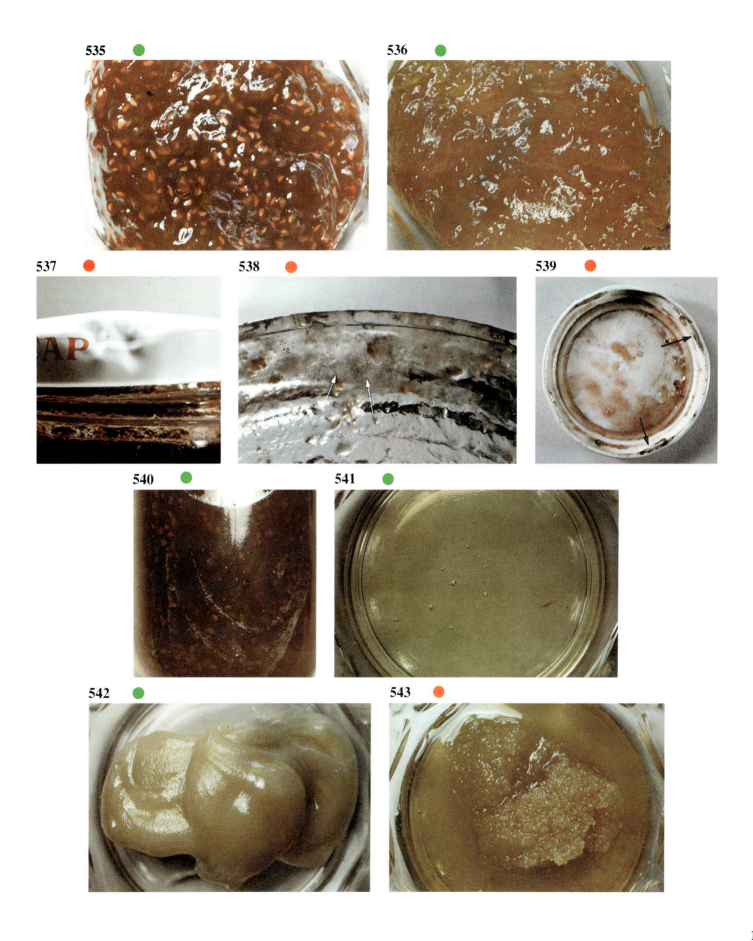

535

536

537

538

539

540

541

542

543

141

Confectionery

Chocolate (544–546)

Chocolate, like cocoa, is derived from the cacao tree *Theobroma cacao* and the initial stages of processing are similar. In chocolate manufacture cocoa nibs are milled which, together with the heat produced by the milling, releases 51%–56% of intracellular fat. The cocoa is then mixed to produce an homogeneous mass and refined to the correct particle size. For good quality eating chocolate not more than 20% of the mass should have a particle size greater than 22 microns.

The final stage in manufacture (excluding moulding and packing) is 'conching' – a holding period where time, temperature, agitation and aeration are balanced to produce the final desired characteristics. During manufacture milk solids may be added to produce milk (white) chocolate as opposed to plain (black).

Chocolate is subject to contamination at source and the heating applied during manufacture may not be sufficient to destroy vegetative micro-organisms. It has consequently been implicated in a number of outbreaks of salmonellosis, the most recent involving *Salmonella napoli* in Italian chocolates exported to the U.K. and West Germany.

544. Black (A) and milk (B) chocolate. These two examples are of good quality although the line of crystallisation (arrowed) on the black chocolate, probably due to uneven cooling, is technically a fault.

545 and **546. Poor quality chocolate,** irregular in shape and of unappetising appearance (**545**). A 'bloom' of sugar crystals resembling dust is also present (**546**) (arrowed).

Boiled Sugar Confectionery (547–548)

Boiled sugar confectionery is produced by boiling a solution of sugar, flavouring and colour until the required concentration is reached and then setting the mix in moulds without crystallisation. Fillings such as sherbet powder or semi-solids may be incorporated. Quality is largely determined by personal preferences, for example sweets produced for children are likely to be of a garish colour unlikely to appeal to adults. High quality sweets of this type are also likely to be more clearly moulded and to have fewer broken pieces.

Specific problems with boiled sugar confections are few but importers should take care to ensure that colourings and flavourings conform to appropriate regulations.

547. Sugar confectionery as an art form. The Italian town of Sulmona specialises in sugar coated fruit, which forms a relatively large part of the local diet.

548. Boiled sugar sweets. Two good quality examples, butter mints (A) and 'rocks' (B). Individual wrapping of sweets improves hygiene when sold loose and prevents coalescing.

Miscellaneous Confectionery (549–553)

In addition to the well defined groups of chocolate and boiled sugar confectionery there exists a multitude of types based on gums, soft sugar mixtures and plastic sugar mixtures such as toffees. The range is too varied for the technology to be discussed even in outline and reference should be made to specialist publications.

There are no significant quality control problems unique to this category although some low sugar content fondant type mixtures may be spoiled by osmotolerant yeast sometimes with spectacular consequences.

An unusual type of food associated illness has resulted from excessive consumption of **crackling candy,** a sweet which rapidly evolves large quantities of carbon dioxide into the mouth. In a typical case (Anon, 1984) severe mouth soreness resulted after eating a type of crackling candy 5 days in succession.

549. Many sweets are based on edible gums such as gum arabic (Acacia) together with appropriate flavouring and colouring. Sample A is of mediocre quality wine gums. Contrary to the expectations of children (and some adults) wine gums contain no alcohol. Sample B is of good quality 'coconut mushrooms' topped with desiccated coconut. Persons responsible for procurement of desiccated coconut should be aware that it has previously been implicated in outbreaks of salmonellosis and ensure that supplies are from a reputable source. The organism implicated, *Salmonella ferlac,* was atypical in fermenting lactose and could therefore not be detected by conventional tests.

550 and **551. Poor quality 'jelly beans'** (**550**) and spearmint gum (**551**). The sweets, which had obviously been handled, were sold from open containers, placed close to the floor and unprotected from contamination. The degree of handling as children select their sweets is considerable and those retailed in this way have been implicated in the transmission of childhood disease. The sweets illustrated were imported into the U.K. from Spain and care must be taken to ensure that the ingredients conform to national regulations and that potentially dangerous substances, particularly colouring, are not present.

552 and **553. Asian confectionery.** A growth of the Asian community in the U.K. has resulted not only in the Asian cuisine becoming widely accepted but also in Asian confectionery being generally available and eaten by children irrespective of racial origin. Such a product, sugar coated fennel seeds, is shown here (**552**). Persons involved in the importation of such sweets should ensure that colouring materials used conform to the appropriate national standards. A further example of the increasingly cosmopolitan nature of the U.K. food industry is given by these Asian sweets (**553**), which consist of a flour, water and sugar paste moulded and coloured to resemble other foods such as potatoes, cheese etc. Production is often on a small scale and it is necessary to ensure that the appropriate hygiene regulations are met.

544 ● ●   A   B

545 ●

546 ●

547

548 ● ●   A   B

549 ● ●   A   B

550 ●

551 ●

552 ●

553 ● *

143

Snack products and nuts

Potato Based Snacks (554–558)

The market for savoury snacks is immense and highly competitive. A large part of it is taken up by children and a constant stream of new shapes and flavours is produced in an attempt by manufacturers to consolidate and increase their share of the market. This results in products which appear, to the adult at least, bizarre in nature, a popular (1982/3) example being a snack which resembles in shape and lurid colour the 'Space Invaders' of the electronic arcade game. The basic technology of all such products is similar and

quality control problems are usually of a cosmetic nature.

Potato based snacks may be made from potato pieces or from reformed potato powder. In either case cooking is by frying. Potatoes of the correct sugar content must be used to obtain proper coloration.

Cereal Based Snacks (559–560)

The commercial background to cereal based snacks and the technology involved in production is in most cases similar to that of potato based snacks.

554. Good quality potato based snack products. Potato sticks (A) are made from whole sliced potatoes whereas potato rings (B) are reformed from powdered potatoes. Some sticks and crisps are also made in this way.

555 and 556. Poor quality potato crisps (chips). Note the patches of extreme blackening on the edge of some of the crisps (arrowed) (555). The cause is failure to remove the 'eyes' from potatoes after peeling. **556** shows in contrast good quality crisps.

557. Cheese flavoured potato snacks with a pink powder (arrowed) on the surface of two pieces. This is the flavouring compound which should be evenly distributed over all pieces. As the powder is normally applied when the snacks are warm after cooking it tends to aggregate in discrete areas leading to uneven colouring

and flavour. No solution has yet been found for this technological problem.

558. Potato and rice flakes (Far-Far). A savoury product analogous and similar in flavour to potato crisps (chips). This product illustrates the effect of colour on perception of quality since in Europe and the U.S.A. bright colours such as reds and greens are not normally associated with savoury products and the concept would produce an adverse customer reaction.

559. Wheat and maize flour based snacks. Two good quality examples which are basically similar although that on the left (A) is prawn flavoured while that on the right (B) is cheese flavoured.

560. Croutons – deep fried bread with garlic flavouring represents a more sophisticated snack than most.

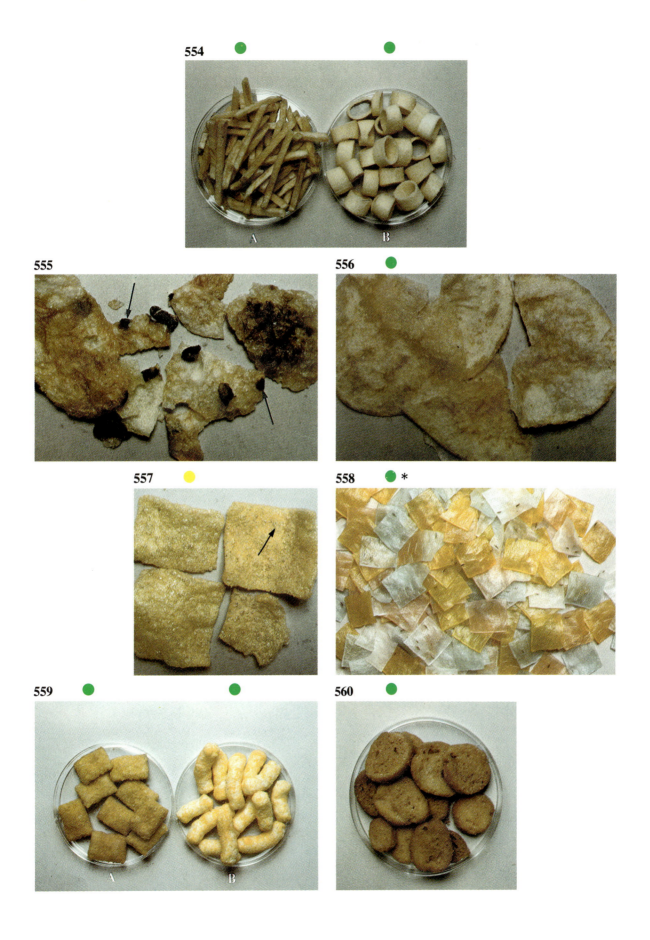

554 🟢 🟢

555

556 🟢

557 🟡

558 🟢 *

559 🟢 🟢

560 🟢

145

Nuts (561–569)

For convenience nuts are discussed in the present section concerned with snacks although it is recognised that many nuts are purchased for other purposes. Nuts in shell or shelled for consumer convenience receive no processing other than pre-packing or, in some cases, chopping or flaking. The predominant nuts for the snack market, peanuts, are cooked either in oil with salt or dry roasted. In the former case particularly, life is limited by oxidative rancidity, and gas (nitrogen) packing is often used to permit extended storage.

561.   Poor quality roasted peanuts. The skin covering the nuts has been improperly removed and parts have remained throughout processing. Ideally such nuts should be removed prior to packing.

562 and 563.   Nuts in shell are sold primarily for the Christmas trade and as such poor quality is likely to lead to considerable loss of goodwill. **562** illustrates Brazil nuts (*Bertholettia excelsa*) of generally poor quality with one nut showing signs of rodent attack during storage. The group of mixed nuts (**563**) are also of poor quality and contain a foreign body (arrowed). This was probably a piece of dried fruit which had developed mould and subsequently dehydrated.

564 and 565.   Pre-packed almonds (*Prunus amygdalus*) showing rodent damage (**564**). Close examination of the almonds (**565**) shows the damage to be too extensive to be caused by shelling machines and to be distinguishable from boring by insects.

Damage probably occurred before pre-packing as debris is absent. The possibility of damage in the consumer's home cannot be totally excluded.

566 and 567.   Rodent damage to whole almonds (**566**). Rodent, particularly rat, jaws may be sufficiently powerful to snap nuts such as the almond (**567**).

568 and 569.   Examination of walnuts (*Juglans resia*) allegedly mould contaminated showed no evidence to substantiate the claim (**568**). A more detailed examination (**569**) showed evidence of rodent damage (arrowed).

Examples of foreign bodies in nuts may also be found in Chapter 17, Foreign Bodies and Infestations as follows:

P. 248 Spider leg in chopped nut cake topping

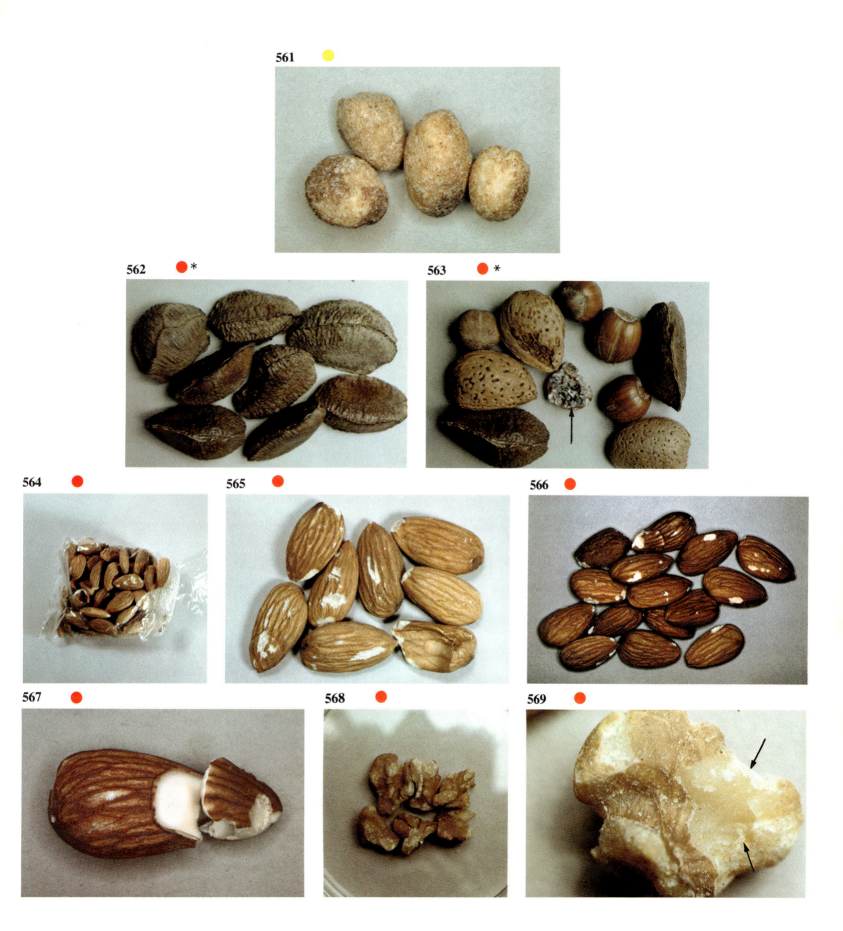

561 ●

562 ● *

563 ● *

564 ●

565 ●

566 ●

567 ●

568 ●

569 ●

Dried formulated products

Dried Soups, Sauces etc. (570–573)

Production of dried soups and sauces is a relatively simple operation involving mechanised weighing of ingredients, mixing and packaging. Ingredients are chosen to impart the desired flavour, colour and texture to the product when rehydrated. No specific quality control problems exist.

Whole Dried Meals (574–576)

There are two basic approaches to whole dried meals. The first is the snack meal concept of a complete light meal prepared in its packaging (plastic pot) by addition of hot water. Such products are usually based on rice or noodles and manufacture is similar to dried soups.

The second approach is a more substantial meal intended, when cooked, to resemble a conventionally prepared meal. The components are individually packed with the exception of the meat or fish component which is a mixture of dried ingredients. Western foods do not lend themselves to this treatment and the products are usually based on Oriental or Chinese dishes. There are no specific problems of quality associated with either type of product.

570. The Brabender Amylograph. While the properties of flour proteins are the major determinants of the suitability of flour for baking (see p. 115) the **properties of the starch** are of greatest importance in such products as **dried soups** where the correct degree of **gelation of the starch in the flour constituent** is necessary to obtain the required 'body'. The gelling properties may be determined physically using the Brabender Amylograph which continuously measures the resistance to stirring of a 10% suspension of flour in water during heating from room temperature to 95°C at a rate of 1.5°C/min. The illustration shows a flour suitable for a soup of moderate viscosity such as tomato. 'Instant' dried soups which do not require prolonged heating usually contain flour or starch which has been pre-gelatinised.

The gelling properties of starch are also important in formulations or canning when gelation at low temperatures may impede heat penetration into the can and possibly result in underprocessing (see p. 222).

571. A good example of a dried soup mix. The mix is dry and powdery, without solid lumps which would have indicated inadequate drying or exposure to moisture after drying. The peas are entire and have not been damaged during processing.

572. Poor quality soup mix which has gained moisture during storage. This is likely to affect flavour adversely as well as the solubility.

573. A good example of a dried sauce mix (onion). Such products are largely flour based (often with partially gelatinised starch) together with appropriate flavouring and dried constituents. As with dried soups the gelling properties of the starch in the flour are of major importance in producing a satisfactory product (see **570**).

574. Complete dried convenience foods represented here by **beef chow mein** attempt to exploit the requirements of consumers who, in western Europe and the U.S.A. have both increasingly adventurous palates and a desire for convenience. This example contains two types of noodle, the chow mein and soya sauce (not shown) in separate packs. The chunks of meat (arrowed) are chopped and reformed although in some varieties constituents such as prawns may be freeze dried. The flavouring of such products is bland, the cooked meal often bearing only incidental resemblance to the original on which it was based.

575 and 576. Instant rice or noodle based meat flavoured snack meals are an advanced formulated product (**575**). The ingredients are prepared to allow virtually instant rehydration in hot water. In this example the rice (**576**) has been 'puffed' to permit this. Note that **textured soya chunks** are used as the 'meat'.

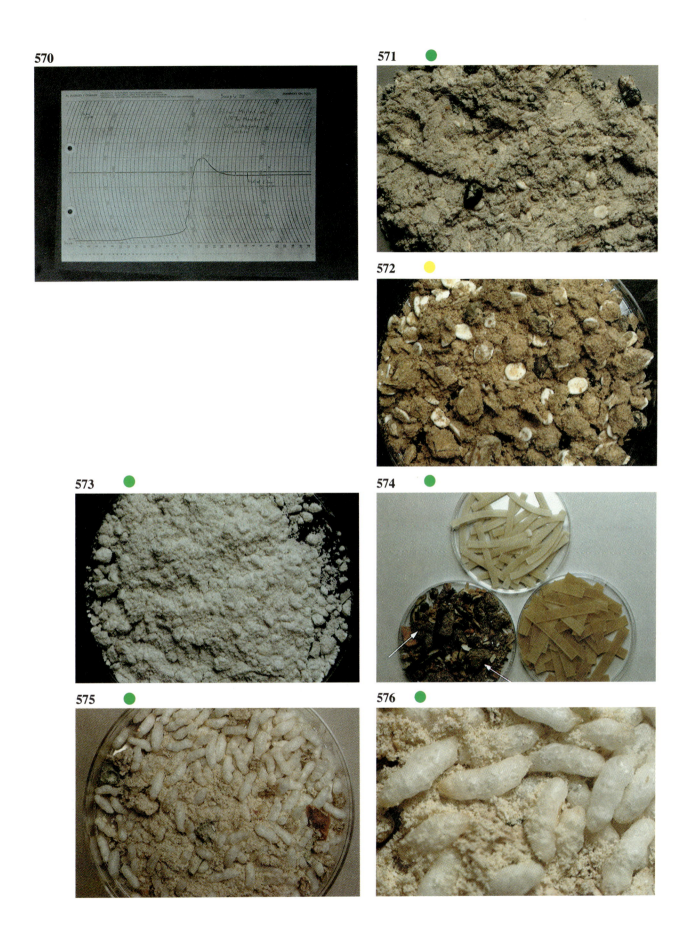

570

571 ●

572 ●

573 ●

574 ●

575 ●

576 ●

Soft Drinks (577–588)

Although a wide variety of soft drinks (non-alcoholic) are produced on a world wide basis they may be considered to be of three basic types.

1. Fruit and vegetable juices based wholly or primarily on the natural product.
2. Drinks, usually carbonated, produced largely from artificial colourings and flavours. For the present purpose this includes the cola drinks.
3. Bottled water.

Fruit and Vegetable Juices

The majority of products in this category are fruit juices since production of vegetable juice is relatively small scale. Fruit juices vary widely in terms of detail and may or may not contain added flavouring and colouring. Juices may be prepared as ready for consumption or as concentrates and may be preserved by freezing, pasteurisation or UHT treatment. Preservatives such as sodium benzoate may also be added. There is an increasing trend towards fresh fruit juices requiring refrigeration, while in-store preparation of citrus juices, usually orange juice, is increasingly common.

The low pH value of fruit juices (typically <4.0) means that spoilage is usually by yeast while mould may grow on the surface of juices. Lactic acid bacteria and acetic acid bacteria are the only bacteria able to develop in fruit juices. Spoilage by lactic acid bacteria (usually *Lactobacillus* or *Leuconostoc*) involves gas production by heterofermentative species, off-flavours and slime production while acetic acid bacteria (usually *Gluconobacter*) produce turbidity (in clarified juices) and off–flavours.

Fruit Juices and Mycotoxins

The possibility of mould damaged and thus potentially mycotoxin containing fruit being processed for juice exists and has given cause for concern. Jarvis (1976) notes that this is of particular relevance to apple juice manufacture. Scott (1972) detected significant quantities (20–100µg/kg) **patulin** in 25 commercial apple juices and other workers have made similar findings. Patulin is destroyed during alcoholic fermentation and is thus not a potential problem in cider. (N.B. Confusion has arisen due to dialectal differences between Standard and American English. The American term 'cider' refers to unfermented apple juice, cider in the Standard English sense being referred to as apple wine or 'hard' cider.)

Artificially Flavoured Drinks

Such products are typically of low pH values and also of low nutrient content. In most cases colourings and flavourings, which are typically aldehydes such as **ethylmethylglycidate** (strawberry flavour), are not prone to microbial attack. Care must be taken with imported products to ensure that ingredients conform to the appropriate regulations. Cola drinks such as Coca-Cola and Pepsi-Cola which probably represent the most widely distributed drinks in the world do not, contrary to tradition, contain cocaine.

Bottled Water

Bottled high purity water is widely used for drinking purposes in countries where mains water is prone to contamination, while in many European countries water from the famous spas such as Vichy has long been bottled and drunk for their supposedly remarkable curative properties. It should be noted that the mineral contents of some waters such as Vichy make them unsuitable for kidney disease sufferers. Bottled water became popular on a large scale in the U.K. during the drought of 1976 when the nitrate levels in some areas became sufficiently high as to make the water unsuitable for babies. Sales have continued at a high level since and several domestic sources are now bottled as well as a wide range of imported waters.

Water for bottling should be of potable quality at source and should be free from both actual and potential chemical or microbiological contamination. The management of the source should be in accordance with such guidelines as The Bacteriological Examination of Water Supplies (Report, 1982).

Carbonation to a level of 3–5 volumes of CO_2 also has a bactericidal effect. During the 1974 cholera epidemic in Portugal some 3,000 persons contracted the disease as a consequence of drinking non-carbonated bottled water but none were affected by carbonated water from the same source. Carbonation should not however be considered to be a substitute for the initial abstraction of high quality water or for good hygienic practice during processing.

Despite all precautions microbial numbers of up to 10^6 colony forming units/ml have been recovered from bottled waters, a considerably higher number than is present in directly drawn mains water. The presence of high numbers has been attributed to **cryptic growth** by **autochthonous** bacteria (ICMSF, 1980) but it is considered that the situation has still not been adequately explained and cases are known to the Authors where high numbers of bacteria were due to growth in the bottling plant and not to multiplication in the bottle. The bacteria present are typically aquatic bacteria such as *Flavobacterium* and *Cytophaga* and are of no direct public health significance. It is considered however that the use of non-carbonated bottled water for making up baby foods, and possibly, foods for invalids or compromised patients should be avoided where reliable mains water is available.

Chemical taints may also be a problem with bottled water. Many of these have been associated with polyethyleneterephthalate (PET) bottles and are attributed to the leaching of plasticisers or their thermal degradation products into the water.

Various types of soft drink based on the same fruit (lemon) are illustrated in the following sequence of illustrations.

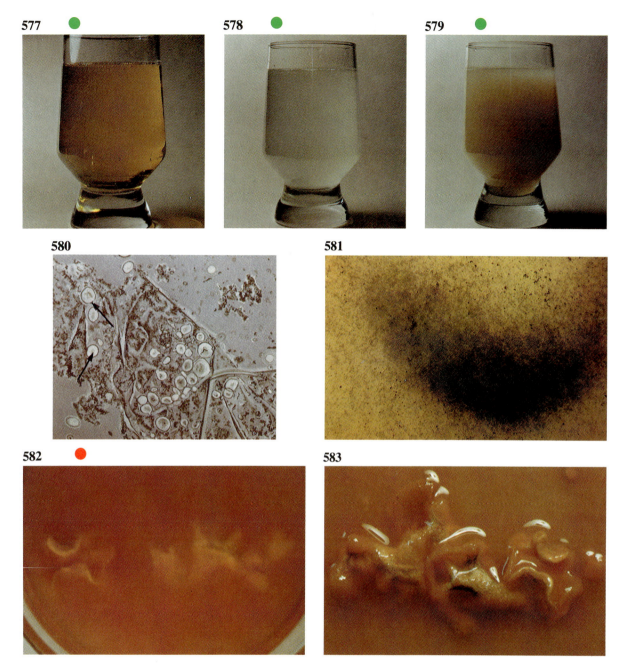

577.   Carbonated lime and lemon drink. This contains no natural fruit juice whatsoever but is based on synthetic colour, flavour and carbon dioxide as a carbonating agent. Legislative details vary from country to country but in every case labelling must clearly distinguish such a drink from a natural fruit juice.

578.   Carbonated 'bitter' lemon drink which contains some natural fruit in addition to flavouring, colouring, carbon dioxide and quinine to impart a bitter aftertaste.

579.   Whole natural lemon juice containing no additive except sulphur dioxide to retard browning. The example shown is concentrated and would require the addition of water (and probably sugar) before drinking. Similar pure juices are sold in various forms of packaging and may be unclarified as in the example illustrated or with insoluble material removed by centrifugation or filtration.

580 and 581.   In clarified fruit juices haze or sediment is a fault. Where the juice is not spoiled by growth of a micro-organism such as yeast **starch haze** is a common cause particularly where unripe fruit has been used. **Starch cells** are easily recognised microscopically. **580** shows starch cells magnified 500 times (arrowed) seen within the granular mass of cells and cell walls in apple juice. The presence of starch may be confirmed by adding diluted iodine solution to the juice which results in a blue or violet red coloration (**581**).

582 and 583.   A mould pellicle derived from fruit drink illustrated in general (**582**) and in detail (**583**). This particular fruit drink is a blend of juices from 10 fruits, UHT treated and then bottled. The mould would not survive UHT treatment and either entered the bottle at filling or through a faulty closure or, most probably, after the bottle was opened.

151

584

585

586

587

588

584 and **585.** **'Blowing' of the lid of a can** of soda-water packed for sale as a mixer drink illustrated from the side (**584**) and from above (**585**). Blowing is not due to microbial growth (pH value and nutrient level too low to support microbial growth) or to chemical activity but to over-carbonation which in this example was exacerbated by sudden pressure changes as a high speed train left the Severn tunnel. Similar effects occur on aircraft.

586 and **587.** **Cans used as containers for soft drinks** suffer the disadvantage of being non-resealable. **586** illustrates a newly developed lightweight aluminium can with resealable closure (**587**). Note (arrowed) the dents in the can due to finger pressure when resealing.

588. **Mains tap water (A) and carbonated spring water (B).** Despite the implicit purity of spring water microbial numbers often exceed 10^5 colony forming units/ml as opposed to direct mains water which in the U.K. typically contains less than that number. (Carbonated water usually has significantly lower microbial counts than non-carbonated due to the low pH value.)

Examples of foreign bodies in soft drinks may also be found in Chapter 17, Foreign Bodies and Infestations as follows:

P. 248 Insect larva in spring water
P. 251 Mould pellicle in orange juice
Pp. 252–253 Paper in lemon and lime drink

589 **590**

589 and **590.** **Two examples of good quality tea,** Earl Grey (**589**) and jasmine (**590**).

Hot beverages

Tea (589–590)

Tea (*Thea sinensis*) is largely grown in developing nations such as Sri Lanka and often forms an important source of export currency. The market, however, is controlled by developed countries, primarily the U.K., and prices which are set in London frequently bear little relationship to the economic needs of the producers. A number of large U.K. companies have in recent years received considerable adverse publicity over low payment to overseas employees. Attempts by producing countries to market tea directly to Europe have not been successful largely due to the lack of a marketing structure and to the power of existing brand names.

Tea may be **fermented (black), unfermented (green)** or **semi-fermented (oolong)**. Most exported tea is fermented. 'Fermentation' of the tea leaves is the result of intrinsic enzyme activity rather than that of micro-organisms. Microbial activity, particularly that of moulds of the genera *Aspergillus*, *Penicillium* and *Rhizopus*, may harm the flavour and reduce the quality of tea, which is blended to give standard flavours and aromas. Blending is carried out by expert tasters who are also responsible for determining the quality which will be selected for blending. As far as is possible the judgement of blenders is objective rather than subjective.

Instant teas – dried solubles of infused teas – may be used particularly in catering but lack the more subtle flavours of tea.

Tea is not usually subject to spoilage although flavour deteriorates with prolonged storage. Many foods readily pick up taint during storage but tea presents a particular problem. Precautions must be taken to avoid taint pick up during transport and storage. The problem is most acute with bulk tea but packaged tea may also be affected. Care must be taken to ensure that suitable packaging materials are used since 'cardboard' taints have been reported in the past due to use of poor quality paper.

Special reference must be made to **herbal** teas. Originally used as cheap rural substitutes for real tea and consisting of dried dock, comfrey etc., such products are now marketed in their own right and frequently have ill defined curative properties bestowed upon them. Production is small but has grown with the increased interest in 'health foods'.

During 1983 publicity was given to **comfrey tea** containing **deadly nightshade** (*Atropa belladonna*) imported into the U.K. from West Germany. This incident highlights the risks associated with 'natural' products where, because of the small scale of the operation, quality control procedures are inadequate.

Coffee (591–594)

Like tea, coffee (*Coffea spp*) is grown in tropical and sub-tropical climates and is of considerable importance in the economies of producing nations such as Brazil.

The first stage in the processing of coffee involves removal of an outer layer of pulpy material from the bean. Traditionally this has been by natural development of a pectinolytic microflora predominated by members of the *Enterobacteriaceae* although a pectinolytic enzyme derived from moulds may be used to avoid the possible production of off-flavours. Following removal of the pulpy layer a lactic fermentation occurs typically involving *Leuc. mesenteroides*, *L. brevis* and *Streptococcus faecalis*. The beans are subsequently dried and hulled before roasting and grinding. The choice of coffee is largely a matter of personal taste and ground coffee is often blended to give consistent taste and aroma. Much is sold as instant coffee where the hot water soluble solids are extracted and subsequently dried using spray, drum or freeze drying. Liquid coffee extracts are also available.

Miscellaneous Hot Beverages (595)

In addition to tea and coffee a number of other hot beverages are drunk in significant quantities. The nature of these varies and in some examples such as the malted milk drink Horlicks the process is specific to the product. Basically however there are three types – **cocoa and chocolate** based; **milk** based and **beef** based (see p. 159). Some of the milk based drinks are vitamin supplemented and the group as a whole is popularly considered to have nutritionally beneficial properties. There are no specific quality problems associated with such beverages.

591 ●

592 ●

593 ●

594 ●

595 ●

596 ●

597 ●

598 ●

599 ●

Four types of good quality coffee.

591. Coffee beans after roasting. The flavour of beans from different geographic locations is characteristic although beans are often blended to obtain a consistent flavour after grinding.

592. Freeze dried coffee granules. Such coffee has a more natural flavour than that of conventional instant coffee. The high energy costs of freeze drying mean a higher price.

593. Conventional instant coffee. The structure of conventional roller or spray dried instant coffee granules differs from that of freeze dried and the changes in flavour during drying are greater.

594. Liquid coffee extract, an early form of instant coffee. The flavour is characteristic and, while unlike that made from beans or instant coffee, may be preferred particularly by older people. The liquid extract is also widely used for flavouring.

Examples of foreign bodies in coffee may also be found in Chapter 17, Foreign Bodies and Infestations as follows:

Pp. 252–253 Child's toy in coffee powder
Pp. 252–253 Hessian in coffee beans

595. Cocoa powder – a good quality example of a hot beverage. (See also Chocolate p. 142).

Miscellaneous products

Herbs and Spices (596–599)

Herbs and spices represent historically one of the most important areas of long distance international trade. It is still significant today although overshadowed by bulk trade in staple commodities in terms of both volume and value. The increasingly cosmopolitan nature of many Western European countries and generally more adventurous palates have both contributed to an increase in importation of herbs and spices to these areas.

The most important factors in determining the quality of dried herbs and spices are the initial quality of the source crop and subsequent storage conditions. The desired properties of many spices are lost if storage, particularly of the ground commodity, is for too long a period or under poor conditions.

Herbs and spices are prone to **infestation** and **foreign bodies** such as small stones may also be present. **Adulteration** can be a major problem particularly in the ground form. For example the addition of finely ground sawdust or spice wastes would be difficult to detect in visual examination but may be detected by laboratory determination of the crude fibre content.

Adventitious contamination of spices may also occur from other plants growing in the vicinity of the spice plant. Such contamination is usually detectable in unground herbs and spices but may not be so in the ground product. A particularly serious example concerned oregano powder (*Origanum vulgare*) contaminated with datura.

The nature of dried herbs and spices precludes major problems of microbial spoilage although moulds may sometimes develop if the moisture content is allowed to rise sufficiently.

Herbs and spices often contain large numbers of **bacterial endospores** and while the proportion of spice in recipes is usually low under some conditions their contribution can be significant. Pepper powder for example may contain large numbers of highly heat resistant spores which cause problems in canning while spices have frequently been found to be the original source of *B. cereus* in outbreaks of *B.cereus* food poisoning associated with meat products (e.g. Ormay and Novotny, 1969).

596. Good quality dried bay leaves (*Laurus nobilis*). Note the absence of both physical damage and indications of disease.

597. A good example of caraway seeds (*Carum carvi*). Note the absence of significant numbers of damaged seeds or extraneous matter and the uniform size of the seeds.

598. A good quality example of cloves (*Eugenia caryophyllata*). Cloves have powerful antimicrobial and antioxidant properties and their use for those purposes (in addition to organoleptic) has been suggested in ground pork products.

599. The drug store beetle, *Stegobium paniceum,* although a relatively common contaminant of foods appears to be particularly attracted to spices. The illustration shows the insect infesting cayenne pepper. (See also Chapter 17 pp. 244–245.)

Examples of foreign bodies in herbs and spices may also be found in Chapter 17, Foreign Bodies and Infestations as follows:

Pp. 244–245 Drugstore beetle from cayenne pepper
Pp. 244–245 Leaf beetle in rosemary

Jellies and Blancmanges (600)

Jellies, blancmanges etc. are early examples of convenience sweet courses. When made up the sweets form a stable gel based on gelatin (jellies), cornflour starch (blancmange) and compounds such as carrageen and carboxymethylcellulose. Flavouring (usually fruit) and colouring is added. Products of this nature have few specific quality problems but where gelatin is an ingredient care must be taken to ensure its freedom from *Salmonella* (see p. 159).

600. A good example of a dessert jelly.

Examples of foreign bodies in jellies may also be found in Chapter 17, Foreign Bodies and Infestations as follows:

Pp. 244–245 Insect larva case in fruit jelly
P. 251 Insect (identity unknown) in 'quick setting jel'

Oriental Fermented Products (601–608)

Oriental fermented products form a diverse group of commodities differing from other fermented products in that moulds are the prime micro-organism used rather than yeasts or bacteria. World wide production in 1976 amounted to a value in the order of £1,500,000,000 with production concentrated in China, Japan and other Far Eastern countries, with lesser production in India. Imitations of some products, e.g. soya sauce formulated to suit Western tastes, are also produced in Europe and the U.S.A. These products are not usually fermented but produced using chemical or enzymic hydrolysis of soya.

Principal oriental fermented products are as follows:

Soya sauce (see below).

Tamari sauce, a Japanese product similar to soya sauce made with *A. tamarii.*

Miso, a Japanese product produced from soya beans using *A.oryzae,* yeasts (*Sacch. rouxii* and *Zygosaccharomyces spp*) and lactic acid bacteria. The final product is in the form of a paste.

Tempeh, an Indonesian product produced from soya beans and fermented with species of *Rhizopus.*

Ang-Khak, Chinese red rice produced by growth of *Monascus purpureus* on autoclaved rice.

Natto, a Japanese soya bean product differing from others in that the main micro-organism involved in the fermentation is *B. subtilis.*

Tofu (tou-fo-ru), see below.

Minchin, a Chinese product made from starch-free raw wheat gluten. The fermentation depends on a natural inoculum and typical examples contain up to seven species of moulds, nine of bacteria and three of yeasts.

Idli, is an Indian product made from ground rice and black gram beans (*Vigna mungo*) fermented overnight. *Leuc. mesenteroides* leavens the idli and its growth is followed in succession by *Str. faecalis* and *Pediococcus damnosus.*

The extensive use of moulds in production of Oriental fermented products has led to concern over the possibility of mycotoxin formation. As far as is known mycotoxin production by any species involved in manufacture has not been demonstrated although in an industry where production methods are often primitive control is obviously difficult. Soya beans may be affected by growth of spoilage moulds such as *Aspergillus* before processing and the temptation to convert such beans into fermented products is strong particularly in poorer or less developed economies. Such a practice is potentially hazardous since Jarvis (1976) states that aflatoxins would not be destroyed in the fermentation process used in soya sauce manufacture.

601. Traditional soya sauce containing chillis. Strongly flavoured, this is common in Oriental cooking. Soya sauce is produced by fermenting a mash of soya beans and crushed wheat with a starter (joji in Japan, chou in China) containing *A. oryzae* (*A.soyae*), various lactic acid bacteria such as 'P. soyae' and *L. delbrueckii,* and yeasts such as *Sacch. rouxii.* The inoculated mash is held for 3 days at ca. 30°C and then soaked in sterile 24% NaCl brine for a period of 2½ months to a year. During the first stage of the process hydrolysis of proteins and starches occurs by the action of *A. oryzae* enzymes. This activity continues throughout manufacture but during the maturation in brine acid development by lactic acid bacteria and alcohol production by the yeasts occurs together with the development of the characteristic odours and flavours.

602. Soya beans in soya sauce. A Japanese product normally sold fresh but bottled for long distance export. The beans are part fermented by mould (usually *A. oryzae*) and then steeped in finished soya sauce (see **601**). Products such as this which are unfamiliar in nature present considerable problems in determining what parameters constitute good quality.

603 to 608. Tofu (tou-fo-ru; soya bean cheese) is a product of Chinese origin now becoming more common in the west due to interest in so-called health and ethnic foods. Soluble protein in a crude filtrate of soya bean extract is curdled by a magnesium or calcium salt and the curd then fermented with moulds, probably *Mucor spp,* for a month at ca. 15°C and finally ripened in brine. Fresh tofu is illustrated (**603**). It is highly perishable and for export is either preserved (see p. 136) or UHT treated (**604**). Note the leakage of 'whey' due to poor quality seaming. The leakage has

spread over the pack (**605**) which should not be retailed in this condition.

Spoilage of tofu is largely by moulds and yeasts. **606** shows extensive colonisation which occurred within 2 days at ambient temperature and ca. 4 days at 4°C. The black mould was identified as *Mucor spp*, the blue as *Penicillium spp* and the red colonies as a yeast (**607**).

In addition to visible microbial growth considerable amounts of 'whey' are produced (arrowed) (**608**) probably as a consequence of breakdown of structural proteins.

606 ●

607 ●

608 ●

609 ●

610 ● *

611

158

Fillers (609–611)

A variety of products are used for their structural rather than their nutritional or organoleptic properties. The three major types are **starches**; **gums** and **mucilages** which are complex polysaccharides traditionally derived from trees, seaweeds or seeds but now also from the bacterium *Xanthomonas campestris* which is reported to yield a product of consistently high quality (Bull and Solomons, 1983); and **gelatin,** a partially degraded protein derived from collagen by acid hydrolysis. Gums and mucilages if unrefined may contain debris and while rarely supporting microbial growth may contribute large numbers of micro-organisms to products in which they are used.

Gelatin is of animal origin and can be an important source of micro-organisms including *Salmonella;* stringent quality control standards must therefore be applied and the International Commission on Microbiological Specifications for Foods (ICMSF) recommend the absence of salmonellas in $10 \times 25g$ samples taken from each batch (ICMSF, 1974). Gelatin also supports rapid microbial growth and if incorrectly handled may contaminate products containing gelatin which are not cooked after filling (see pp. 56, 76).

Extracts and Hydrolysates

Meat, vegetable and yeast extracts are prepared by acid hydrolysis followed by neutralisation, concentration, and in some cases, drying. Dried meat extract is usually combined with other ingredients including dried hot water extract of meat, flavourings and flavour enhancers such as monosodium glutamate and nucleotides to produce stock (bouillon) cubes, gravy powder etc., while the concentrates are used as sandwich spread or form the basis of hot drinks (see p. 153). Extracts and hydrolysates are of low water and high salt content and are therefore shelf stable. However, although heating during manufacture kills vegetative bacteria such products may contain large numbers of heat resistant endospores which may be of importance if the extract is used as an ingredient in, for example, a canned product.

Laboratory Tests

The diversity of products discussed in this chapter is such that laboratory tests cannot be listed. Those wishing to apply laboratory tests should consider the product nature and the common spoilage pattern.

609. Seaweed polysaccharides such as agar and alginic acid are used in large quantities as stabilisers and gelling agents in compounded foods. Seaweed itself is rarely consumed as food in the Western Hemisphere, an exception being the lava bread of Wales. In Japan seaweed is widely used as a food usually after a crude extraction. The nutritive value, however, is minimal. This illustration shows a dried crude extract of seaweed (probably the giant kelp *Macrocystis pyrifera*). This is of good quality containing little extraneous material. Poorer quality seaweed can contain large quantities of sand which must be removed before consumption.

610 and 611. Gums from trees such as the acacia are used in the food industry as thickeners. The Asian domestic cuisine also makes extensive use of gums. **610** shows Goonder (a type of Acacia, *Robinia sp*), an Indian edible gum containing a high percentage of polysaccharides. The collection of such gums is based on small scale rural economy and large quantities of extraneous material are present (**611**).

Chapter 10

Bakery

For the purpose of the present discussion Bakery is considered in two main areas, first **bread and cakes** and second, **biscuits.** Breakfast cereals are discussed with grocery in Chapter 9.

In terms of volume bread is the most important bakery item although, in the U.K. and U.S.A. at least, sales are declining overall. The marketing situation in the U.K. and the U.S.A. differs from that in Europe in that in the former countries the market for bread is dominated by large plant bakeries producing a uniform product and sold either under a brand name or under a supermarket own label. In continental Europe the market continues to be dominated by small local bakeries. In recent years in the U.K. there has been a reaction against the somewhat bland, if convenient, 'plant' bread and traditional varieties of loaf have regained popularity, a trend aided by the development of in-store bakeries in large supermarkets and the growth of specialist hot bread shops.

It should be noted that this trend has revived the argument concerning the nutritional status of white *vs* wholemeal bread while the recent fashion for high fibre diets has proved a further stimulus to sales of wholemeal and bran containing breads.

The principles of bread baking are straightforward although quality may be adversely affected at any stage. In practice many different bakery techniques are used. It is beyond the scope of this book to detail these but two basic techniques are described. These are the conventional long fermentation batch process and the more modern high speed mix process (see **612**). The bread bakery process and potential quality problems at each stage are discussed below.

Bread (612–642)

The Ingredients

The basic ingredients of flour, yeast, water and salt are common to all types of bread baking. The choice of a correct type of flour is essential for production of good quality. The flour must be from a **strong** wheat of high protein content. Classically the Canadian Manitoba wheats are considered most suitable for bread baking, and most flours are compounded by the miller to give the desired properties for the particular bread. A number of physical and chemical tests are available to the miller. The characteristics of strong wheat suitable for bread making determined by an 'Extensometer' are shown on p. 115.

Yeast (*Saccharomyces cerevisiae*) may adversely affect the quality of bread if stale and inactive since insufficient gas to inflate the loaf may be produced. Similarly the correct strain of *Sacch. cerevisiae* must be used to ensure sufficient gas is produced rather than alcohol. Where yeast is cultured in the bakery and where cleanliness is poor the yeast may become contaminated with 'wild' yeasts and possibly bacteria which can produce off-taints in the dough before baking which are detectable in the finished product.

Oxidising improvers are added to the high speed mix dough in order to help produce –S:S– bonds in the dough and hence enable it to attain its correct rheological properties during mixing. Improvers may also be added to flour for traditional baking but considerably greater quantities are required for high speed mixing.

Fat is usually added to a high speed mix, as it helps to give an even crumb structure in the loaf and to retard staling.

Mixing

Conventional dough is batch mixed using a low speed mixer with dough hook. The prime function of the mix is the even distribution of the dough constituents. In contrast the high speed mix has the function of imparting to the dough the physico-chemical and rheological properties necessary to produce a satisfactory loaf. To this end a considerable expenditure of work, typically 40 joules per gram, must occur. High speed mixing may be continuous as in the Wallace and Tiernan Do-maker process or batch as in the U.K. Chorleywood process. Failure to expend the necessary work results in a low quality loaf.

Fermentation

During fermentation 'maturation' or 'ripening' of the dough occurs, the changes taking place producing the same effect as the high speed mixing. The classical long 12 hour or overnight fermentation is still used in some areas of Continental Europe but elsewhere a 3 hour fermentation is now almost universally accepted. The quoted temperature (26°C) is an average; lower or higher temperatures may be used. A shorter fermentation period of 1½ hours is normally used for wholemeal bread.

The fermentation stage is of considerable importance in determining the quality of conventionally baked bread, either under- or over-fermentation producing a poor quality loaf. Dough produced by high speed mixing has either no fermentation stage or a very short one of ca. 15 minutes.

See also bread mixes, pp. 114–115.

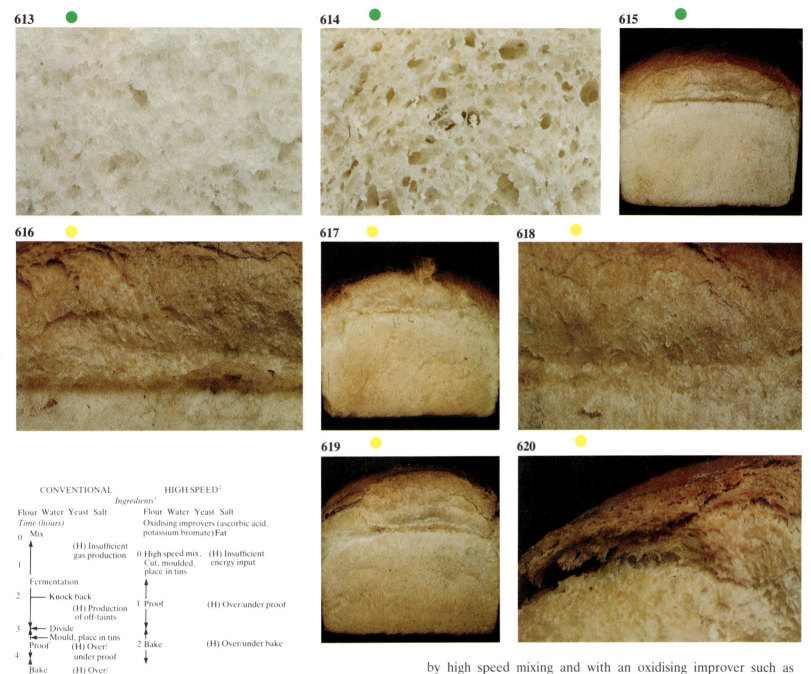

613 •   **614** •   **615** •

616 ○   **617** ○   **618** ○

619 ○   **620** ○

CONVENTIONAL		HIGH SPEED[2]
Ingredients[1]		
Flour Water Yeast Salt		Flour Water Yeast Salt
Time (hours)		Oxidising improvers (ascorbic acid, potassium bromate) Fat
0 Mix		
	(H) Insufficient gas production	0 High speed mix. (H) Insufficient Cut, moulded, energy input place in tins
1		
Fermentation		
2 Knock back		1 Proof (H) Over/under proof
	(H) Production of off-taints	
3 Divide		
Mould, place in tins		2 Bake (H) Over/under bake
Proof (H) Over/	under proof	
4		
Bake (H) Over/	under bake	
5 232°–260°C		3 (H) = Specific quality hazard

[1] Other ingredients such as anti-mould agents may be added
[2] High speed processes vary considerably in detail
[3] A 3 hour fermentation is normal today

612. A flow diagram for the manufacture of white bread comparing the traditional fermentation process with the more modern high speed process. Times shown for the stages of each process may vary according to technological requirements or to local practice.

613 and **614.** **The differences in crumb structures** produced by high speed and low speed mixing processes are considerable. **613** shows the crumb structure of a typical 'factory' white loaf produced by high speed mixing and with an oxidising improver such as potassium bromate added to the mix. **614** shows, in contrast, the crumb structure of a traditional batch white loaf produced by low speed mixing and without oxidising improver. The open texture is often preferred to the soft close structure of the high speed mix loaf.

615 to **620.** **The effect of the proof stage on bread quality.** Correctly proven batch bread shown in general view (**615**) and in close-up (**616**) is compared with bread 20% underproven (**617** and **618**) and bread 20% overproven (**619** and **620**). The underproven loaf is seen to be smaller while in the overproven the crust tends to lift away from the body of the loaf. The flexibility of modern bread flours is illustrated by the fact that despite large differences in proof time the effect in the finished loaves is relatively small.

Proof

The proof stage is essentially common to both high speed and conventional processes although in the former continuous provers are likely to be used in a large scale plant.

During the proof stage gas produced by the yeast is retained by the mature dough, the loaf swelling to near its final size and the characteristic crumb structure being produced. Both overproving or underproving can lead to serious quality defects.

Baking

The baking stage is also common to each process. Small batch ovens usually have less even temperature distribution than the larger reel ovens or the continuous ovens used for large scale bread manufacture. These latter are frequently over 100 feet in length and, properly operated, permit a highly uniform level of baking.

621 ● * 622 ● * 623 ●
624 ● 625 ● 626 ●

621. **Two loaves showing slight burning** and splitting due to the oven being overheated. Although technically a fault these loaves would be fully acceptable to most consumers.

622. **Pan dirt on the side of a loaf** (arrowed). This is derived from grease used to lubricate the baking tins. Unless excessive it is not a serious fault.

623 and **624.** **Staling of bread.** No staling retarder is added to traditional bread and the crumb rapidly begins to take on a 'crystalline' appearance and stale texture (**623**). Staling is not due to drying out but to a reversible change in the starch involving crystallisation and various additives retard staling. **624** shows high speed mix 'factory' bread which has had a staling retarder added.

This bread is of the same age as that shown (**623**) and gives little sign of staling. The retardant usually added is fat which also aids in production of the even soft crumb. Staling is temperature dependent, the rate becoming faster as the storage temperature is lowered to reach a maximum at $-2°C$ to $-3°C$. This can be of importance where pre-prepared sandwiches are stored in refrigerated display cabinets for extended periods. Staling does not occur below $-18°C$ and bread may thus be stored deep frozen.

625 and **626.** **Mould on bread.** Despite the addition of mould inhibitors such as calcium sorbate to bread, mould will ultimately develop. **625** shows *Mucor sp* growing on a bread loaf and **626** shows the growth of *Penicillium sp* in addition to *Mucor*.

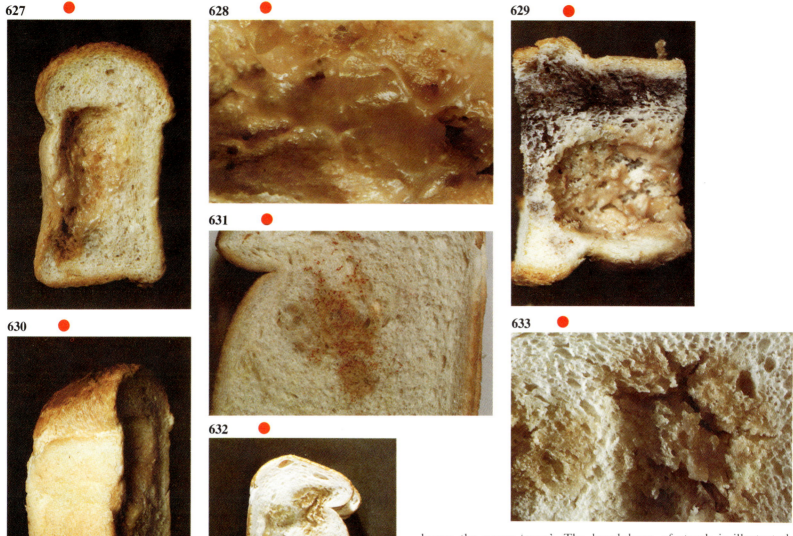

627 to 630. **A general view of advanced 'rope' in bread (627).** Ropiness is caused by the growth of *Bacillus spp* particularly *B. subtilis* or *B. licheniformis*. Large bakeries control rope by addition of calcium sorbate which prevents outgrowth of the bacterial spore after germination. Rope, however, is still a problem in bread produced by small traditional bakeries particularly in hot weather or in large bakeries where poor housekeeping has permitted a build up of spores of *Bacillus spp* to such an extent that sorbate is no longer effective.

The close-up view (**628**) shows clearly the area of stickiness. This is produced by hydrolysis of flour proteins by bacterial proteolytic enzymes and hydrolysis of starch to sugars by bacterial amylases. The bacteria responsible are able to utilise the simple sugars to produce a polysaccharide capsule. At a certain stage of development the slimy material may be pulled into long threads:

hence the name 'rope'. The breakdown of starch is illustrated (**629**). The bread around the affected area no longer gives the characteristic purple reaction with iodine, the structure of the bread in this area remained relatively intact due to the proteolytic enzymes being slower acting than the amylases.

It should be noted that rope only affects the crumb of the loaf and hence is difficult to detect from the outside. This angled view (**630**) shows that while the interior is heavily affected by rope the exterior remains intact.

631, 632 and **633.** **Serratia marcescens** is able to grow on bread when ambient temperatures are high and classically produces copious quantities of red or pink pigment (see p. 60). When heavily affected the bread appears to bleed and early examples were often attributed to miraculous interventions. The condition is rare today in bread and only infrequently as spectacular as classically described. **631** shows development of colonies of *S. marcescens* on sliced bread. The slicing machine is often considered to be a source of the organism. More extensive growth of *S. marcescens* is shown (**632** and **633**) (cf. ropey bread). Note that in this example the organism is a non-pigmented strain and no red pigment is produced.

Examples of various types of bread other than white bread. Many of the faults described in particular relation to white bread such as staling, rope, mould etc are common to all bread but eating quality must be assessed individually.

634. Pitta bread. Common in Eastern Europe, Asia Minor and North Africa. Pitta is baked in a thin sheet from a relatively wet flour dough.

635. Wheaten loaf. A mass produced batch loaf with 81% wheatmeal flour and 19% cooked whole wheat. Wheatmeal flour has a higher enzymic activity than white flour and short fermentation times are usually employed.

636. Hovis bread. A mass produced loaf baked with a proprietary flour consisting of 90% white flour and 10% wheat germ (embryo). The germ is an important source of vitamins which are lost in the milling of white flour although in many countries white flour is supplemented with thiamine, niacin and iron.

637 and 638. Turkestan loaf illustrated from above (**637**). The crumb is illustrated (**638**). Turkestan loaf is typical of bread produced in Central Asia. Its popularity in the U.K. stems from its high fibre content. The flour used is a mixture of wholemeal wheat and rye, the crust being topped with wholewheat and sesame seeds. The dough is supplemented with honey, and rope formation is particularly rapid in breads of this type.

639 and 640. Pumpernickel. A soft German rye bread made from coarsely ground rye. The coarse pieces of rye are shown in close-up (**640**) (arrowed) and may be mistaken for foreign bodies, including maggots, by consumers unfamiliar with the product.

641. Crispbread. Originally of Scandinavian origin it is now sold widely elsewhere and is often eaten as part of a calorie controlled diet. Crispbread may be produced from rye or wheat flour and may be fermented or unfermented. Traditional manufacturing methods involve the use of snow or powdered ice as a leavening agent, the expansion of small bubbles in the ice during baking causing the dough to rise.

642. Parisian barm or sour dough bread is produced using a continuous barm rather than a pure yeast culture. The barm usually contains a yeast and associated lactic acid bacteria giving bread of an acid flavour and closer texture than the conventional.

634

635

636

637

638

639

640

641

642

PARISIAN HEAT and SERVE

$2.75

Rye Bread and Ergotism

Ergotism is a disease of humans caused by consumption of rye bread produced from grain infected by the growth pathogen *Claviceps purpurea*. The disease was recognised in the Middle Ages (St. Anthony's Fire; Holy Fire) and the connection with diseased rye was noted in the mid 16th century. Ergot poisoning is due to the production by the sclerotia of **ergotamine**, a pharmacologically active compound related to the hallucinogen LSD.

The symptoms of ergotism vary widely and include hallucinations, violent burning feelings, extreme fatigue, giddiness and convulsions. There are basically two types – **gangrenous** ergotism where constriction of blood vessels leads to gangrene of extremities together with violent burning feelings and formication; and **convulsive** ergotism with fatigue, giddiness and convulsions. The latter form is usually coupled with a Vitamin A deficiency. The last known outbreak of ergotism in the U.K. occurred in 1927 amongst Jewish immigrants in Manchester who had consumed rye bread made from flour containing 0.1%–0.3% ergot. The most recent major epidemic in Europe occurred in France in the early 1950s. The risk of ergotism however remains wherever economic circumstances make necessary the cultivation of rye in poor soils which provide suitable conditions for *Claviceps*.

A further clinical condition associated with mycological invasion of grain is **Alimentary Toxic Aleukia**. This syndrome, which is due to toxin production by a species of *Fusarium*, became epidemic in Western Siberia during the early 1930s but is rare today. A full description of the syndrome is provided by Joffe (1971).

Cakes (643–645)

Cakes and biscuits, while representing a far more diverse group of products than bread, are technically simpler. Use of correct flour type is essential and guidance may be drawn from the classification of Kent-Jones and Amos (1957) shown in Table 19. Failure to adhere to these recommendations results in products lacking the characteristic properties expected.

Although cakes normally present no problems of a public health nature special attention must be given to cakes topped with, or containing, fresh cream, which is an excellent medium for microbial growth. The degree of handling necessary in production of cream cakes means a high level of risk of contamination, in addition to which the need to refrigerate fresh cream cakes is still not fully appreciated. (See also p. 40.)

Biscuits (646–647)

Although produced in other parts of the world, the U.K. probably has the greatest diversity of type. Consumption has been falling for several years although the per capita consumption is still the highest in the world. In the U.K. biscuit manufacture is dominated by two large combines, United Biscuits and Associated Biscuits each manufacturing under a variety of brand names. Biscuit manufacture has long been organised on a large scale and is an early example of fairly advanced automation in the food industry.

Table 19 Recommended Flour Types for Cakes and Other Flour Based Confections

Confection	Flour Type
Fermented e.g. buns	Strong bread making
Chemical aerated	Self raising flour (weak flour plus sodium carbonate and acid calcium phosphate)
Sponges	Weak flour of fine particle size
Puff pastry	Strong bread making
Short crust	Weak
Cakes – low fruit – low sugar	Medium strength
Cakes – high fruit – high sugar	Weak flour of fine particle size treated with chlorine to modify starch

Modified from Kent-Jones and Amos (1957)

Technically, manufacture is straightforward. The flour used is a weak flour of low protein content which permits the dough to be rolled to desired thickness but which does not alter shape when the biscuits are stamped out.

Under good manufacturing conditions there are few quality control problems, the greatest economic loss being broken biscuits. Similarly they are shelf stable but unless correctly packaged will pick up moisture and lose crispness while in high fat biscuits rancidity may develop.

Laboratory Tests

Laboratory tests are not normally considered relevant for products discussed in this chapter.

Examples of foreign bodies in bakery may also be found in Chapter 17, Foreign Bodies and Infestations as follows:
Pp. 252–253 Glass in bread
Pp. 252–253 Plastic fragments in cake
Pp. 252–253 Seed from cake

643 and **644.** **High sugar content cakes** such as the fruit cakes illustrated (**643**) have a low water activity level and are resistant to microbial spoilage. Storage in damp conditions or condensation under wrapping film may permit mould growth as shown (arrowed). Detailed examination (**644**) shows the mould growth to be restricted by the low water activity and to have an atypical appearance without aerial hyphae.

645. **Ovenbake.** The material removed from the top of Chelsea buns was originally thought by the consumer to be mouse droppings. It was in fact ovenbake, a carbonaceous material which collects on the interior walls and fittings of ovens as a result of splashes of dough mix etc. which become burnt on to the walls occasionally falling onto the product. Although difficult to eliminate entirely regular cleaning of oven interiors markedly reduces the incidence of ovenbake. (See also pork pies, p. 64.)

646 and **647.** **Insect (fur beetle, *Attagenus sp*) damage** to sesame seed biscuits (**646**) is indicated by irregular shaped holes (A) on the surface of the biscuit. These should not be confused with the normal round regularly shaped holes (B) which occur as a result of air bubbles bursting during baking. The irregular holes are shown in detail (**647**).

Chapter 11

Flans and Pizzas

Flans and pizzas are technically similar, each consisting of a pastry base on which a topping is placed. Flans (and quiches) typically have a base of short pastry and the topping is usually based on egg together with other ingredients such as cheese, bacon and vegetables. Flans may be eaten hot or cold.

Pizzas are traditionally based on bread dough although lighter pastry doughs may be used. The toppings may vary considerably according to locality but the Italian pizza topping is usually cheese and tomato based whereas that of the French equivalent, the Pissaladière, does not contain cheese and is commonly based on onion and anchovies.

Pizzas, which are generally eaten hot, represent one of the earliest forms of fast food being popular not only in Italy and Southern France but also in the U.S.A. Their popularity has recently spread to the U.K. both with respect to catering and home cooking. A wide variety of fresh and frozen pizzas are available and while these are generally based on the Neapolitan type, derivatives such as those based on a slice of French bread have appeared.

Examples of laboratory tests are shown in Table·20, p. 170.

648 to 652. Spoilage of pre-prepared fresh pizzas is usually by moulds. The moulds are generally derived from the cheese but grow well on the tomato sauce component of the topping. **648** shows a totally spoiled pizza on which four types of mould and visible yeast colonies have developed. The mould *Geotrichum candidum* is illustrated in detail (**649**). Note that the development of aerial hyphae has been restricted by the overwrapping film. Blue mould (*Penicillium sp*) is present (arrowed **648**) and this is illustrated in greater detail (**650**). *Rhizopus* (**651**) and *Aspergillus* (**652**) are also present.

653. Visible yeast colonies developing on the topping of a pizza. Note the 'doughnut' shape colony characteristic of many yeasts (arrowed A). Mould colonies are also present (arrowed B) and the pizza had been subjected to excessively long storage.

654, 655 and 656. French bread pizzas are a U.K. development from the traditional Italian style, and consist of a filling piled on top of a split piece of French bread. Such products are often made by small bakeries or by the in-store bakeries of large supermarkets. In the example shown (**654**) processed cheese has been used. Blue mould (*Penicillium sp*) is developing (arrowed) on some of the cheese and this is illustrated in detail (**655**). Mould colonies (arrowed) are also developing on the bread base (**656**). The pizza illustrated was on sale in this condition but bore no 'sell by' or 'use by' date.

657 and 658. A very poor quality pizza illustrated in overall view (**657**) with the discoloured ham shown in greater detail (**658**). The pizza was probably fresh chill, frozen by the retailer to 'save' the product at or beyond the end of the shelf life.

Table 20 Laboratory Tests Flans and Pizzas

I. Microbiological
Total viable count
Enterobacteriaceae (Pizza only)
Yeast (Pizza only)
Staphylococcus aureus (Pizza only)

II. Chemical/physical
Analysis is not usually required on products of this type.
Foreign body/infestation examination not normally required.

Notes:
(1) These tests are examples only and are not inclusive.
(2) Examples are for finished products only.
(3) Techniques are detailed in specialist publications such as ICMSF (1978) – Microbiology: Egan *et al* (1981) – Chemistry.
(4) Instrumental techniques such as Impediometry may be substituted for some classical microbiological methods.

659, 660 and **661** **A cheese and onion quiche** with wholemeal base and of apparently good quality (**659**). A cross section through the quiche (**660**) shows (arrowed) a layer of blackening at the boundary between the wholemeal base and the topping, shown in detail (**661**). The blackening is thought to have been caused by a reaction between **iron in the flour and sulphite** probably used in excess as a preservative for the (dried) onions but also possibly from egg powder used in the formulation.

662 to **665.** **Mould growth in cheese and onion quiche (662).** The upper surface of products such as quiche is dry with a low water activity level. Microbial spoilage is almost invariably by mould, illustrated in detail (**663**). Growth is, however, restricted and photographed after a further week at ambient temperature (**664**) shows little further progress.

It is more rapid on the moist interior filling after removal of a slice for sale (**665**). Initiation of mould growth on flans is usually a consequence of a high storage temperature or of a fluctuating storage temperature causing moisture to condense on the upper surface.

666 and **667.** **Pre-prepared par-baked potatoes.** In Western Europe and the U.S.A. the growth of the market for fresh chill convenience foods continues and allows the introduction of value added products of this type, illustrated before final cooking (**666**) and after final cooking (**667**).

659

660

661

662

663

664

665

666

667

Prepared Salads

Prepared salads typically consist of chopped vegetables in either a mayonnaise or oil and vinegar base. Such products are retailed in pre-packed containers or from bulk. The best known is coleslaw (basically cabbage in mayonnaise) which also forms the base for many variants including those containing meat or fish. The market for prepared salads has increased vastly in recent years largely to meet the demands of calorie controlled diets, and the types available have increased accordingly. Care is needed when formulating salads because interaction between ingredients may create conditions suitable for the growth of potentially pathogenic micro-organisms. It should also be appreciated that while traditional coleslaw type salads are relatively shelf stable some types, particularly those based on dried legumes, have only a very short shelf life and are fundamentally unsuitable to the long shelf lives required for large scale retailing. (See also fresh salads: Produce, Chapter 13.)

Coleslaw has recently been implicated as the origin of an outbreak of adult and perinatal disease caused by *Listeria monocytogenes* (Schlech *et al*, 1983, see p. 175). Although the cabbage used in the preparation of the coleslaw was the source of the organism rather than mishandling during preparation, manufacturers of coleslaw and other salads should ensure that raw materials are obtained from growers whose practices preclude the possibility of contamination with *Listeria monocytogenes* or other pathogenic micro-organisms.

Examples of laboratory tests are shown in Table 21, p. 174.

668, 669 and **670.   Gas production blowing the lid of a pre-pack** of mayonnaise based (celery, apples and onion) salad (**668**). In this example yeasts were responsible for the gas production but gas in salads may also be produced by fermentative Gram-negative rods of the family *Enterobacteriaceae* (**671**) or by heterofermentative lactobacilli. Salads of high carbohydrate content are particularly prone to this type of spoilage. Examination of the contents (**669**) shows gas bubbles in the mayonnaise and a generally unappetising appearance. The deterioration is illustrated in detail (**670**).

671.   Gas production in coleslaw. In this example gas was produced by members of the family *Enterobacteriaceae* and the coleslaw had a strong faecal smell. Members of the *Enterobacteriaceae* involved in food spoilage typically comprise such genera as *Klebsiella*, *Enterobacter*, *Citrobacter* and *Hafnia*. They normally have no public health significance and are common in soil and on plant material probably, in this example, being derived from the cabbage component of the coleslaw.

672 to **675.   Taramosalata** is a dish of Greek origin made from cod's roe with breadcrumbs, seasoning and colouring (**672**). The intrinsically highly perishable nature of the product causes it to be prone to fermentation (usually by yeast) and to spoilage by surface colonies of mould (**673**). Colonies of *Penicillium sp* are illustrated in detail (**674**). Fermentation by yeast is illustrated (**675**). The body of the product has been totally disrupted by gas production by ca. 10^7 colony forming units/g of yeast (*Saccharomyces sp*). In the example illustrated sufficient gas was produced to blow the lid completely off the container.

676 to **679.   Mayonnaise based salads.** The trend in the U.K. towards eating only a light midday meal has led to the development of a number of products to meet the demand for a 'substantial snack'. Rice and chicken salad (**676**) and pasta and ham salad (**677**) are perishable products based on mayonnaise sauce containing the appropriate ingredients. Some combinations of ingredients in a perishable base may be ecologically unstable and be prone to microbiological problems. **Recipe Hazard Analysis (RHAS)** should therefore be applied in development of products of this type.

Deterioration and loss of quality of mayonnaise based products may be of a physico-chemical nature. Examination of the rice and chicken salad (**678**) and pasta and ham salad (**679**) after 10 days' storage at 6°C shows (arrowed) separation of the mayonnaise leading to accumulation of free liquid and a general thinning of the sauce.

680.   Salad with aspic topping. A further example of the growth in pre-packed salad products. This type of product consists of a base layer of mayonnaise topped with aspic containing various ingredients (in this example boiled egg, peppers and olives). The degree of component handling involved in production is high and aspic provides a good substrate for microbial growth. The use of RHAS is again indicated to quantify the risks presented by products of this nature.

668

669

670

671

COLESLAW

672

673

674

675

676

677

678

679

680

681 ●

682 ●

681.   Cous-cous salad, basically a North African semolina-based product, is illustrative both of increasingly adventurous tastes in Western Europe and the growth of the market for exotic pre-prepared salads. It is frequently not recognised that such salads may present a good substrate for growth of micro-organisms including those of food poisoning potential. This has been of little importance in the past where product batches have been small and storage lives very short (<1 day) but large scale production with attendant long life may lead to problems. **Recipe Hazard Analysis (RHAS)** should be used to check the safety of new formulations.

682.   A Turkish fresh salad typifies the Middle Eastern product. The vegetables are vinegar brine pickled and additional colour and flavour is derived from turmeric and paprika powder. In countries such as Egypt and the Lebanon the colour is derived from colouring added to the brine in which the vegetables are pickled. The ethanoic acid and salt concentration of such products is low and the storage life correspondingly short.

Table 21 Laboratory Tests Prepared Salads

I. Microbiological
Total viable count
Yeasts
Escherichia coli
Staphylococcus aureus (some types only)
Lactic acid bacteria (some types only)

II. Chemical/physical
pH value
Foreign body/infestation examination not normally required

Notes:
(1) These tests are examples only and are not inclusive.
(2) Examples are for finished products only.
(3) Techniques are detailed in specialist publications such as ICMSF (1978)–Microbiology: Egan *et al* (1981)–Chemistry.
(4) Instrumental techniques such as Impediometry may be substituted for some classical microbiological methods.

Produce (Fresh Fruit and Vegetables)

Fruit and vegetables represent one of the most complex areas of food production, distribution and marketing. The seasonal nature of production, the frequent dependence on suitable weather and the highly perishable nature of produce leads to economic uncertainties. In the U.K. and many parts of Europe the trade has in the past been concentrated through small specialist shopkeepers and market stall holders, and until fairly recently the large scale retailers such as the then emergent supermarket chains were reluctant to be involved with produce. This thinking was changed largely by the experiences of American supermarket chains who overcame the problems associated with long distance haulage by developing sophisticated means of transport and storage; insisting on high initial quality standards and, to attract custom, a high standard of presentation and packaging. Large scale retailers have probably also benefited to a greater extent than small retailers from the wider availability of 'out of season' fruit and vegetables. Income from such crops is often of considerable economic importance for producing nations such as Morocco, Egypt and Kenya.

Although supermarkets and other large scale retailing operations are now well established outlets small shopkeepers and market traders continue to command a large percentage of the total market. In many cases their continued significant role is based on the perishable nature of produce where **freshness** is an all-important quality parameter and where poor overall quality is readily discerned by the untrained eye. It is thus recognised that the higher level of individual service and inspection required in the handling of produce benefits small retail outlets and their wholesale suppliers.

In one sense the ultimate realisation of consumer determination of quality lies in the **pick your own** concept although there are also fairly significant benefits to the consumer in terms of cost. In 'pick your own' operations the consumers harvest their own fruit or, less commonly, vegetables. The producer risks a higher wastage rate but benefits from greatly reduced labour costs and, as a consequence of eliminating the profit share due to wholesalers and/or retailers, a higher profit margin and a more favourable cash flow.

In spite of the above factors the increasing role of large scale retailing with its ever growing market share has radically altered produce marketing and has necessitated the introduction of formal quality assurance at every stage. This in turn has led to a greater emphasis on quality assurance within the whole of produce marketing.

Unless grown for canning (p. 212), freezing (p. 209) or drying (p. 126) fruit and vegetables usually receive no processing beyond cleaning and possibly trimming between harvest and retail sale (for exceptions see cooked beetroot pp. 182–183 and salad trays pp. 194–195). Although the life of produce may be prolonged by refrigeration, gas storage or treatment with preservatives as appropriate, in many cases deterioration continues unchecked from harvest to the end of life and the highly perishable nature of produce leads to a high level of actual spoilage with consequent economic loss.

While the primary and/or secondary causes of spoilage are likely to be of little concern to the small retailer or wholesaler they are of major importance in large scale post-harvest operations (wholesaling, distribution, retailing) where storage losses are of prime concern as well as growing operations which have the added problem of pre-harvest losses. Diagnosis of the cause of produce spoilage is thus essential for large scale handlers in order that future losses may be obviated. Quality control of produce requires a high level of inspection together with an acceptance of a high level of rejection due to quality defects as well as a sufficient degree of knowledge to recognise the underlying causes of spoilage.

The practice of storing vegetables such as cabbage for long periods at low (ca. 2°C) temperatures has been linked with food-borne **listeriosis** (Schlech *et al*, 1983). A large outbreak of adult and perinatal infection due to *Listeria monocytogenes* was positively associated with the cabbage component of coleslaw. The cabbage had been grown in fields fertilised with sheep manure from a flock with known cases of 'circling disease' (listeriosis). Subsequent cold storage of the cabbages had probably led to multiplication of a small initial inoculum of *List. monocytogenes* to a significant population size.

Lettuce or other raw vegetables have also been linked (without experimental evidence) to listeriosis in hospitalised, immuno-suppressed patients (Ho *et al*, 1981).

Schlech *et al* (1983) suggest that while the role of vegetables in sporadic and epidemic listeriosis requires further investigation growers should be cautious in the use of raw manure to fertilise crops that are subject to prolonged cold storage and subsequently consumed uncooked. Where listeriosis is known to be present in cattle or sheep the manure from these animals should not be used on crops designated for human consumption.

The Onion Tribe (Genus *Allium*) (683–689)

The onion tribe are important food plants in many parts of the world. The major species are bulb and salad onions (*Allium cepa* in the Western Hemisphere, *A. fistulosum* in Japan and China); leeks (*A. porrum*) and the herb, garlic (*A. sativum*). Shallots which are similar to bulb onions may be placed in a further species *A. ascalonicum*.

All members of the tribe are subject to pre-harvest disease and infestations, some of which affect post-harvest quality. **Stem eelworm** (*Ditylenchus spp*) for example attacks both onions and leeks and results in plants of a bloated and twisted appearance having poor storage properties. Correct crop rotation is essential to prevent such infestations. Problems of quality deterioration affecting the consumer are more usually due to post-harvest spoilage. Onions, garlic and to a lesser extent leeks exert an antibacterial effect and spoilage is normally caused by mould growth (N.B. The pathogenic bacterium *Staphylococcus aureus* grows well on peeled onions. See p. 132).

683. **Neck rot** caused by the mould *Botrytis* is a common form of spoilage of both onions and leeks. This illustration shows a healthy onion (arrowed A) compared with an infected onion (arrowed B). The mould superficially resembles *Penicillium spp* but is usually one of three species of *Botrytis*, *B. allii*, *B. byssoidea* and *B. squamosa*. Primary invasion of the onion leaves occurs, the disease then spreading downwards to the bulb. Onions are particularly prone to neck rot if stored under damp conditions or if the bulbs have not been properly dried after harvesting. **Downy mildew** caused by *Peronospora destructor* often predisposes onions to neck rot since bulbs with this condition are small, soft and fail to mature. Equally damage to leek by **leaf moth** (*Acrolepia assectella*) may lead to rapid development of *Botrytis*.

684. **Despite the prevalence of mould spoilage** of onion a slight but significant anti-fungal effect is demonstrated by the close-up illustration which shows the lack of aerial mycelia production by *Botrytis sp*.

685 and **686.** **Spread of invasive mould growth** is preceded by tissue damage due to the mould's exo-cellular enzyme activity. An area of tissue so affected is shown (arrowed) (**685**). The enzymes are usually pectinolytic or cellulolytic in nature and such damage which may be seen in many types of vegetable, develops with particular rapidity when preceded by physical damage such as bruising. Secondary infections like that illustrated (**686**) where a second mould species is present often occur in the region of such lesions and develop as a result of the metabolic activities of the 'primary' invader *Botrytis*. In older and physiologically damaged onions (and other vegetables) different moulds or indeed bacteria are not infrequently found growing independently on the same product.

687. **The invasive nature of the fungal enzymes** is shown by the section through an affected onion. Although the inner layers are seen to be healthy the tissue damage extending inwards from the exterior is clearly illustrated by the brown discoloration of the outer layer; an enzyme reaction associated with tissue damage. Where it is difficult to assess the extent of damage by inspection of the whole vegetables it is essential that a section across one or more should be cut to enable an interior examination to be made.

688 and **689.** **Garlic.** Like all members of the onion family garlic is prone to attack by moulds. **688** shows a poor quality bulb of garlic with patches of black mould, a condition referred to as 'smut'. A cross section through the bulb (**689**) shows the condition of the individual cloves to be entirely satisfactory with no significant penetration by the mould into the interior.

The Cabbage Family (Brassicas) (690–698)

The cabbage family contains a large number of food plants, the majority of which will withstand moderately severe climates such as the U.K. winter. They are of particular importance as winter vegetables in such regions and are of dietary significance as providers of ascorbic acid (Vitamin C). Some members of the family, kale for instance, are of importance as a fodder plant for cattle. Most widely used for human consumption are cabbage (*Brassica oleracea var. bullata* and *var. capitata*); cauliflower (*Br. oleracea var. botrytis cauliflora*); Brussels sprouts (*Br. oleracea var. gemmifera*); broccoli (*Br. oleracea var. botrytis cymosa*) and curly kale (*Br. oleracea var. acephala*).

Despite the diverse appearance of these plants they share a common ancestor *Brassica oleracea*, the wild cabbage, and are in fact closely related.

Like all vegetables the brassicas are subject to a number of pre-harvest diseases and infestations which lower the quality of the vegetable if the affected plants are not rejected before sale. These include attack by insects such as cabbage root flies (*Delia brassicae*) and flea beetles (*Phyllotreta spp*); caterpillars, the larval stage of the cabbage white butterfly (*Pieris brassicae*) and the cabbage moth (*Plutella xylostella*); physical damage by birds and the classical fungal disease of the brassicas, club root (*Plasmodiophora brassicae*). Plants affected by the latter rarely reach maturity and retail sale. Brassicas are also affected by storage diseases usually bacterial rots.

Turnip and swede which are also brassicas are discussed in the section concerned with root vegetables p. 182.

690. Larva of a pre-harvest infestant, probably cabbage root fly (arrowed) revealed during preparation of a cabbage for cooking.

691. Pre-sale halving of cabbages. Like most other horticultural products cabbages are sold basically 'as taken from the earth' with no more than a preliminary removal of outer leaves and excess soil. In industrialised nations a fairly basic form of processing such as manually halving the cabbage may be applied. The effective display life of such products is less than for the whole item. The illustration shows withering and dehydration of the leaves and also a new outgrowth of leaves from the stem. This latter growth may be caused by light mediated induction of growth hormones.

692 and 693. Cauliflower is often affected by patches of dark brown or black discoloration on the curds which is frequently mistaken for mould growth or rots. The usual cause is a deficiency of the micro-nutrient **boron** or unusually dry growing conditions. There is no effect on either odour or taste and unless grossly affected such cauliflowers are suitable for sale.

694 and 695. A floret affected by browning. A close-up illustration (**694**) shows that while some localised tissue damage has occurred

(arrowed) there is no deep penetration or extensive superficial spread as would be encountered with bacterial or mould spoilage. A section across a whole affected cauliflower (**695**) shows deep tissues to be unaffected. This is in contrast to a superficially similar condition caused by the plant pathogenic bacterium *Pseudomonas syringeae* pv *maculicola* (black rot). Such an infection is accompanied by a strong rotting odour. Other brassicas may be infected by this organism and affected plants are not fit for sale.

696. Attempts may be made to save cauliflowers seriously affected by browning by cutting out the damaged curd. Such attempts are largely futile leading to poor appearance due to cuts (arrowed) and subsequent rapid dehydration.

697 and 698. Deterioration of calabrese; a type of broccoli. **697** shows discoloration of some of the florets (arrowed). Deterioration of this nature can occur through ageing of the product or through inappropriate storage conditions. **698** shows a cross section through a stem of calabrese with a lesion due to a rot, probably of bacterial origin, clearly seen (arrowed). Providing the rot is confined to the stem interior the saleability of the product is unlikely to be significantly affected.

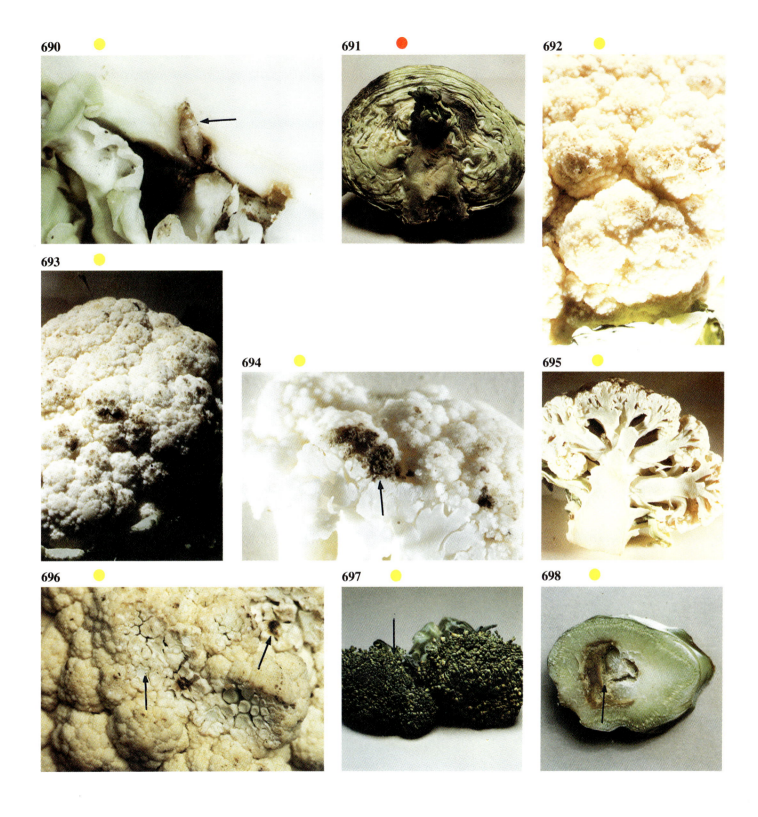

690   •   691   •   692   •

693   •

694   •   695   •

696   •   697   •   698   •

Peas and Beans (Legumes) (699–710)

This section is concerned with fresh peas and beans. Other legumes which are merchandised primarily in dried form are discussed elsewhere (p. 124) as are products derived from legumes such as soya sauce (p. 156) and oil (p. 126).

In addition to their importance as food plants legumes are of considerable importance with respect to soil fertility. They form a symbiotic relationship with the atmospheric nitrogen fixing bacterium *Rhizobium* which develop in nodules in the roots. The practical consequence is a reduced need for nitrogenous fertiliser during growth of subsequent crops.

The most commonly grown fresh legumes are the garden pea (*Pisum sativum hortense*); green pea (*P. sativum macrocarpum*); sugar pea (mange tout); broad bean (*Vicia faba*); French bean (*Phaseolus vulgaris*) and scarlet runner bean (*Ph. multiflorus*). Pre-harvest field diseases can lead to reduced quality in all cases. These include insect damage such as that caused to peas by **thrips** (thunderflies, *Parthenothrips spp*), fungal diseases such as *Botrytis*, and fungal **anthracnose** in beans. Viral diseases mostly spread by **aphids** (*Aphis fabae*) or thrips may also affect legumes; leaves are commonly affected but the disease may spread to pods.

Fresh legumes are also subject to a variety of bacterial and fungal infections post-harvest. In most cases the disease affects first the pod and subsequently the seeds.

It should be noted that large quantities of fresh legumes, particularly peas, are canned or frozen. The acreage involved in the U.K. alone totalled more than 100,000 acres (40,000 hectares) during 1982, a figure far greater than that involved in commercial production for the fresh unprocessed market. The canning and freezing of peas represents one of the few areas where objective testing is applied to determine quality, a device known as a 'Tenderometer' being used to physically determine tenderness of peas.

699.   A general view of poor quality pea pods. Some are over-ripe and desiccated and some show pathological lesions such as pod spot. Pea pods of such obviously inferior quality should be removed by quality control staff or by the retailer before being displayed for sale.

700.   A detailed example of pod spot of peas. The three possible causes are all moulds and are:
(a) *Ascochyta pinodella* (the likely cause in this example) which gives rise to very small brown or black lesions.
(b) *Ascochyta pisi* which causes round tan to brown spots of 9 mm diameter.
(c) *Mycosphaerella pinodes* which causes purple to black lesions.

A. pinodella and *M. pinodes* are the commonest causes of pod spot in the U.K. and the disease can result in considerable economic loss.

701 to 704.   Damage to peas in pod by pea thrips (701) and, in closer view (**702**). Eggs of the pea thrip are laid in the flower and the larvae tunnel into the developing pod and hence the young pea. Typically, as in this example, the infestation and resultant damage is evident only on shelling the pod. Note in the close-up illustration insect excreta (arrowed A) and secondary mould growth (arrowed B). A larva in an affected pod is illustrated (**703**) (arrowed) and is shown in close-up (**704**).

705 and 706.   Mould spoilage of peas and other legumes may develop first within the pod. **705** shows mould growth (*Pythium butleri*) within a pea pod. The extent cannot be assessed from external examination and while in this example it is obvious that the pod was diseased and should have been rejected (**706**) less fully developed internal mould growth may show no external evidence. In examples such as this the mould may gain access with insects such as thrips (see also **701** to **704**).

707 and 708.   Runner beans of general poor quality showing excessive curvature, discoloration and being over-ripe (**707**). The pods are limp and flaccid rather than firm as in the good quality beans (**708**) which are relatively straight and of good colour and correct degree of ripeness.

709 and 710.   Anthracnose. A seed borne fungal (*Colletotrichum spp*) disease which affects all types of bean (**709**). First appearing as sunken black spots on leaves and stems the pods are ultimately affected as shown. Seeds of affected plants have a brownish black discoloration. In some cases the affected areas are brown rather than black (**710**).

See also **Frozen peas. Frozen broad beans** Chapter 14, pp. 210–211.

699

700

701

702

703

704

705

706

707

708

709

710

Roots and Tubers (711–721)

Root vegetables are derived from four families of plants. The term 'root' as applied is technically incorrect as turnip, swede and celeriac have swollen parts of the hypocotyl and/or stem included in what is familiarly termed the root. Some root vegetables such as beetroot and carrot have a dual role as a salad vegetable eaten cold and as a hot winter vegetable. They are all subject to a number of pre-harvest diseases that affect quality. These include **heart rot** of beetroot (*Beta vulgaris*) due to **boron deficiency**, **carrot fly** (*Psila rosae*) infestations of carrots (*Daucus carota sativus*) and parsnips (*Pastinaca sativa*) and the fungal **violet root rot** (*Helicobasidium purpureum*) of swedes and turnips (*Br. napus: Br. rapa*). Root vegetables particularly carrots are also prone to damage by slugs (*Agriolimax reticulatus*) and snails (*Helix spp*). Storage diseases are predominantly rots which may be either bacterial (usually *Erwinia carotovora* or soft rot pseudomonads similar to *Ps. fluorescens* biovar II, Palleroni, 1984) or fungal (e.g. *Fusarium spp*) in origin.

Radishes (*Raphanus sativus*) are normally eaten as spring or summer salads although the winter radish produces large solid roots similar to beetroot which may be grated for salads. They are cruciferous plants and suffer from similar pests and diseases to other root brassicas such as turnips. **Flea beetle**, **club root** and **downy mildew** (*Peronosporaceae*) may be troublesome while a fungal scab disease of radish produces sunken areas in the roots. Poorly stored winter radish may be affected by bacterial soft-rots. They are also prone to damage by chewing pests such as slugs while the upper part of the root may be attacked by birds. Any of these factors may adversely affect the market quality, together with delays in harvesting producing 'woody' roots and dehydration if storage or display periods are excessive.

Potatoes (*Solanum tuberosum*) are the most important tuber cultivated as a food plant and are of considerable economic importance. Consumption tends to rise during economic depression. Current (1981) U.K. production involves ca. 500,000 acres (200,000 hectares) the bulk of which is for fresh consumption although considerable quantities are processed in various forms such as crisps (see p. 144), potato flour and alcohol.

The best known pre-harvest disease of potatoes is, of course, potato blight caused by the fungus *Phytophthora infestans* which is illustrated historically by the 19th century Irish famine. Blight is readily controlled by **Bordeaux mixture** (copper sulphate) sprays and by use of resistant cultivars. Blighted potatoes rot easily and can serve as a focus for large scale rotting during storage. Potatoes are also subject to attack by insects such as **aphids** and the **Colorado beetle** (*Leptinotarsa decemlineata*). This latter pest is rare in the U.K. and its presence must be reported to the police. Eelworm is an important cause of crop loss and cysts of the eelworm remain in infected ground for several years. Secondary damage to potatoes arises from chewing insects such as slugs, mechanical damage and physiological damage due to storage at too low temperatures.

The major storage disease of potatoes is bacterial rots. These cause massive economic loss with in some cases hundreds of tons of potatoes being lost in a matter of days (Wilson, 1969). Sub species of *E. carotovora* are the most common cause. These include *E. carotovora* subsp *atroseptica* which also causes the pre-harvest disease 'blackleg', and *E. carotovora* subsp *carotovora*. Other bacteria including *Clostridium spp* are also involved in the rotting of potatoes.

This section discusses roots and tubers common in temperate climates. Other root vegetables such as sweet potato are discussed in the section concerned with exotic vegetables p. 196.

711, 712 and **713.   A comparison of good quality** (711) **and poor quality** (712) carrots. The poor quality examples are unwashed, desiccated and have bent, deformed roots. Such deformations may be due to cultivation in unsuitable stony ground or to uneven availability of water. A comparison of cross sections (713) of the carrots shows that while the poor quality (arrowed) have shrunk due to dehydration there is no internal tissue damage and no evidence of bacterial rots or other disease.

714.   Mould growth (arrowed) **on a poor quality carrot.** The mould growth is superficial only and is a consequence of poor storage conditions.

715 and **716.   Sclerotinia disease.** A comparison of a carrot affected by Sclerotinia disease (A) with a good quality example (B) (715). A close-up of the lesions is illustrated (716). Sclerotinia disease is soil borne, caused by the mould (*Sclerotinia sclerotiorum*). Affected carrots are either of lowered quality as in this example or, where the mould has fully developed, unfit for sale. The establishment of the mould in soils intended for carrot growing may thus cause considerable economic loss.

717.   Beetroot is unusual as a vegetable in being retailed cooked as well as raw (and pickled). The illustration is of good quality cooked beetroot. The life of such products is short and is limited by dehydration, mould and occasionally yeast and bacterial growth.

718.   Scorzonera (*Scorzonera hispanica*) – black salsify, is rarely grown in the U.K. although its cultivation presents no problem and the plant grows wild in some coastal regions. It is relatively disease free and a good quality example is shown. It has alleged medicinal properties and was previously used as a source of inulin.

719.   Potato cultivars may be classified according to their time of maturity. Thus in the U.K. 'first earlies' are harvested between June and early August while the foliage is still green; 'second earlies' between August and September and 'main crop', which comprise the bulk of the stored crop, in September to October and possibly early November. This illustration shows good examples of Maris Peer, a second early potato, resistant to **wart disease** (*Synchytrium endobioticum*) which causes brownish-black outgrowths near the 'eyes' (arrowed). The tubers are free from skin diseases such as **common scab** ('*Streptomyces scabies*') and **powdery scab** (*Spongospora subterranea*). When present these diseases are unsightly but do not spread to the interior of the tuber.

720.   A section through a good quality second early potato tuber, Maris Peer. Note the absence of the yellow-brown eelworm cysts, damage or rots.

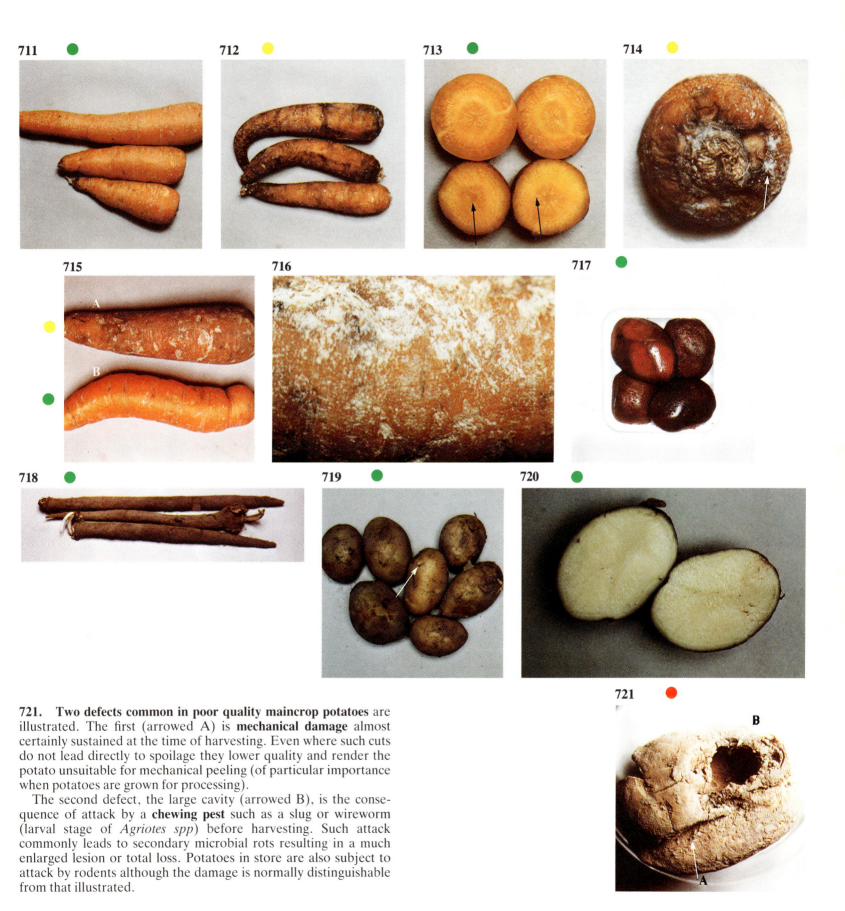

721. Two defects common in poor quality maincrop potatoes are illustrated. The first (arrowed A) is **mechanical damage** almost certainly sustained at the time of harvesting. Even where such cuts do not lead directly to spoilage they lower quality and render the potato unsuitable for mechanical peeling (of particular importance when potatoes are grown for processing).

The second defect, the large cavity (arrowed B), is the consequence of attack by a **chewing pest** such as a slug or wireworm (larval stage of *Agriotes spp*) before harvesting. Such attack commonly leads to secondary microbial rots resulting in a much enlarged lesion or total loss. Potatoes in store are also subject to attack by rodents although the damage is normally distinguishable from that illustrated.

Leaves, Stalks and Stems (722–733)

Vegetables eaten as leaves, stalks and stems are a largely unrelated group from a number of plant families. The culinary uses are varied, some plants being eaten as cooked vegetables, such as spinach, and some cold such as celery. Rhubarb is often thought of as a fruit rather than a vegetable. The major groups are the leaf plants like spinach (*Spinacea oleraceae*); members of the *Umbelliferae* such as celery (*Apium graveolens var. dulce*) and fennel (*Foeniculum vulgare var. dulce*) grown for the swollen and elongated petioles, parsley (*Petroselinum crispum*) grown for leaves and petioles and asparagus (*Asparagus officinale*) grown for its stems. Also grown for their enlarged petioles are rhubarb (*Rheum rhaponticum*) and seakale (*Crambe maritima*) while leaf beets such as spinach beet or chard (*Beta vulgaris*) grown for leaves and/or petioles are closely related to the root vegetable beetroot (see p. 182).

Many of the pre-harvest diseases which affect leaf, stalk and stem vegetables are common to other vegetables and include attack by aphids, moulds and chewing pests such as slug. Club root and violet root rot which normally are associated with brassicas, also affect seakale, while asparagus is affected by a specific pest, the **asparagus beetle** (*Crioceris asparagi*). The storage life of this type of vegetable is normally short and wilting due to moisture loss is a common cause of poor quality. Bacterial rots develop and damage by a variety of pests may occur.

722. Deadly nightshade (*Atropa belladonna*) found in fresh spinach. Although it is good practice to ensure that ground used for cultivation of horticultural produce is free from poisonous plants the risk is greatest with leaf and stem varieties such as spinach where the contaminating plant may not be easily detected.

723 and 724. Fresh parsley rapidly deteriorates in quality. **723** illustrates poor quality pre-packed parsley which is showing wilting and yellowing of the foilage. A close-up (**724**) shows a rot to be present. The development of such rot, which in this example is of bacterial origin, is often enhanced by pre-packing in plastic films.

725 and 726. Bacterial rot of fennel. The blackened affected areas are illustrated in general view (**725**) and in detail (**726**). The rotting is accompanied by a rotting odour. As with other vegetables (see **723**) pre-packing may contribute to the development of such rots.

727. A poor quality bunch of celery which has suffered mechanical damage resulting in the breaking of some petioles, which are flaccid due to dehydration. Lesions on the petioles are shown in detail (**728**). In this example the lesions were due to mechanical damage followed by invasion by secondary rot micro-organisms. A superficially similar condition with yellowing of the leaves and brown cracks in the petioles is caused by boron deficiency. The cracks in such a case would be dry and odour free, unlike those illustrated which were slimy and had a rotting odour.

729. Two basic systems of cultivation exist for celery. Winter grown celery, illustrated, is grown in trenches and earthed up with soil to blanch the petioles. In this example earthing has been inefficient and only the base has been blanched. This leads to a bitter flavour. In contrast to winter grown, summer celery, which is not frost hardy, is grown on the flat in dense blocks to exclude light and is thus self blanching. One cultivar however, **American Green,** has green petioles, is of superb flavour and should not be confused with imperfectly blanched winter grown species.

730, 731 and 732. Asparagus for many years has been considered to be a luxury product in the U.K. It is now more generally available both in specialist greengrocers' shops and in supermarkets. **730** shows good quality asparagus with firm stems and no indication of significant mechanical damage. In contrast that illustrated (**731**) is limp and tired and close examination (**732**) shows a soft rot (arrowed) in the centre stem. The stems on either side are not visibly affected but in some circumstances rots spread rapidly from one stem to others.

722 🔴

723 🟡

724 🟡

727 🟡

725 🔴

726 🔴

730 🟢

731 🟡

728 🟡

729 🟡

732 🟡

185

733. Poor quality rhubarb as demonstrated by the thin, weak stems which have a green coloration. Good quality rhubarb is red along most of the length of the stem which is relatively thick in proportion to the length.

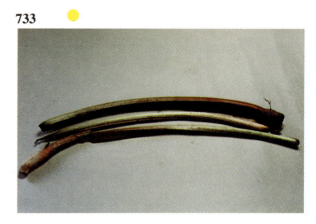

733

Fruiting Vegetables (734–768)

Fruiting vegetables, with the exception of sweet corn, are grown solely for their fruits and are from either the *Cucuberaceae* or the *Solanaceae*. Sweet corn (*Zea mays*) which is grown for the seed head (cob) is a member of the *Gramineae*, the grass family. In this section members of the *Solanaceae* discussed include aubergines (*Solanum melongena ovigerum*) and peppers (*Capsicum spp*); of the *Cucuberaceae* courgettes (*Cucurbita pepo ovifera*) and melons (*Cucumis melo*) which are eaten as fruit. Other fruiting cucurbits are pumpkins and squashes (*Cucurbita maxima* and *Cucurbita moschata*). It should be noted that tomatoes (*Lycopersicon esculentum*), which are members of the *Solanaceae*, and cucumbers (*Cucumis sativus*) which are members of the *Cucuberaceae*, are both fruiting vegetables but are discussed in the section concerned with salads (p. 191) while peppers also have a function as salads. In addition some fruiting vegetables such as aubergines would until recent years have been considered to be exotic vegetables in the U.K. (see p. 196).

The *Solanaceae* and the *Cucuberaceae* are both affected by a wide variety of pre-harvest diseases caused by insects, moulds, viruses and nutrient deficiencies. A number of these affect the post-harvest quality of the fruit. The high moisture content also means that they are prone to post-harvest rots which may be fungal or bacterial in origin.

Sweet corn is relatively little affected by pre-harvest diseases although problems may be caused by the **frit fly** (*Oscinella spp*). Birds however are an important cause of economic loss during growth and cobs are also prone to attack by chewing pests including rodents during storage.

Courgettes occasionally undergo atavistic mutation to an earlier form and produce significant quantities of cucurbitacin E. This has an extremely bitter taste and a small number of fruit containing the compound recently led to rejection of an entire batch of canned courgettes (zucchini). Although cucurbitacin E is toxic it has been previously thought that the extreme bitterness would preclude consumption of affected fruit. Food poisoning by cucurbitacin E containing courgettes has, however, been reported in Australia; symptoms being severe cramps, diarrhoea and collapse. (Rymal *et al*, 1984.)

734

735

736

737

734. A good quality red pepper.

735. Lesions on the exterior of a red pepper. The lesions are probably bacterial and have developed after mechanical damage. The quality is poor and the pepper should not have been retailed in this condition.

736 and **737. A green pepper** from a plant affected by **tobacco mosaic virus (736)**. Affected plants are stunted in growth and have mottled leaves. The interior of the pepper is normal (**737**) with the exception of the brown rot developing from the stem end which is probably not associated with the viral infection.

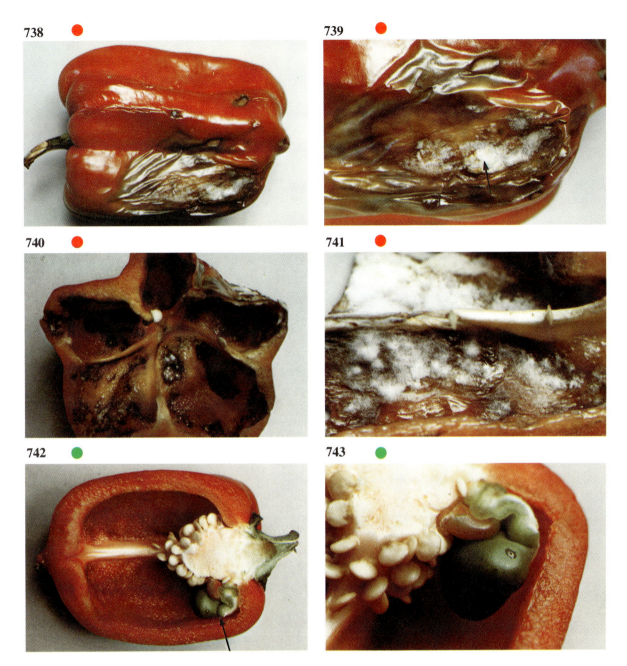

738 ●   **739** ●

740 ●   **741** ●

742 ●   **743** ●

738 to 741.   A red pepper of very poor quality (although still on sale!) showing a large area and some smaller areas of rotting (**738**). **739** shows the lesion in close-up and mould (arrowed) is present on the affected area. A cross section **740** shows the interior to be affected. Note that the seeds have fallen away from the stem end where they normally develop. **741** shows white mould attacking the wall of the pepper from the interior. The mould has probably entered from the stem end and may have been derived from a pre-harvest plant infection.

742 and 743.   Although normal in external appearance a section through this mature pepper (**742**) showed a second pepper (arrowed) developing within the older, shown in detail (**743**). The cause which does not affect eating quality is a fault in cell differentiation in the early stages of fruit development.

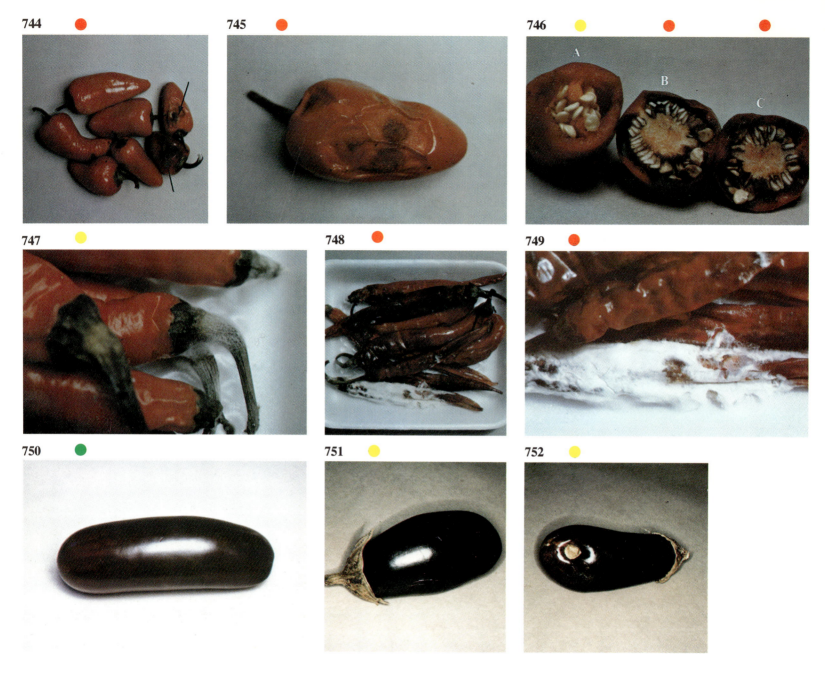

744. **Chilli (hot) peppers** (*Capsicum frutescens*) of general poor quality. The peppers are shrivelled due to dehydration and lesions (arrowed) are visible.

745 and **746.** **Lesions on a chilli pepper** due to rotting caused by *Alternaria sp* (**745**). A cross section through affected peppers (**746**) shows the mould to be present in the interior of the fruit (B & C). Fruit A is not affected by *Alternaria* but is physiologically damaged.

747, 748 and **749.** **Mould development on pre-packed chilli peppers** often commences at the stem end (**747**). After two days' further storage at ambient temperature (and within the 'sell by' date of the pack) the lower chilli has become completely covered with white mould (**748**), illustrated in detail (**749**). As with many

pre-packed vegetables condensation below the overwrapping film can lead to rapid mould growth.

750 and **751.** **A good quality aubergine** of good colour and free from blemishes (**750**). In contrast **751** shows **bruising** round the stem end. This probably is a consequence of post-harvest damage and the aubergine is still suitable for sale although spoilage of the bruised areas is likely to be rapid.

752. **A hole in the flesh of an aubergine** extending some 2 cm into the flesh. The regularity of the hole suggests that it is unlikely to be of insect or animal origin and was probably caused by a sharp object while the aubergine was growing with subsequent deposition of scar tissue around the site of the damage.

188

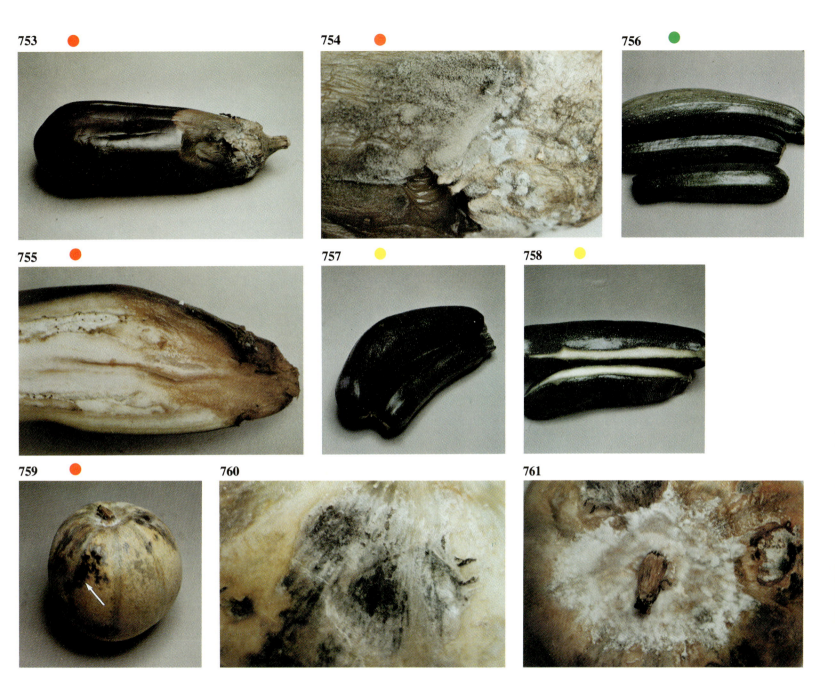

753, 754 and **755.** **Fruit rot (phomopsis blight)** caused by the mould *Phomopsis vexans* is the most important post-harvest disease of aubergines and is characterised by tan coloured lesions which darken with age (**753**). The mould itself is shown in the close-up illustration (**754**). More than one type of mould may be present in advanced cases of rot as secondary invaders. The rot penetrates deep into the tissues of the fruit as shown in the cross section (**755**).

756. **Courgettes** like aubergines were until recently regarded as 'exotic' in the U.K. In recent years the vegetable has become extremely popular and is readily available. The examples shown are of good quality products. Courgettes are in fact baby marrows but those of highest quality are from specially developed cultivars such as the Italian Zucchini.

757 and **758.** **Courgettes occasionally develop as a 'double' fruit** (**757**). When the double is split it is seen that the fruit are, in fact, two separate entities (**758**) and the quality of either half of the double courgette is unimpaired.

759 to **763.** **Melons are prone to mould spoilage** especially if stored under damp conditions. **759** illustrates an Ogen melon with patches of mould (arrowed) on the rind, shown in close-up (**760**). The condition is referred to as **charcoal** or **rhizoctonia** rot (*Rhizoctonia spp*). In other examples several moulds may be present on the same fruit. **761** illustrates a honeydew melon with **stem end rot**, a common fault, caused by *Diplodia spp*. The melon has also been bruised and resulting enzymic breakdown of tissue has encouraged the development of two further types of mould (arrowed) (**762**). At this stage the mould growth is largely superficial with little penetration into the interior of the melon (**763**).

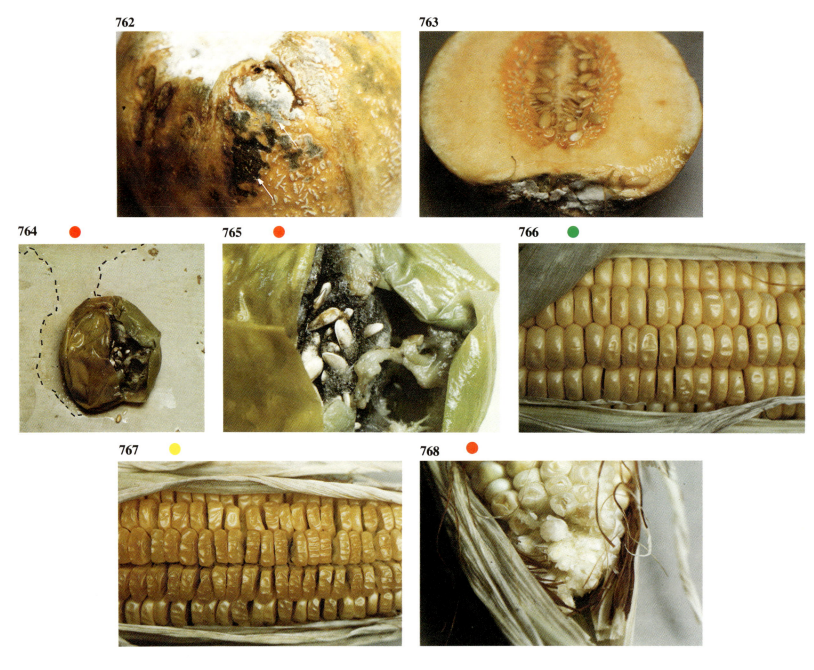

762

763

764 ●

765 ●

766 ●

767 ●

768 ●

764 and 765.   Advanced decay of a melon which has collapsed due to internal rotting (**764**). Apart from the split due to external pressure the rind was largely undamaged. The melon appeared intact when purchased but deteriorated to the condition illustrated within one day. Note the large quantity of liquid (outlined) which has exuded from the melon and which in store would spread the rot over a large number of fruits. The prime cause of the rot in this example was bacterial (*Erwinia sp*) but close examination of the interior of the melon showed a mould (*Rhizopus*) to be present (**765**) in the area of the seeds.

766 and 767.   A good example of sweet corn (766). Note the firm, evenly coloured seeds with no evidence of disease or physical damage. In contrast, the cob illustrated (**767**) is of poor quality although no specific defects are present. The seeds are dehydrated and the surrounding leaves withered.

768.   Damage to sweet corn. Although birds cause damage to sweet corn before harvesting the nature of this damage suggests attack by rodents during storage.

Salad Vegetables (769–786)

For the present purpose salad vegetables are defined as those grown primarily for use as cold eating salads although some, such as tomato, are extensively grown for canning and other processing. Many salad vegetables are members of plant families which have been discussed previously. They include spring onion (*Allium* genus pp. 175–177), tomatoes and cucumbers (Fruiting Vegetables pp. 192–194), radish (*Raphanus sativus,* Roots and Tubers pp. 182–183). The general quality problems associated with salad vegetables are thus the same as for other members of the group. The remainder to be discussed are grown for their leaves.

The major species are lettuce (*Lactuca sativa*); endive (*Cichorium endiva*) and chicory (*Cichorium intybus*). All are members of the *Compositae* (daisy family). (Note that Chinese lettuce, *Pe-tsai,* is a member of the brassicas (*Br. cerua*) and that in China and the Far East similar species are grown primarily for their enlarged root.) In addition to those grown for salad use vegetables grown primarily for other purposes may be used in salads. These include onions (pp. 175–177), carrots (pp. 182–183), cabbage (pp. 178–179) and peppers (pp. 186–187).

Salad vegetables such as tomatoes and cucumber require the protection of glasshouses for extensive cultivation in the U.K. and much of Western, Central and Northern Europe. Until recently the availability of most salad vegetables has been highly seasonal. The development of horticultural industries in developing nations such as Morocco and the availability of cheap air freight (see Exotic Vegetables p. 196) has resulted in a more or less year round availability in Europe not only from countries such as Morocco and Israel but also, in the case of radish and iceberg lettuce, from the U.S.A.

In discussing salad vegetables it is necessary to refer to a technological innovation which may affect both the economics of the industry and the quality of the vegetables. **Nutrient Film Technique (NFT)** is a development of hydroponics whereby plants are cultivated in a mineral salts solution. In NFT plants are supported either in peat blocks or in inert supports in a circulating solution of buffered mineral salts. The initial capital cost required to set up a full scale NFT operation is high and fairly sophisticated equipment is needed to maintain nutrient levels and pH values within correct tolerances. The advantages are claimed to be high yield coupled with rapid growth and uniform quality as well as reduced operating expenses. The quality of NFT vegetables compared with that of traditionally produced is open to dispute with claims being made that such produce is deficient in flavour and, in the case of lettuce, produces a crop with no heart.

In addition to a possible role in transmission of listeriosis (see p. 175) salads have been frequently implicated in the transmission of a number of other bacterial and viral diseases including **shigellosis** and **infectious hepatitis.** The problems are particularly acute in countries where raw human sewage (nightsoil) is used as fertiliser; where highly contaminated water is used for irrigation and where supplies of potable water for washing the crops are inadequate. Leafy salad vegetables such as lettuce which readily entrap contaminants are considered to pose a particular risk as is watercress which may be grown in contaminated water. Watercress also poses a risk with respect to **liver flukes** (*Fasciola hepatica*) and the primary hosts, usually sheep or cattle, should not be allowed to graze in the vicinity of cress beds. A Code of Practice (Anon, 1979) exists for growers in the U.K. and similar Codes are produced elsewhere.

769. Spring (salad) onions such as illustrated, although not totally unfit for sale should have been rejected by a quality controller, produce manager or shopkeeper. In addition to general poor quality as a result of age and handling damage the onions exhibit (arrowed) blackening of the leaves due to physiological lesions.

770. Lettuce are prone to damage by chewing pests such as slugs and to storage diseases like moulds (frequently *Botrytis*) and bacterial rots. The blackening illustrated in this poor quality lettuce is characteristic of the bacterial soft rot caused by *Erwinia spp*. Soft rots are commonly associated with, and encouraged by, poor storage conditions. Damage to the vegetables also encourages the condition.

771. An extreme case of bacterial soft rot in a cucumber by *Erwinia sp*. The cellular organisation of the cucumber has broken down completely. Cucumbers halved before sale are particularly prone to soft rots due to tissue damage at the cut and due to the organism being spread by the cutting knife.

772. In addition to soft rot, fly (Order *Diptera*) maggots (arrowed) were present in the cucumber. The eggs were probably laid in the softened tissue before deterioration was complete.

773. Good quality examples of three types of tomato; English (A); Marmande (B) and cherry (C). These demonstrate the diversity of produce available in large supermarkets. Different quality control criteria are likely to be required for each type of tomato.

774 and 775. Blossom end rot of tomatoes (**774**) is normally due to dry soil conditions or to an excessively high concentration of mineral salts. The rot should be distinguished from the infectious mould and bacterial rots, and penetration into the interior of the fruit is slight (**775**). Although fit for sale the value of such fruit is reduced and in more severe cases of blossom end rot total loss may occur.

776. Damage to a tomato caused by a slug. In this example damage occurred before harvest but similar damage may be incurred during storage. Tomatoes having such an extent of damage are unlikely to reach the retail market but the problem, if widespread, can cause considerable economic loss.

777. Curved scars on a tomato, probably the result of damage by a moth. Such a tomato would be downgraded but still be considered fit for sale.

778 and 779. A stem end infection of a tomato caused by the plant being infected with a pathogen, probably a mould during growth (**778**). A section through the fruit (**779**) shows penetration into the interior to be limited.

780. A tomato from a plant infected by *Alternaria sp* during growth. The lesion is dry and a section through the tomato showed penetration to be superficial only.

781. A tomato from a plant infected by tobacco mosaic virus (TMV). In addition to the area of yellow-bronze discoloration the fruit is soft and of general poor quality. In extreme cases of TMV

769

770

771

infections the entire fruit is bronze in colour. TMV resistant cultivars such as the F1 hybrid 'Cura' have been developed.

782. 'Greenback' disease of tomatoes. The green area on the shoulder of the fruit is hard and the fruit fails to ripen correctly. The cause is attributed to excessively high temperatures during the ripening stage of growth. Some cultivars are said to be resistant to the condition.

783 and 784. A general view of poor quality tomatoes showing splits which are subsequently being colonised by mould (**783**). The mould illustrated in detail (**784**) (identified with *Rhizopus*) infected the tomatoes during post-harvest storage. Despite the poor quality such tomatoes still command a market for home conversion into ketchups and sauces etc.

772 🔴

773 🟢 🟢

🟢

A   B

C

774 🟡

775

776 🔴

777 🟡

778 🔴

779

780 🔴

781 🔴

782 🔴

783 🔴

784 🔴

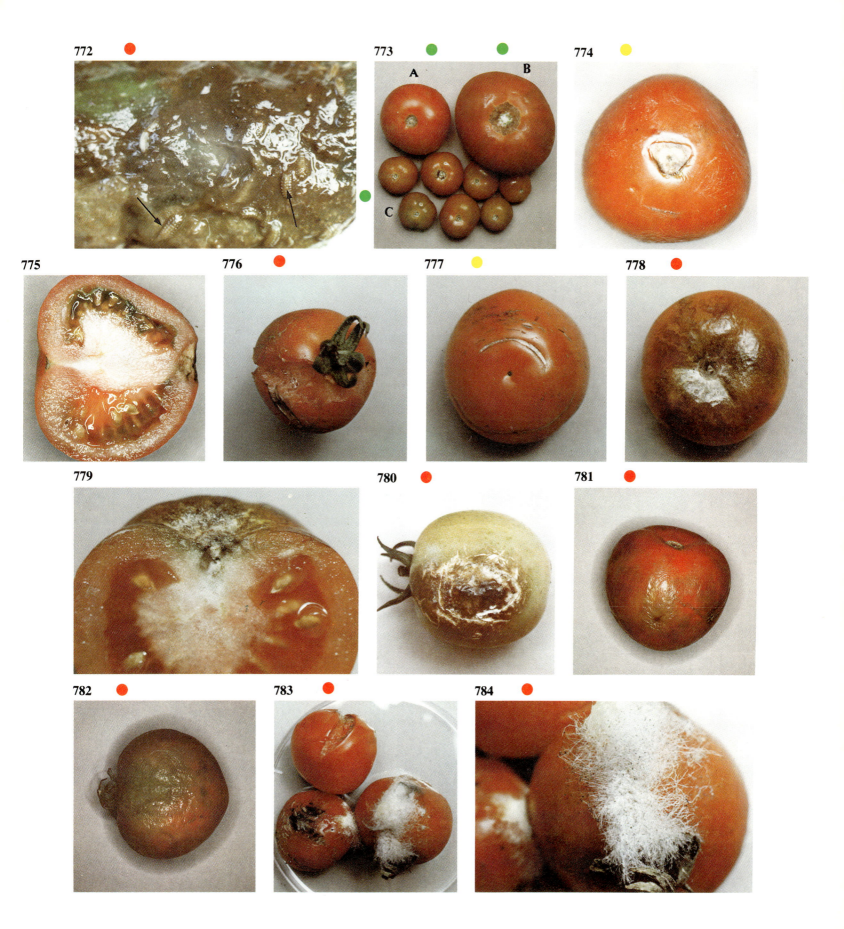

785 ●   **786** ●

787 ●   **788** ●

789 ●   **790** ●

791 ●

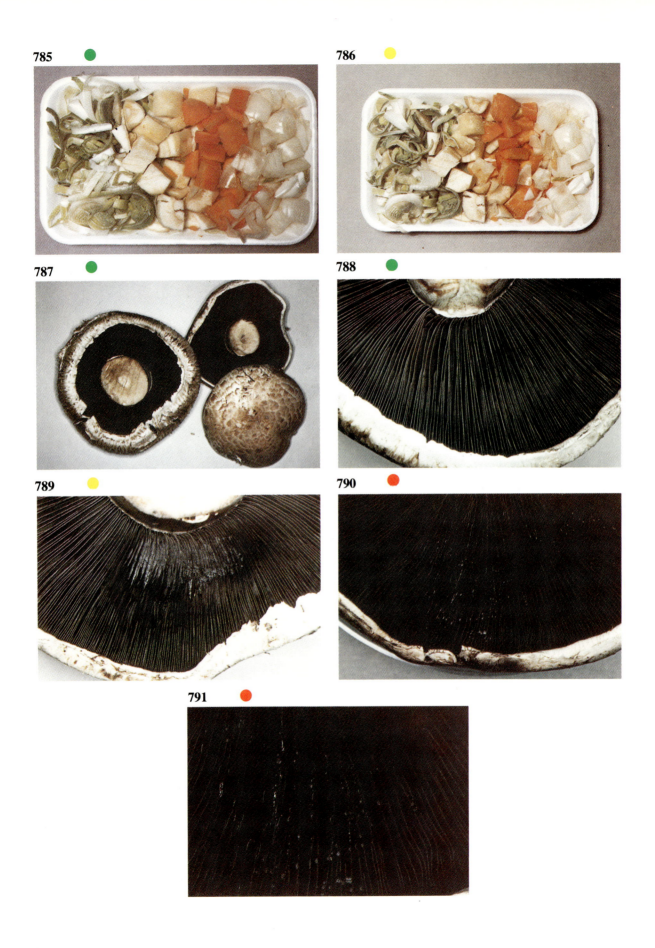

194

785 and 786. Pre-prepared fresh salad trays have become a feature of many supermarkets in recent years in the U.K. A good quality example is illustrated (**785**). Typically, the vegetables are not mixed but grouped on the tray. Note that in this example most are those with a dual cooked/salad usage. Packaging of such products is normally on a polystyrene tray with a permeable film overwrap. The life of cut vegetables is short, usually 1 to 2 days due to tissue damage and dehydration. **786** shows the same salad tray after four days' storage at ambient temperature, and the vegetables can be seen to be limp and wilted. See also Prepared Salads, Chapter 12, p. 172.

Edible Fungi (787–791)

The vast majority of edible fungi retailed in the more developed countries are cultivated and usually derived from the common or wild mushroom *Agaricus campestris*. In less advanced areas wild fungi may be more common while in the U.K. and other European countries a variety of wild fungi are picked for personal consumption and small scale sale. Some of the more exotic varieties such as Chanterelles (*Cantharellus cibarius*) may be dried for export. In addition large numbers of truffles (*Tuber spp*) which grow below the ground on the roots of oak and hazel trees develop naturally and pigs or dogs are used to detect them for harvesting. Truffles for export are usually heat processed (bottled).

Harvest of wild fungi whether for personal consumption or for sale requires a high degree of experience to ensure that the edible species are correctly identified. The Death Cap (*Amanita phalloides*) may be confused with the common edible mushroom by inexperienced pickers, particularly children, and illness and death results. Cases of mushroom poisoning in the U.K. have increased markedly in recent years with the growth of 'back to nature' philosophy.

Illness has also resulted from conscious consumption of species of mushroom known to have hallucinogenic properties due to their containing derivatives of **lysergic acid diethylamide** (LSD). The best known species in Europe is Fly Agaric (*Am. muscaria*); the 'magic mushroom'. It should be noted that sale of hallucinogenic mushrooms is not illegal in the U.K. and several other countries and at least one company openly distributes dried 'magic mushrooms' and other hallucinogenic species.

Cultivated mushrooms are highly susceptible to disease and complete crop failure can occur as a result of pathogens, which include viruses, flies and mites such as **sciarid flies** (fungus gnat; *Sciara spp*) and bacteria. The most important bacterial pathogen which also affects market quality is bacterial **blotch disease** caused by a fluorescent pseudomonad *Ps. tolaasi*. The epidemiology of the disease is complex (Wong and Preece, 1980) but mites and sciarid flies are involved in transmission of the disease in addition to the direct physical damage they cause. A newly recognised bacterial disease **ginger blotch** caused by a member of the *Ps. fluorescens* complex distinct from *Ps, fluorescens per se* (Wong *et al*, 1982) appears to be increasing in importance.

Mushrooms are also subject to post-harvest diseases. These include bacterial rots and mould growth as well as deterioration due to dehydration.

787. Large oyster mushrooms (cèpes). The largest is nearly 6″ in diameter, and they are of generally good quality.

788. The gills of a good quality oyster mushroom, seen in close-up.

789. Mechanical damage, probably during harvesting, to the gills of a mushroom which has become the focus of a bacterial rot infection.

790 and 791. Bacterial colonies developing on the gills of a mushroom illustrated in general view (**790**) and in close-up (**791**). The organism involved is a *Pseudomonas sp*.

Exotic Vegetables (792–798)

For the present purpose exotic vegetables are considered in the context of north west Europe, specifically the U.K. Vegetables in this category would not be considered to be exotic in the producing countries, e.g. sweet potato in the West Indies. The range widely available in the U.K. has increased markedly in recent years. There are two main reasons for this trend. The first is the large number of Commonwealth citizens of non-European ethnic origin who have migrated to the U.K. in post war years. Each group (predominantly Asian or Caribbean with smaller numbers from Africa) has retained to a greater or lesser extent their traditional food habits and thus created a demand for new commodities. At the same time the indigenous population have been exposed to the immigrant cuisine particularly the Asiatic and have adopted some items of it.

The second reason for the growth in exotic vegetables is the expansion of relatively cheap air travel. The immediate effect has been to vastly increase overseas travel with, in many cases, an expansion of the population's culinary outlook. Vegetables such as peppers and aubergines (see pp. 186–189) which are typically part of southern European cuisine are now commonplace in the U.K. and are no longer regarded as exotic. At the same time the building of airports primarily for the tourist trade and the availability, particularly during the winter months, of spare aircraft capacity, permits the air freighting of high quality perishable vegetables. (Note that more traditional fruit and vegetables are imported out of season from tropical countries in this way.)

The value of the trade in exotic vegetables is difficult to assess. The majority of trade is controlled by small firms and sole traders and while large supermarkets, particularly in areas of high immigrant population, are carrying stocks most sales are made through small outlets usually owned by immigrants, primarily Asians. Estimates for examples of staple carbohydrate commodities (Coursey and Brueton, 1983) were yams (*Colocasia esculenta*) 7,500 tonnes, sweet potatoes (*Ipomoea batatas*) ca. 4,000 tonnes, plantains and cooking bananas (*Musa* cvs) ca. 5,000 tonnes and other root crops ca. 1,700 tonnes. These authors point out that the volume of vegetables for the Asiatic sector of the market is much smaller but probably higher in unit cost.

Exotic vegetables are derived from a number of plant families and their pests and diseases causing loss of market quality are primarily those associated with their commonplace counterparts and discussed previously. Dasheens (*C. esculenta* var *esculenta*) and eddoes (*C. esculenta* var *antiquorum*) are the aroid corms of the plant and do not have a counterpart.

792. **Okra** (*Abelmoschus esculentus*) **or 'Ladies fingers'** originate in the Middle East and Asian countries and play an important role in the cuisine of all the main immigrant groups. These are poor quality examples being limp and dehydrated with skin markings probably caused by damage during harvest or transit. Okra are also imported canned and dried but a feature of the immigrant market is a strong preference for poor quality fresh produce over high quality processed even if the price of the former is higher.

793. **Moulis** (*R. sativus* var *longipinnatus*). A root vegetable of African origin. These vegetables which taste like radish may be eaten raw as a salad or cooked. The example illustrated is of good quality.

794 and 795. **Sweet potatoes** (**794**) are a staple part of the West Indian diet even in the U.K. and are selected in preference to the nutritionally similar potatoes even at a considerably higher cost. The examples illustrated are of poor quality showing dehydration, poor cleaning and mechanical damage with associated necrosis (detailed **795**). Again, quality is a secondary consideration in preference to alternatives.

796. **Dasheen** consists of the aroid corm of the plant. Quality problems are largely the same as those associated with root vegetables. The example illustrated is of good quality.

797 and 798. **Fresh bean sprouts** (**797**) from the soya bean (*Soya max*) are produced in the U.K. but as their use is derived from Oriental cuisine they are considered as exotic vegetables. Although not overtly spoilt by excessive dehydration or mould growth these sprouts are of poor quality showing signs of wilting. A lack of commitment to quality by the retailer is demonstrated by the label (**798**) which carries none of the information required including the 'sell by' date.

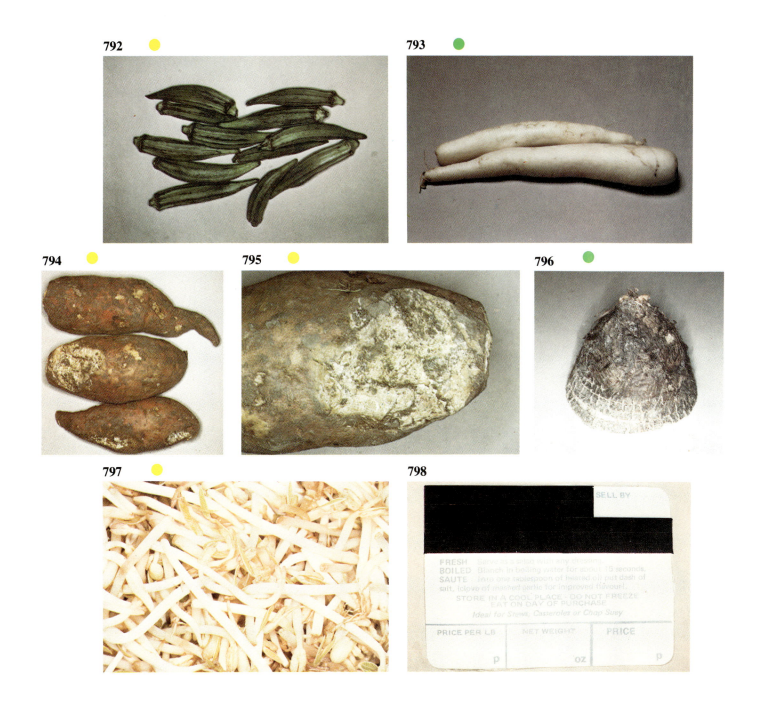

792

793

794

795

796

797

798

SELL BY

FRESH   Serve as a salad with any dressing.
BOILED   Blanch in boiling water for about 15 seconds.
SAUTE   Into one tablespoon of heated oil put dash of
salt, 1clove of mashed garlic for improved flavour.
STORE IN A COOL PLACE - DO NOT FREEZE
EAT ON DAY OF PURCHASE
Ideal for Stews, Casseroles or Chop Suey

PRICE PER LB   NET WEIGHT   PRICE

p   oz   p

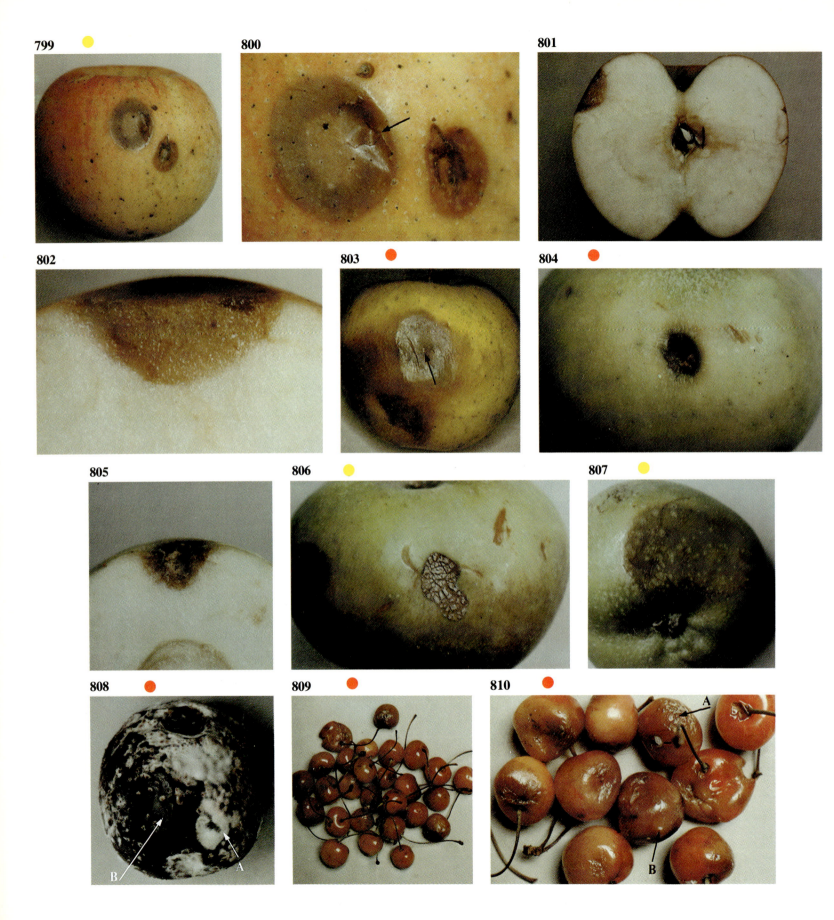

799

800

801

802

803

804

805

806

807

808

B   A

809

810

A

B

198

Apples and Pears (799–807)

Apples (*Pyrus malus*) (and to a lesser extent pears (*P. communis*)) are economically important tree fruits in temperate climates. They may be sold as fresh fruit, processed as dried, canned and frozen or processed for either fresh or fermented juice (cider). Apples in particular are important export crops for countries such as New Zealand, South Africa and France.

Apples for retail sale are broadly classified as **dessert** (eating) for eating raw, and **cooking** which, as the name implies, require cooking to be palatable. Certain varieties of cooking apples may be selected for canning and freezing while special cultivars are grown for cider production. There is at present (1983) considerable controversy over the quality of apples. Many traditional English varieties were low yielding, prone to disease and sometimes of poor appearance and yet had superb flavour. Such varieties have been replaced for commercial growing by high yielding, disease resistant and visually attractive apples such as Golden Delicious which have a flavour described at best as 'acceptable' and often likened to cotton wool!

Apples are subject to a number of pests and diseases which affect the market quality of the fruit. They are also subject to attack by birds although to a lesser extent than pears where birds are often the major cause of pre-harvest damage. Apples and pears are both prone to damage by bruising during handling often followed by secondary mould or bacterial growth. Apart from the deleterious effects on quality of individual apples such fruits serve as a focus for infection of other fruit in store. **Controlled atmosphere storage** in a regulated atmosphere of carbon dioxide and oxygen, usually in conjunction with refrigeration is now widely used to delay post-harvest deterioration. The effect of carbon dioxide in controlling micro-organisms on fruit may lie in improving the physiological status of the fruit rather than in a direct antimicrobial effect.

---

799 to 803. **Bruising** is a major cause of quality loss and spoilage in apples (**799**), shown in detail (**800**). Note the tear in the skin (arrowed) above the bruise. Tough skins are an important factor in choice of variety for commercial growers. A cross section through the apple showing penetration into the interior is shown (**801**). The close-up (**802**) shows cell breakdown providing ready access for spoilage micro-organisms. In a more severely damaged example (**803**) white spots of mould growth (arrowed) are beginning to develop on the damaged area.

804 and **805.** **The codlin moth** (*Cydia pomonella*) remains an important cause of economic loss. The moth lays its eggs under the skin of the apple and on hatching the larvae burrow toward the centre of the fruit. **804** illustrates the entry hole on the surface of an apple and (**805**) shows the penetration in a cross section. In this example penetration is only limited but in more extreme cases penetration throughout the interior can occur. Effective control may be achieved by spraying with **derris** before and after the trees blossom.

806. **Scab** which affects both apples and pears is a further cause of lowered quality and economic loss. The causative organism is the mould *Venturia inequalis*. The disease initially affects the leaves appearing as spots at blossom time. Dark patches subsequently appear on the fruit. Control is by spraying at blossom time with a fungicide such as **captan (Benomyl).**

807. **Damage to an apple** probably due to its having rubbed against an adjacent branch. Modern cultural and pruning techniques have largely obviated damage of this nature.

Plums and Cherries (808–810)

The plum commonly grown in the U.K. *Prunus domestica* is derived from a natural hybrid of the wild cherry plum and the sloe. At least 13 other species of *Prunus* are grown elsewhere in the world for varying processing purposes such as canning or jam making. Closely related species also grown in the U.K. are damsons and greengage. Commercial production of plums has fallen in the U.K. and indeed other countries since the second world war. Contributing causes are decline in demand for canned fruit and jams and difficulties with cultivation. Plums are prone to erratic cropping if springs are cold, suffer considerable bird damage and from an, as yet uncontrollable fungal disease, **silver-leaf** (*Stereum purpureum*). In addition the industry has been severely affected by a new aphid-borne viral disease **Plum-pox** or 'sharka' disease which is incurable and requires complete eradication of affected trees for control. Plums also have a short shelf life after harvest and are prone to rapid mould and yeast growth.

In addition to cultivated species of *Prunus* it should be noted that wild sloe is harvested in many parts of the Northern Hemisphere often as a flavouring for spirits such as gin, vodka and schnapps.

Cherries are also members of the genus *Prunus*, sweet cherries being *Pr. avium* and sour cherries typified by the Morello being *Pr. cerasus*, hybrids between the two types being referred to as 'dukes'. Sweet cherries are commonly grown for dessert use while sour or dukes are more generally used for processing.

The two species of cherry differ in size and in cultivation associated problems. The problems affecting market quality of cherries are however similar, the major one being damage by birds. Post-harvest cherries are prone to similar problems as other fruits notably mould and, to a lesser extent, bacterial and yeast infections.

808. **Mould** (arrowed A) **and associated yeast** (arrowed B) developing on the surface of a plum. The rapid development may have been due to bruising. Fruit affected in this way should be removed from store or display as soon as possible to prevent rapid spread.

809. **Deterioration of cherries** shown in general (**809**) and in close-up (**810**). The cherries are over-ripe and bruised. Fungal rots are becoming established (arrowed A) and yeast is developing on one cherry (arrowed B).

Apricots and Peaches (811–813)

Apricots and peaches (which for this purpose include nectarines) are both members of the genus *Prunus* (*Prunus armeniaca* and *Prunus persica* respectively). Although both species may be cultivated in climates such as the U.K. commercial production is in the warmer areas of southern U.S.A., Australia and South Africa. In addition to production for fresh consumption apricots and peaches may be canned or dried and, particularly with apricots, converted to jam.

Apart from damage by birds and insects pre-harvest diseases of apricot and peach trees tend not to directly affect fruit quality although a tree infected with, for example, the fungal disease **leaf curl** (*Taphrina deformans*) is likely to produce fruit of general poor quality.

Storage diseases of apricots and peaches are largely mould rots and in common with other species damaged fruit are most susceptible to attack.

Vines (814)

The grape (*Vitis spp*) is economically one of the world's most important fruit. Total world production involves in excess of 9,000,000 hectares spread over all continents. While the majority are grown for winemaking considerable quantities are produced as dessert grapes, for drying and, to a much lesser extent, for canning.

Grapes are subject to attack by mould notably *Botrytis* (see p. 239) during growth and, if over-ripe may be subject to yeast fermentations on the vine. Moulds and to a lesser extent yeast are the major specific causes of quality deterioration during storage. As with other fruit birds are a problem both during growth and, if inadequately protected, after harvest.

Strawberries (815–817)

Although strawberries (*Fragaria spp*) are both cultivated and grow wild in many parts of the world, strawberries and cream for tea is thought of as being quintessentially English; symbolic, together with Wimbledon tennis and Henley regatta, of an England most of the population has never known.

Strawberries are an important horticultural crop grown not only for the fresh dessert market but also for canning, freezing and jam manufacture. They are essentially seasonal, the bulk of the U.K. crop being harvested in June. The season is however extended by use of late ripening varieties, and out of season imports air freighted from countries such as Kenya means all year availability although out of season strawberries are a luxury. The seasonal nature of the crop together with dependence upon favourable climatic conditions brings problems for both growers and retailers. A cold damp spring followed by good weather, for example, is likely to result first in shortage with poor quality strawberries commanding high prices followed by a glut where the price of good quality fruit falls below profitable levels. In some years large quantities of good quality strawberries may be dumped.

Birds and to a lesser extent slugs cause damage to strawberries pre-harvest and plants are also affected by a number of aphid borne viral diseases which, if not controlled, result in poor quality fruit being produced. The most important field disease which directly affects market quality is **grey mould** rot caused by *Botrytis spp*. *Botrytis* and other moulds also attack strawberries after harvest, particularly when harvested under damp conditions. See also frozen strawberries, Chapter 14, p. 209.

811, 812 and **813. A poor quality peach** showing severe dehydration and a patch (arrowed) of mould growth (**811**). A cross section (**812**) shows tissue and stone deterioration and a close-up of the flesh (**813**) indicates that tissue degradation is advanced. There is no evidence of mould penetration and the cause of deterioration is physiological.

814. Grapes of general poor quality. The presence of damaged grapes (arrowed) is likely to lead to rapid mould growth.

815 and **816. Botrytis (grey mould) rot of strawberries.** The grey mould (arrowed A) developed to this extent within 2 days of picking during wet weather conditions and the strawberries were probably infected in the field. A second mould (arrowed B) probably *Rhizopus* (see below) is also developing (**815**). The grey mould is shown in detail (**816**). The rotting of the strawberry tissue is visible where arrowed below the mycelium.

817. Rhizopus rot of a strawberry. Note the characteristic 'cotton wool' appearance of the mould on which the black sporangia are visible and the collapse of the fruit due to rotting.

Cane Fruit

Cane fruit, which are all members of the genus *Rubus*, are cultivated for fresh dessert use and for canning, freezing and jam making. Some species such as blackberry are common wild plants in the U.K. although hedge clearing and stubble burning associated with large scale agriculture have reduced the numbers of wild blackberries. It should be noted that the flavour of cultivated cane fruits, particularly the blackberry, is generally considered to be inferior to that of the wild.

In the U.K. raspberry, blackberry and the natural hybrid the loganberry are grown most extensively although the latter is not commonly grown for fresh dessert consumption. The genus *Rubus*

readily hybridises and in addition to the above a number of unusual varieties such as the boysenberry, the wineberry (Chinese blackberry), the veitch berry and the King's Acre berry exist. These are largely grown as garden curiosities although the boysenberry is cultivated on a commercial scale in the U.S.A. The major pre-harvest disease of cane fruits which affects market quality is infestation by **raspberry beetle** (*Byturus tomentosus*) which results in maggoty fruits. Control is by **derris** sprays. Post-harvest spoilage is largely due to mould growth by genera such as *Rhizopus* and *Mucor*, and is particularly rapid when berries have been picked under damp conditions. See also canned raspberries, Chapter 15, p. 212.

Currants and Gooseberries

Currants are all members of the genus *Ribes*. The most common commercially grown species is the **blackcurrant** (*Ribes nigrum*) although lesser quantities of redcurrants and whitecurrants are produced. The blackcurrant is grown as a fresh dessert fruit and for canning, freezing and jam making, as well as for processing into juice. Blackcurrants, particularly varieties such as Baldwin's Black are of high ascorbic acid (Vitamin C) content and the juice is frequently used as a dietary supplement for children and invalids. It is also used to produce an alcoholic drink 'Cassis' (see p. 238).

Blackcurrants are affected by a large number of pests and diseases, the most important being the viral disease **reversion**; infestation by **mites** (big bud, *Cecidophyopsis ribis*) and by **aphids.** These diseases can cause heavy economic loss. Damage to fruit by birds is a problem with all currants, the redcurrant being particularly prone to attack.

Fresh currants have only a very short storage life and are subject to rapid physiological deterioration followed by mould growths.

Gooseberries (*R. grossularia*), a close relative of currants, are grown on both a domestic and commercial scale in the U.K. and other temperate countries. In the U.K. a gooseberry cult previously existed whereby growing clubs held competitions for the heaviest berry. A few such clubs still exist in the North of England but a more important legacy is the modern fast growing varieties of gooseberry which may be harvested not only when fully ripened but also prematurely as early green gooseberries. Special varieties are grown for processing and commercially may be canned, frozen, converted into jam and, on a small scale, used for wine making. Typical examples of domestic varieties are the early harvested 'Careless' and 'May Duke' while dessert varieties include the red skinned 'Whinham's Industry' and the favourite yellow-green skinned 'Leveller'.

Like all fruit crops gooseberries are prone to severe damage by birds and require protection both before and after harvest. Viral diseases, a variety of which affect gooseberries, may be avoided by selection of good quality rootstock while greenfly may be controlled by tar-oil during winter or by malathion during spring. Two diseases specific to gooseberries, the **gooseberry sawfly** (*Nematus ribesii*) and **American gooseberry mildew** (*Sphaerotheca morsuvae*) may cause considerable damage and corresponding economic loss. Attack by gooseberry sawfly is erratic and unpredictable but the larvae rapidly defoliate affected bushes. Control is by **malathion** or **derris**. American gooseberry mildew affects not only the plant shoots but also the berries which become coated with a felt-like growth. Control is by a fungicide such as **benomyl**, **pirimicarb** or **thiophanate-methyl.**

Citrus Fruits (818–826)

Citrus fruits (*Citrus spp*) are typically the products of warm climates and cannot be grown successfully on a commercial scale in temperate countries. There is thus a considerable export trade in citrus fruits which are of major economic importance to producing nations such as Spain, Morocco, Israel, South Africa and parts of the U.S.A. The extent of their economic importance may be judged by the political attempts to damage the economies of South Africa and Israel by boycotting of South African oranges (and other produce) by opponents of that country's apartheid policies and the attempt to discredit Israeli citrus fruit by claiming it had been injected with mercury.

In fiscal terms total U.K. imports during 1980 (United Kingdom Trade Statistics, H.M. Customs and Excise) amounted to over £129,000,000 worth. The largest tonnage of citrus (or any imported fruit) is oranges (*C. sinensis*) although that has declined slightly in favour of other citrus fruits and juices. Sour oranges (*C. aurantium*) are imported for marmalade manufacture but now account for only 1% of imports.

Imports of grapefruit (*C. paradisi*) continue to increase as do other citrus lines. The once popular tangerine (*C. reticulata*) has largely been replaced by other varieties of the same species such as satsumas and clementines while hybrids such as uglis (*C. reticulata* × *C. paradisi*) and ortaniques (*C. reticulata* × *C. sinensis*) are now more common.

818 to 821. **Fungal rots** are the most common storage diseases of citrus fruits. Typically mycelia are only present on the outside of the fruit if storage has been in humid conditions such as a closed jar in the home. **Navel end rot** caused by *Alternaria citri* is illustrated (**818** and **819**). The example shown is atypical in that the rot is normally more pronounced at the navel end. Affected fruits ripen early and may be detected by the unusually deep colour of the peel. **Brown rot** caused by *Phytophthora sp* is illustrated (**820**). The rot is accompanied by a characteristic odour and usually develops after wet growing conditions. It only develops on the outside of the fruit under very humid conditions (**821**).

822, 823 and **824.** **Anthracnose disease** of citrus fruits produces scabs on the outside of the peel with rotting extending inwards when the infection is severe. Examples of the disease in oranges (**822**), limes (**823**) and lemons (**824**) are shown.

825 and **826.** **Mechanical damage** causes loss of quality in citrus fruits and may act as a focus for mould growth. **825** illustrates a poor quality grapefruit with mould growth beginning at the site of damage while (**826**) illustrates bruising on a tangerine possibly by guide bars on a conveyer belt.

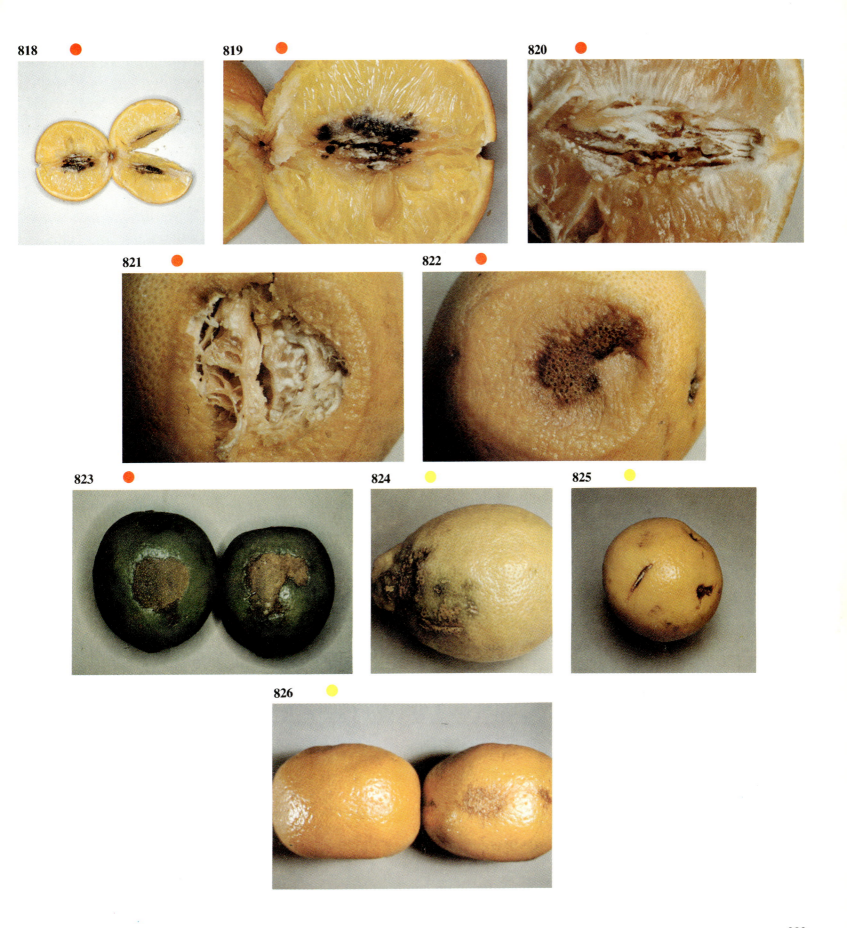

Tropical Products (827–835)

For the present purpose tropical fruits are restricted to those, primarily banana and pineapple, which have traditionally been imported into the U.K. and North Western Europe in significant quantities. Other tropical fruit such as mangoes and avocados are discussed with 'Exotic fruit' (see p. 206).

Bananas (*Musa cvs*) are the most important tropical fruit crop in international trade. As with some other tropical and sub-tropical products (cf. citrus fruits p. 202; sugar p. 138) bananas form a highly important part of the external economics of many producing nations exemplified by the description 'Banana Republics' and attempts to destabilise economies by sabotaging banana crops have been made. Major exporters to Europe include the Canary Islands, the Caribbean and tropical America, Cameroon and Surinam.

Imports to the U.K. in 1980 (United Kingdom Trade Statistics, H.M. Customs and Excise) totalled over 300,000 tonnes at a value of almost £75,000,000. Small quantities of dried and canned bananas are also imported, the latter largely as a purée for use in flavouring and as baby food, as well as green cooking bananas and plantains for the West Indian immigrant trade. Coursey and Brueton (1983) note that consumer acceptance of bananas is based largely on cosmetic factors rather than organoleptic quality, taste, texture and aroma. These authors state that the change from the 'Gros Michel' cultivar shipped on the stem to 'Cavendish' cultivars shipped in box together with other changes in handling have led to substantial improvements in ripe appearance and shelf life of bananas on the U.K. market.

Note that the antibiotic nisin permitted in products such as processed cheese to prevent bacterial spore germination and out-growth (see p. 36) is also used to prevent mould growth on bananas.

Pineapples (*Ananas comosus*) are highly perishable and until the introduction of air freight, and of more sophisticated container ships, most pineapples imported were canned. Although substantial quantities of canned pineapple are still imported consumption has recently declined partly as a consequence of the availability of the fresh product. Substantial quantities of pineapple juice are also imported because of rapidly increasing juice consumption and a diversification of the products available. U.K. imports of fresh pineapples totalled nearly 12,000 tonnes in 1980 (United Kingdom Trade Statistics, H.M. Customs and Excise), an increase of over 4,000 tonnes since 1970 and equivalent to a value of £5,233,000.

Countries supplying significant quantities to the European fresh market include South Africa, Kenya and the Ivory Coast. Although historically pineapples have been grown in hot-houses in the U.K. to supply the most affluent sectors of the community cultivation has, for practical purposes, ceased for many years. The era of the pineapple is remembered by the stylised stone pineapples still commonly found on the gateposts of some country houses and occurring as further stylised imitations elsewhere.

One disease of pineapples, '**pineapple pink**' disease, caused by acetic acid bacteria or *Erwinia herbicola* is noteworthy in that the symptoms, a pink or brown discoloration of the flesh only become apparent when the flesh is heated. (See page 212.)

827 to 831. Banana freckle (Black spot) caused by the mould *Macrophoma musae* (**827 and 828**) is a common condition affecting the exterior of the peel only (**829**) unlike the superficially similar '**pitting disease**' (*Piricularia grisea*) in which the flesh below the peel is rotted (**830 and 831**). This latter condition is rare. The incidence of banana freckle at retail level has increased with the substitution of the Cavendish cultivar for the highly resistant Gros Michel.

832. Banana ripeness is a subjective topic. Most consumers in the U.K. and Europe would consider bunch A to be over-ripe and unfit to eat. Other persons, particularly those from producing countries, would consider such a degree of ripening to be desirable. Bunch B would be considered ideal for eating by most U.K. and European consumers while bunch C would be considered to be

under-ripe and unpleasant to eat. Considerable quantities of green bananas and unripe plantains (*Musa* AAB group) are imported for cooking as carbohydrate staples by West Indian and to a lesser extent, other immigrant groups.

833, 834 and 835. Fresh pineapples previously imported into the U.K. were often very large fruit used for decorative as well as culinary purposes at banquets etc. Small pineapples for family consumption now make up the bulk of the trade. In some cases these are halved before sale (**833**). The cut surface is usually protected by a plastic film and while overall appearance may remain unchanged for up to 4 days (**834**) physiological lesions develop on the flesh (**835**) (arrowed). There is also an increased possibility of mould growth.

827 🟢 *

828 🟢 *

829 🟢 *

830 🔴

831 🔴

832 🟢 *   🟢   🟢

A   B   C

833 🟢

834 🟡

835 🟡

Exotic Fruit (836–849)

As with exotic vegetables (p. 196) exotic fruit are defined in terms of the North West European market, specifically the U.K. The growth of trade in exotic fruit parallels that of the vegetable trade e.g. the demands of an immigrant market, widening of the tastes of the indigenous population and increased availability of cheap air freight, and sophisticated refrigerated container ships.

Acceptance of exotic fruit by the native population is probably greater than that of exotic vegetables (if courgettes and aubergines are excluded) partly because they are seen as luxury items as opposed to staple vegetables like sweet potato. Indeed with examples such as avocados (*Persea americana*) the demand comes from the indigenous population and lies not with spontaneous demand but from vigorous promotion by the major producing countries, Israel and South Africa. Such an 'artificial' creation of demand has an interesting effect on perceived quality of avocados in that the use of sea transport (at ca. 5°C) necessitates the use of cultivars such as Fuerte which is firm textured, low in oil content and chill resistant. This is seen by the British public as the typical (and by implication high quality) avocado (Coursey and Brueton, 1983) whereas cultivars of softer texture and higher oil content but less easily transportable are actually of far better eating quality.

Avocados comprise the highest tonnage of exotic fruit imported into the U.K., 1980 imports being in excess of 6,000 tonnes (Coursey and Brueton, 1983). Second to avocados were mangoes (*Mangifera indica*) with an imported tonnage in excess of 3,000 tonnes. Unlike avocados mangoes were first imported to satisfy the immigrant market but have been widely accepted by the indigenous population. Other tropical fruit such as paw-paw (papaya, *Carica papaya*) have not been widely accepted and imports of 'novelty' fruit such as kiwi fruit (*Actinidia chinensis*) have remained at low levels despite fairly intensive promotion.

In addition to the tropical exotic fruit this section includes the subtropical fruits dates (*Phoenix dactylifera*) and figs (*Ficus carica*). Both are imported in significant quantities in dried and semi-preserved form but fresh fruit are now appearing, in small quantities, in the U.K. market.

The fig is native to the Mediterranean where historically it played an important part in subsistence economies, although it is grown on a commercial scale in other parts of the world with a suitable climate such as California, while elsewhere other species of *Ficus* such as *F. roxburghii* in India also serve as food plants. Like dates most figs in international commerce are partially dried but there is a small international trade in fresh figs for the luxury market. These rapidly deteriorate in quality and this factor limits their availability. The demand for figs in non-producing countries is restricted, Coursey and Brueton (1983) noting that 'figs with their strong flavour and profuse seeds are liked by only a limited proportion of the population, and tend to be consumed by the health conscious for their well known purgative effect, rather than for pleasure'. The purgative effect of figs is further exploited in that the extract of 'syrup' figs is a widely used laxative particularly for children.

836. A good quality but under-ripe mango. One of the problems with exotic fruit is that quality defects are not recognised as readily as with more familiar fruit.

837. Fully ripened mango. Although mangoes as sold in European markets are usually under-ripe to provide sufficient shelf life for the retailer, they are ideally eaten fully ripened. The appearance of the ripe fruit, however, may not be acceptable to a consuming public basically unfamiliar with the nature of the product.

838. Over-ripening of mangoes leads to a major enzymatic breakdown of structural tissue. The illustration shows the interior of this over-ripe mango to be semi-liquid. In this state, however, it may still be considered edible by those familiar with the fruit.

839. Paw-paw, like mangoes, is a relatively new type of fruit on the U.K., European and North American markets and the same problems of recognising poor quality exist. The example shown is of good quality but under-ripe.

840 and 841. Fully ripened paw-paws are prone to mould growth as a consequence of tissue breakdown. **840** illustrates an unacceptable degree of mould development (arrowed) on a ripe fruit. The extent of penetration of enzymes from the mould is illustrated (arrowed) (**841**). The mycelia have not yet penetrated the damaged tissue.

842. The flesh of over-ripe paw-paw rapidly deteriorates as with mangoes. Such an example would, however, be acceptable to some consumers.

843 to 847. Fresh dates (as opposed to dried or sugared) are a new departure in U.K. produce retailing. Good quality dates (**843**) have a shelf life of only a few days and rapidly deteriorate (**844**). Deterioration is primarily physiological and the effects on the internal tissues are illustrated by comparing the cross section of a good quality date (**845**) with that of a poor quality (**846**). Deteriorating dates are prone to secondary mould invasion, and mould growth (arrowed) on the stem end of a physiologically damaged date is illustrated (**847**).

836 ● *

837 ● *

838 ● *

839 ●

840 ●

841 ●

842 ● *

843 ●

844 ●

845 ●

846 ●

847 ●

848

1cm

849

1cm

Laboratory Tests

Laboratory tests are not normally considered relevant for products discussed in this chapter.

Examples of foreign bodies in produce may also be found in Chapter 17, Foreign Bodies and Infestations as follows:

Pp. 244–245 Locust from salad
Pp. 246–247 Moth in lettuce
Pp. 246–247 Moth from bananas

See also:

Pp. 244–245 Weevil in peaches (canned)
Pp. 244–245 *Tenebrionidae* in peas (canned)
Pp. 246–247 Myriapods from fruit (dried)
P. 251 Cigarette end in green beans (frozen)
P. 251 Human hair in peas (frozen)
Pp. 252–253 Masonry nail in peas (frozen)
Pp. 252–253 Plaster dust in peas (frozen)

848 and **849.   A secondary problem** with the quality control of exotic fruit is that infestants may be unfamiliar species. **848** and **849** illustrate a species of weevil (*Curculionidae*) from a batch of Brazilian mangoes imported into the U.K. As far as the authors are aware there are no weevils of this type indigenous to the U.K. At the time the insects were found there was considerable publicity over Colorado beetles and a number of people finding the weevils mistook them for Colorado beetles and were caused considerable anxiety. (Note that one of the dangers of increased import trade of fruit and vegetables is the possibility of introducing not only insect pests but also bacterial and viral pathogens.)

Chapter 14

Frozen Foods

Ice and its attendant low temperatures have been associated with long-term storage of food since antiquity. Ice-houses for example were a feature of English country houses from the 18th century onwards and references are made to the use of snow and ice in earlier civilisations. The full potential for frozen storage began to be developed with the advent of mechanical refrigeration at the end of the 19th century and a large trade in frozen meat from countries such as Australia and Argentina into the U.K. was developed in the early 20th century. Such meat was rarely more than crust frozen and spoilage due to moulds such as *Cladosporium, Sporotrichum, Thamnidium, Mucor* etc. was common.

The modern frozen food industry originated in the 1930s with the well publicised observations of Clarence Birdseye into the rapid freezing of foods. Today the range of products available in frozen form is enormous and encompasses most types of food although it should be noted that some commodities such as tomatoes are inherently unsuited to preservation by freezing. It is obvious that the technology of frozen food manufacture will vary according to the product nature. There are, however, basic principles common to all frozen products, which relate to quality. These are discussed briefly below.

Selection of Raw Materials

Raw materials must be of good quality. It is not possible to improve the quality of any product by freezing and equally no attempt should be made to 'save' products that are known to be deteriorating by means of freezing.

Large scale producers of frozen fruit and vegetables usually grow cultivars specially selected for their suitability. These cultivars would often not be adjudged to be the highest quality if consumed without processing but retain the desired properties after freezing and subsequent defrosting.

Where meat and fish products are concerned it is necessary for manufacturers to ensure that the correct species is in fact being used and that in the case of meat it is fit for human consumption (see p. 47).

Maintenance of Raw Material in Good Condition Before Freezing

The nature of processing operations may entail a considerable delay between receipt at the factory and freezing. During this period deteriorative changes may occur. These include incipient or actual microbial spoilage and post-harvest enzymic changes in fruit and vegetables which may cause softening, loss of flavour or development of off-flavours and colour changes. Blanching in hot water or steam is usually employed to arrest these changes and also aids in colour retention for vegetables such as peas. Blanching reduces the number of vegetative micro-organisms by up to 99%. With any food, delays before freezing must be minimised and where appropriate, refrigeration should be employed to reduce the effect of any unavoidable delays. In planning production of multi-component foods it is necessary to ensure that all components are processed rapidly before final assembly and freezing.

With respect to public health aspects it is particularly important to ensure that holding stages which permit the growth of food poisoning bacteria are eliminated. *Staphylococcus aureus* in particular grows rapidly on many foods in the absence of a saprophytic spoilage microflora and appears able to grow and elaborate toxin particularly rapidly on vegetables after blanching (authors' unpublished observations). Examples of outbreaks have involved vegetables such as frozen peas, potato chips (French fried potatoes) and meat products such as burgers.

Freezing

Most frozen foods are now produced using quick freezing techniques which imply a freezing time not greater than 30 minutes. The means by which quick freezing is obtained vary but in all cases the product quality is higher than that obtained by slow freezing, with freezing times of 36 to 72 hours. Economic benefits are also involved in the choice of quick freezing. Quality of some products is improved by direct freezing in a refrigerant, examples being the freezing of berry fruits in syrups and fish in brine.

Post-freezing Storage

Correct storage conditions are an essential factor in maintaining quality of frozen foods. This implies correct temperatures at all stages. The recommended storage temperatures vary but in commercial situations something in the order of −30°C is common. The principle that the lower the temperature the better is not necessarily correct (at least above the temperature of liquid nitrogen −186°C) because there is some evidence that products highly prone to development of rancidity such as frozen kippers maintain quality as well and possibly better at −20°C than at −30°C.

Temperatures should never be allowed to rise above −20°C except for short periods e.g. during loading into transport, when −12°C should not be exceeded. It is most important that foods defrosted due to refrigeration failure are not refrozen although it is accepted that correctly defrosted bulk ingredients may be refrozen in the complete formulated product.

It is necessary to realise that chemical changes such as onset of rancidity continue in foods during frozen storage, the rate depending on the food and the particular chemical reaction involved. Storage times cannot be allowed to exceed manufacturers' recommendations without risk of quality loss.

In defining the scope of this chapter most of the examples illustrated refer to problems and factors associated with the freezing process. Factors implicit in the product whether frozen or not are discussed in the appropriate commodity section, cross referred where appropriate. Similarly while ice cream is implicitly a frozen food it is discussed with Dairy Products (Chapter 2) as most of the problems are related to the dairy products component.

Examples of laboratory tests are shown in Table 22.

See also **Frozen cream and ice cream**   Chapter 1, pp. 42–43, 45–46
            Frozen meat   Chapter 3, p. 50
            Frozen poultry   Chapter 7, pp. 90–91
            Frozen fish   Chapter 8, pp. 106–107

Examples of foreign bodies in frozen foods may also be found in Chapter 17, Foreign Bodies and Infestations as follows:

P. 248   Rodent parts from frozen mixed vegetables
P. 251   Cigarette end in green beans
P. 251   General debris from frozen peas
P. 251   Human hair in peas
Pp. 252–253   Masonry nail in peas
Pp. 252–253   Plaster dust in peas

Table 22 Laboratory Tests Frozen Foods

I. Microbiological

Total viable count is the basis for assessing the microbial-status of most frozen foods (yeast for fruits and fruit products). Choice of further tests depends on the nature of the product and is usually similar to the fresh equivalent.

Recommended methods for examination of frozen foods may be found in Hall (1982).

II. Chemical/physical

Choice of analyses is dependent on the nature of the product.

Infestation examination not normally required but some products, e.g. vegetables, should be examined for foreign material such as pods which should have been removed during processing.

Notes:
(1) These tests are examples only and are not inclusive.
(2) Examples are for finished products only.
(3) Techniques are detailed in specialist publications such as ICMSF (1978) – Microbiology: Egan *et al* (1981) – Chemistry.
(4) Instrumental techniques such as Impediometry may be substituted for some classical microbiological methods.

850 and **851.** **Breakdown of refrigeration** equipment such as frozen food display cabinets can lead to considerable stock loss and the temptation exists to refreeze. **Refreezing must not take place** under any circumstances and product which has fully defrosted such as the peas illustrated (**850**) must be discarded. Where the product remains partially frozen with remaining ice crystals (arrowed) (**851**) it may be used for immediate consumption in catering etc. It must however be appreciated that **microbial growth rates** on defrosted frozen products are usually more rapid than on fresh and that the decision to use partially defrosted foods should be taken only by experienced and responsible personnel.

852 to **855.** **Correct storage of frozen foods** is essential to maintain quality. Incorrect storage conditions may occur during production, at the retailer's, during transport or in the home. Most problems up to point of sale probably occur during transport or at retailers. **852** illustrates correctly stored frozen peas, which are recognised as 'individuals' rather than being clumped together, and flow freely when shaken from their container. In contrast the peas illustrated (**853**) are encased in a block of ice and it is obvious that at some stage they have been defrosted and refrozen. Frozen broad beans present similar examples, those illustrated (**854**) having been correctly stored while those illustrated (**855**) have defrosted and subsequently been refrozen.

856. **Frozen broad beans** (*Vicia faba*) showing loss of quality due to dehydration during storage. This probably occurred in bulk storage prior to pre-packing.

857. **Frozen broad beans** showing brown spot as a result of pre-harvest disease. Only good quality raw materials should be selected for freezing, and a correctly operated visual inspection stage should be incorporated in vegetable processing for freezing to alleviate such problems. (See also Legumes, Chapter 13, pp. 180–181.)

858. **Many fruit such as the strawberries (*Fragaria sp*) illustrated** are suitable for freezing by commercial, but not necessarily domestic techniques. The strawberries shown are of good quality, in poorer quality samples loss of texture and excessive drip is common on defrosting. Varieties chosen for freezing are not those normally chosen for the fresh market. (See also Strawberries, Chapter 13, pp. 200–201.)

859. **A frozen prawn curry.** Early varieties of complete frozen meals included the ancillary vegetables (see dried meals pp 148–149). These largely comprised traditional meals such as roast beef with vegetable, were poor quality and designed for easy preparation while viewing television. The modern frozen meals are usually the major meal component only and are of considerably higher quality.

850

851

852

853

854

855

856

857

858

859

211

Chapter 15

Canned Foods

The market for canned (and bottled) goods although still large is declining. This to a large extent is due to competition from the newer frozen form of preservation coupled with the fact that canned foods are generally considered to be of inferior quality. At the same time changing patterns of food consumption, in the U.K. for example the demise of the traditional high tea, have led to a general trend of a reduced consumption of canned goods. In recent years, canners have responded by improving quality and extending product range particularly into up-market products such as 'gourmet' soups.

At the same time the increasing influx of immigrant communities in the U.K. and much of Western Europe has led to increased importation of exotic and ethnic canned products.

In addition to economic and socio-economic factors underlying falling consumption of canned goods the industry has recently been affected by renewed concern for the safety of canned materials. Provided that cans are adequately processed and that re-contamination is prevented, canned foods are extremely safe. The effect of failure at any stage is potentially cataclysmic both with respect to product safety and the attendant publicity. This is illustrated by the consequence of the pre-war botulism outbreak at Loch Maree, the Aberdeen typhoid outbreak of 1963 and, latterly, the problems of botulism associated with North American canned salmon.

A high level of inspection is thus required at every stage of production and during post-production handling. Canning technology is complex and for a full discussion reference should be made to one of the specialist textbooks such as Hersom and Hulland (1980) or to the technical literature produced by can and canning machinery manufacturers such as The Metal Box Company. A flow diagram for a typical canning process is shown (**1**, repeated on p. 214).

A brief discussion of the various factors involved is given below and is related to a Hazard Analysis of a typical canning process, control and inspection procedures and illustrations of various problems.

Examples of laboratory tests are shown in Table 23, p. 227.

The Canning Process

Stage 1 – Raw (and Pre-processed) Materials

To some extent raw and partially pre-processed (e.g. boned) materials to be used in canning present no problem particularly associated with the canning process. In most cases no precautions other than those of good manufacturing practice (GMP), always necessary in food manufacture, are required. There are, however, exceptions usually relating to choice of ingredient. It is often thought that canning may be used to disguise poor quality ingredients. With fruit and vegetables the reverse is true and the potentially severe effect of thermal processing is such that ingredients must be of a higher quality for canning than for the fresh market.

Canners, where possible, use varieties specifically developed to impart the desired qualities after processing while contract planting and harvesting by growers working to strict timetables ensures as far as possible a continuing supply of raw material of optimal quality over the desired period. Objective measuring devices to assess quality may be used in some cases such as the Tenderometer with peas. (See also Chapter 13, p. 180.)

Ingredient choice with some fruits such as tomato impinges on safety in that traditionally the tomato as a low pH (high acid) product does not require a full thermal process. However in recent years canned tomatoes with pH values significantly above 4.7 (the accepted lower limit for growth of *Clostridium botulinum*) have been produced. Measurement of pH value is therefore of prime importance in selection of tomatoes for canning and this consideration also applies to related fruit such as sweet peppers.

Where meats are canned, the situation is somewhat different in that canning is often used as a means of upgrading low quality meat and the heating period may even be extended beyond that required for safety to produce adequate tenderisation. After such lengthy thermal processing it is usually difficult to recognise the initial nature and quality. This has resulted in the substitution of horse meat for beef and in the use of unfit or condemned (knacker) meat (see also Chapter 4, p. 55). Relatively simple laboratory tests exist for detection of horse meat before heating but in the case of condemned meat it is necessary for canners to buy from reputable suppliers and ensure that where wholesalers are used, the wholesaler is aware of the ultimate source of the meat.

In some cases the appearance of the raw material does not relate to the appearance of the canned. Pineapples affected by '**pineapple pink**' disease for example, are of normal appearance when raw but develop a pink to brown discoloration when heated during canning. This is thought to be due to the formation of **2,5-diketogluconic** acid by the causative organisms; the acetic acid bacteria, *Acetobacter* or *Gluconobacter* (DeLey *et al*, 1984) or *Erwinia herbicola* (Cho *et al*, 1980).

In addition to the major constituents it is necessary in some cases to apply special consideration for the selection of minor ingredients to be used in canning. Sugar, starch, spices etc. may contain large numbers of bacterial endospores of particularly marked resistance. In some cases such as tomato soup adverse organoleptic changes preclude 'total' elimination of endospores if initial numbers are high and special 'cannery' ingredients with low

spore counts are used. Many canners carry out routine microbiological examination of such products. Small numbers of spores of highly heat resistant thermophilic bacteria are still likely to survive and while such bacteria are unable to develop at normal ambient temperatures in temperate climates, they can cause rapid spoilage in tropical and sub-tropical countries. The permitted (U.K.) food antibiotic nisin is often included in formulation of canned products to be used in such areas.

With respect to formulation it is also important to be aware that changes in the type of ingredient as well as the quantity of ingredients may affect heat-penetration during processing. This is discussed in more detail (Stage 6).

Some canned products which appear to be fully processed are in fact chemically preserved and receive no thermal processing after the can is sealed. Normal canned goods quality control parameters concerning sterility obviously do not apply and persons involved with these products should be aware of any particular problems.

860. **Canned sweetened condensed milk** is stable at room temperature by virtue of its high sugar concentration and thus low water activity level (see pp. 37–39).

861. **Canned anchovies** are preserved by high salt concentration. Before packing in cans the anchovies are packed with salt and held in large barrels. A 'fermentation' occurs softening the flesh and reducing the pH value. This is thought to be due to autolytic

enzyme activity but extremely halophilic bacteria may be involved. The packed product is not fully shelf stable at ambient temperatures and storage in cool, preferably refrigerated, places is required. The rate of spoilage is low since the causative organisms, extremely halophilic bacteria such as *Halobacterium salinarium*, have only limited metabolic activities.

862, 863 and **864.** **Very poor quality canned sardines** turned straight from the can onto a plate shown in general view (**862**) and in close-up (**863**). Note particularly the almost total disintegration of the sardines. This may be caused by use of poor quality raw material, or to over-processing. Good quality sardines are shown in comparison (**864**).

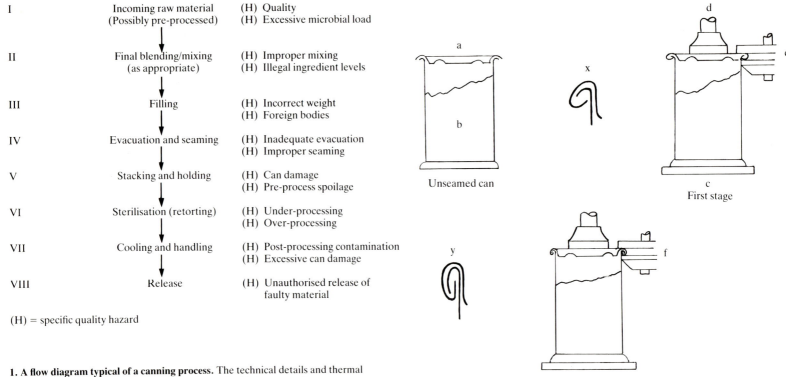

| I | Incoming raw material (Possibly pre-processed) | (H) Quality |
| | | (H) Excessive microbial load |

I   Incoming raw material   (H) Quality
        (Possibly pre-processed)   (H) Excessive microbial load

II   Final blending/mixing   (H) Improper mixing
       (as appropriate)   (H) Illegal ingredient levels

III   Filling   (H) Incorrect weight
                             (H) Foreign bodies

IV   Evacuation and seaming   (H) Inadequate evacuation
                             (H) Improper seaming

V   Stacking and holding   (H) Can damage
                             (H) Pre-process spoilage

VI   Sterilisation (retorting)   (H) Under-processing
                             (H) Over-processing

VII   Cooling and handling   (H) Post-processing contamination
                             (H) Excessive can damage

VIII   Release   (H) Unauthorised release of
                                 faulty material

(H) = specific quality hazard

1. A flow diagram typical of a canning process. The technical details and thermal processing applied will vary according to the product being canned.

865. Can seaming operation. The lid or end (a) is placed on the can body (b). During the first stage of the seaming operation the body is placed on a base plate (c) which lifts pressing the can end against a chuck (d). The first seaming roll (e) then comes into operation and forms a partially completed seam shown in cross-section (x). The second stage involves a seaming roller (f) which has a shallower groove than the first and completes the seam. (Shown in cross section y). After Hersom & Hulland (1980).

Stage 2 – Final Blending/Mixing

This stage involves no problems unique to canning, control lying primarily in ensuring correct formulation, consistency or mix (where appropriate). It is necessary to ensure that any legislative requirements such as level of sodium nitrate are complied with and laboratory analysis may be required for this purpose.

Stage 3 – Filling

Filling is an important aspect of many food packaging operations and in most countries is controlled by legislation concerned with filled weight.

Of unique importance to the canning industry is the fact that incorrect filling may affect the sterilisation process. Underfilling may lead to an excessive amount of air remaining in the can after evacuation, possibly leading to a reduction of temperature in some areas of the can. At the same time underprocessing can result from overfilling which may mean impaired convection within the can and thus inadequate heating.

The filling stage also represents the last stage at which the product may be inspected for presence of foreign bodies. This aspect again is of importance in many filling operations but the totally opaque nature of cans can introduce particular problems. If possible metal detectors should be fitted to the filling line to check the product before it enters the can and a regular check schedule should be introduced to ensure the efficient operation of the detectors. A final visual inspection is also desirable at this stage to detect non-metallic contaminants.

Stage 4 – Evacuation and Seaming

Evacuation and seaming refer to two different technological processes but are considered together since, in most modern canneries, the two processes are carried out concurrently by the same equipment.

Evacuation is required to remove air from the can prior to sealing and retorting because the presence of air pockets can lead to underprocessing.

Correct seaming is a factor of prime importance in ensuring that the possibility of post-processing contamination is minimal. Modern 'open-top' or 'sanitary' cans are manufactured either as two- or three-piece cans. In the first case the can body and bottom end

piece are extruded in a single operation and the lid is added by the canner after filling. The three-piece can consists of a body (usually formed from sheet metal and having a soldered side seam) and two end pieces (lids). The lower of these is usually applied by the can manufacturer and unless cans are physically damaged faults in this part of the manufacturer's seam are rare.

As with the two-piece can the upper lid is seamed onto the can body after filling. Production of a correct seam at this, the canner's seam, is of prime importance. The manner in which the modern double seam is formed is shown (865). The seam dimensions shown are of prime importance and must be checked at regular intervals (usually every 30 minutes). (See also Appendix 2.)

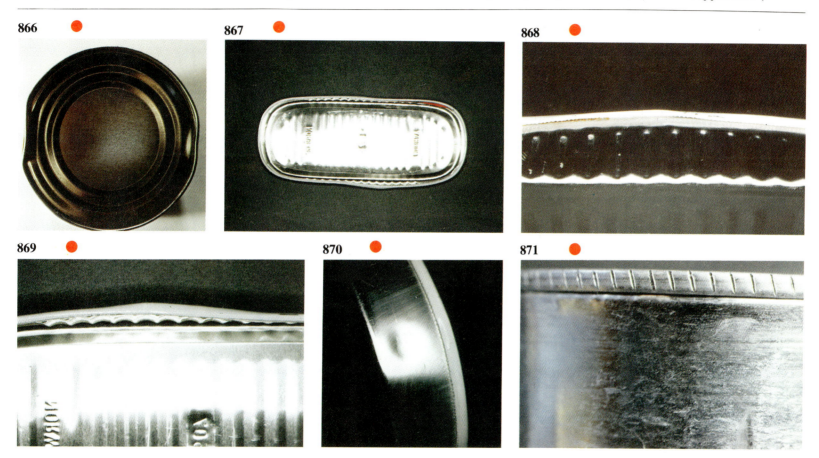

866.   **A defective seam** is shown by the flattening seen in the accompanying illustration. The lack of an indented code indicates that this is the manufacturer's seam where defective seaming is rare. Such flattening is indicative of faulty adjustment of the roller used to finish the seam-making procedure.

867, 868 and **869.   Seam damage in the canner's seam** of a flat, rectangular two-piece can. An allegation of food poisoning arose from consumption of the can contents (cod roe). Damage of this nature is compatible with damage by a can opener as well as with a seamer fault but as identical damage was found in other cans the

fault may be attributed to a faulty seamer, organisms being able to gain admittance through the faulty seam.

870.   **Secondary damage to the cod roe can** referred to above consisting of a sharp dent on one corner. Aluminium cans such as these are particularly prone to this kind of damage which can result in a minute puncture being made which is too small to be easily seen by visual inspection but which readily admits bacteria.

871.   **A seaming fault** manifested as damage to the outer part of the seam. The cause is probably an incorrectly set seamer and in such cases a large number of cans are likely to be affected.

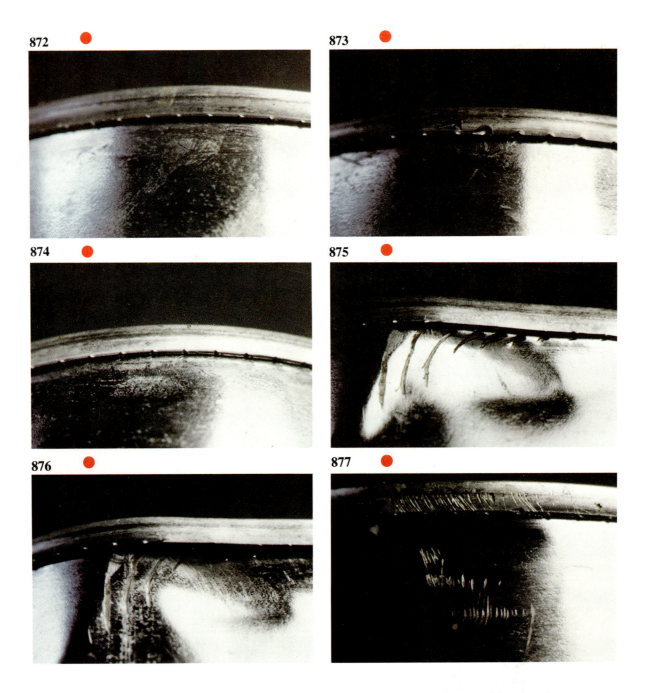

872 to 877. A sequence of plates showing the development of a seaming fault during a production run at a cannery. The illustrations cover a 5 hour period and it is obvious that the necessary regular seam checks have not been made. It should be noted that seam checking should not be used merely to check the historical performance of the seamers but also to pre-empt future problems. A simple statistical record such as a mean and range chart (**878**) may be of help in achieving this.

872. 11.15 hours. Minor seam damage only.

873. 12.45 hours. The damage worsens as the seamer slips out of register. Note also that there is some damage to the seam outer.

874. 15.00 hours. The fault continues to worsen and minor damage to the body of the can becomes apparent.

875 and 876. 15.45 hours. Major damage to the body of the can is now apparent.

877. 16.15 hours. The fault has lessened although there is still some body damage. This may be due to adjustment having finally been made or may have occurred spontaneously.

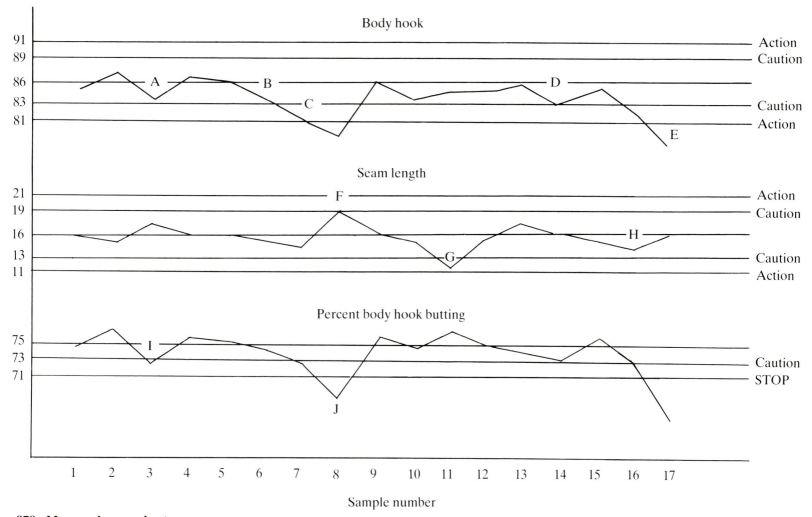

878. Mean and range chart.

The Use of Mean and Range Charts in Canning Quality Assurance
As already indicated an essential part of canning quality assurance lies in ensuring that can seams are properly made. Hersom and Hulland (1980) note that in control of seam production, '. . . . some form of graphical control procedure is preferred. . . . The use of such a statistical chart will indicate clearly when a machine is drifting out of control and will avoid unnecessary adjustments being made for minor fluctuations'. Such a graphical control procedure is the mean and range chart which plots the mean of measurements made at each sampling occasion in relation to pre-set tolerances, the limits of which are marked on the chart. Mean and range charts have many applications in quality assurance, examples being automated filling or weighing operations, and their application is not specific to canning nor indeed to the food industry. **878** is intended to depict the application of mean and range charts to canning quality assurance rather than to illustrate canning quality assurance *per se*. For this purpose the data is simulated but relates to actual measured parameters and realistic values.

The mean and range chart depicts variation within two measured variables which contribute to determination of the important parameter **percent body hook butting.** Details of the important can seam parameters are given in Appendix 2 and fuller details together with practical procedures may be obtained from Hersom and Hulland (1980); Anon (1973).

$$\text{Percent body hook butting (PB)} = \left(\frac{BH - 1.1\,tb}{SL - 1.1\,(2te + tb)} \right) \times 100$$

Where BH = body hook
SL = seam length
tb = body plate thickness
te = end plate thickness

On the third sample occasion the body hook has shortened (A) (in practice this may be due to insufficient base pressure or excessive clearance between rolls and chuck – see **865**) and the PB has fallen into the 'caution' zone (I). Corrective action is taken and the situation remedied but the body hook length drifts down until the 'caution' limit is approached with sample 6 (B). A re-sample 7 (C) is taken and shows immediate action to be required. On this occasion the wrong corrective action is taken and the seam length increased (F). The body hook, however, is continuing to decrease and at sample 8 the PB has fallen to the point at which the canning line must be stopped and affected cans removed (J). Corrective remedial action is taken and at sample 9 both PB and body hook are satisfactory.

The PB and body hook dimensions continue to be satisfactory over a number of sample occasions although remedial action is required for the seam length after measurement of samples taken at number 11 (G). Note that the short seam length has not affected the PB adversely. No further corrective action is taken until sample 16 (although the body hook dimension decreased to almost the 'caution' level at sample 14 (D) this was correctly adjudged to be part of normal variation and not a continuing trend). At sample 16, however, the body hook measurement decreases markedly and corrective action taken at this stage (E) has no effect. As a consequence the PB falls below permissible levels and production is stopped. Major seamer repair such as replacement of seamer rolls (see **865**) would be required at this stage. In addition to the major problem involving the body hook, corrective action was also taken to the seam length (H) which had been steadily decreasing since sample 13.

A sequence depicting the consequence of a complete failure of canning quality control (879–898). All of the cans shown are of Canadian salmon. All were spoilt and each can showed major defects. Canadian salmon was responsible for fatal botulism in Europe in 1981 and was withdrawn from the U.K. (and other European) markets for 9 months in 1982. The cans shown were produced after its re-introduction. The two most common types of damage shown are improperly made seams and lid damage including corrosion. The cans had been stored for a short period only after processing and it is obvious that corrosion was at least initiated before canning.

879, 880 and **881. Damage to the lid.** The first two illustrations show a general (**879**) and close-up view (**880**) of the damage from the exterior and the third (**881**) shows that the interior of the lid has been penetrated. Although apparently minor such damage would permit the ingress of bacteria. Rust around the damaged site suggests that the damage occurred some time previously and cannot be attributed to, for example, a Stanley knife for opening the outer carton.

882 and **883. A dent in the lid.** The first illustration (**882**) shows the damage to the exterior and the second (**883**) the corresponding interior damage, which was probably caused by a semi-sharp object. In this example the interior of the can has been penetrated but even if this had not occurred the dent weakens the metal and serves as a focal point for corrosion.

884 and **885. Apparently superficial damage to a lid (884)** which has, however, penetrated to the interior of the lid (**885**) with attendant risk of contamination.

886 and **887. Pitting due to corrosion** on the lid (**886**) penetrates to the interior (**887**). To a certain extent any can is subject to corrosion, which is enhanced by mechanical damage, failure to dry cans adequately after cooling, overwrapping in plastic film before cans are dried and cooled and adverse storage conditions. If can bodies and lids are stored under adverse conditions corrosion may occur before the cans are filled and processed. Good factory management should ensure that corroded and otherwise damaged can components are never used.

888 and **889. Massive damage to the lid** seen from the exterior (**888**) and interior (**889**). This led to wholesale contamination of the contents. The can has been cleaned to facilitate examination but the cause was corrosion following initial mechanical damage.

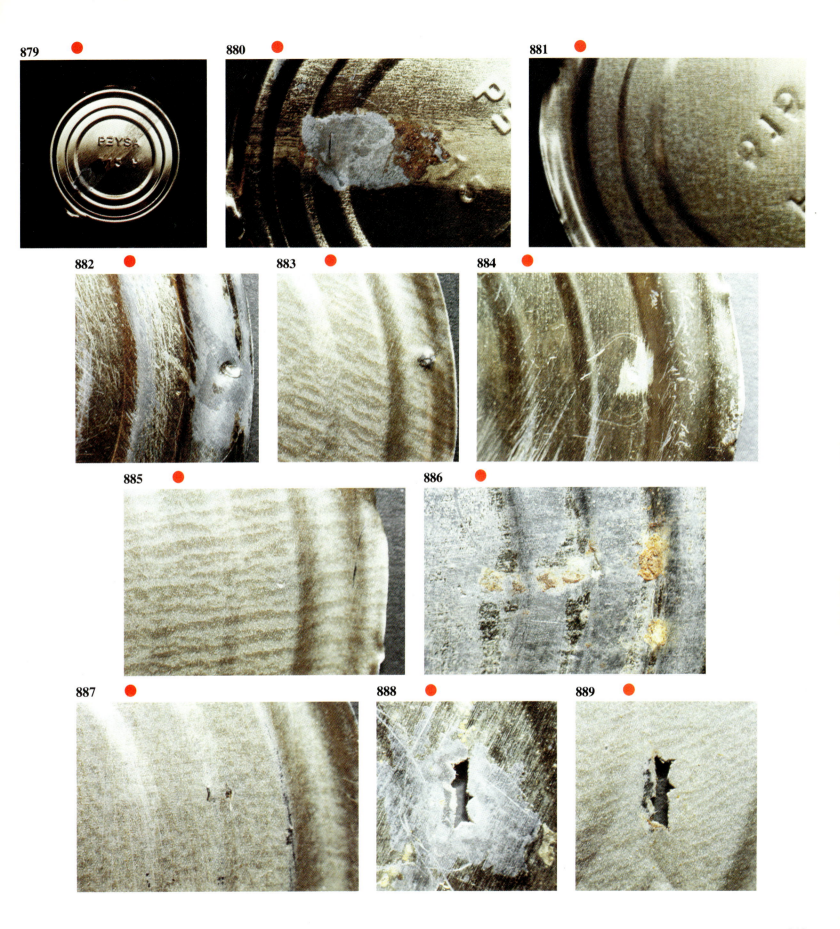

890. **Severe and widespread corrosion** on the lid.

891. **Removal of the protective lacquer** shows pitting of the lid due to corrosion.

892. **The protective lacquer coating** has been mis-applied to this lid resulting in the initiation of widespread corrosion in the unprotected area.

893. **A pin hole in the lid** of a salmon can. The probable cause is a combination of mechanical damage and corrosion.

894. **A top (angled) view of a major seam fault.** The lid at the arrowed position is bent back from the seam leaving a large hole.

895. **A side view of the same can** showing at points A and B failure to complete the seam. A fault of this nature and severity could be caused only by total mis-adjustment of the seamer or to a damaged lid. In either case cannery quality control procedures have been grossly inadequate.

896. **A further example of grossly incorrect seaming** of a can. The indentation in the upper body of the can has prevented the seam being formed. Again this fault was not detected by cannery quality control procedures.

897. **A less obvious but potentially hazardous** seam fault. The seamer has failed to make the seam correctly at Point A. This point is therefore an area at which leaker contamination could readily occur (see **898b**).

898

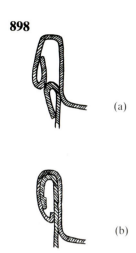

(a)

(b)

899

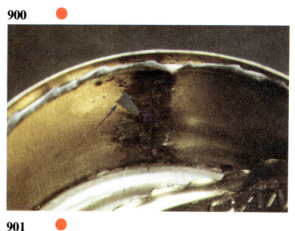

900

901

902

903

898. Cross section of mis-locked can seams. (a) Failure to complete the seam due to a knocked down flange; (b) Failure to complete the seam due to a knocked down curl (see **897**).

899 and **900.** **A large cut in the side panel** of a can illustrated from the outside (**899**) (arrowed) and the interior (**900**) (arrowed). The label had been applied over the cut which was probably present before filling and passed undetected by quality control procedures. The contents were grossly spoilt.

901 and **902.** **A further example of a seaming fault** at the canners' end of a can. (N.B. The label is applied later and its orientation does not relate to canners'/manufacturers' end) (**901**). A close-up view of the fault shows this to have the appearance of a tear (**902**) and it is likely that it arose as a result of the can catching on a broken or incorrectly fitted part of the seamer rather than to maladjustment of the operating parts. It should be noted that visible leakage even from such large tears may be relatively slight as the viscous juice of such products as baked beans may 'seal' the aperture. Although such damage could have occurred after seaming similar damage was found in a number of cans and the label was intact over the damage (before removal for examination).

903. **A close-up of metal dust** on canned pineapple chunks. This was caused by a misadjustment of the seamer causing metal to be scraped from the top of the can. The faulty seam could not be detected by visual inspection but the incorrect seam dimensions would be detected by seam stripping. Regular quality control checks are required to ensure that faults do not go undetected for a long period.

Stage 5 – Stacking and Holding

Stacking and holding stages are not applicable in the large volume continuous canning processes such as the Hydrostat system. Many canners use batch retorts and cans leaving the seamer must be loaded into retort baskets with attendant risk of mechanical damage which may pre-dispose the can to post-retort leakage. More important are the delays which can occur at this stage. In extreme cases spoilage of the contents of the can may take place before sterilisation. The organisms responsible are then destroyed by sterilisation resulting in a spoiled but sterile can.

Avoidance of such delays is primarily a matter of good plant management but if for any reason delays are unavoidable cans should be refrigerated prior to processing. This is not possible with some products e.g. hot dog sausages (frankfurters) where the retort is loaded with hot filled cans – the initial temperature being taken into account in the process calculations. In these cases it is necessary to determine a minimum temperature for cans to be loaded into the retort.

Stage 6 – Sterilisation (Retorting)

The retorting process is the heart of all canning operations and while there are many types of retort and retort processes, the fundamental objective of obtaining 'commercial sterility' is the same in every case.

Calculation or derivation of thermal processing parameters is beyond the scope of this book but a number of fundamental assumptions are made. Foods are usually categorised according to their pH value, four categories being commonly used (see page 8 for details). The most important division is that between **medium acid** foods which have a pH value in the range 4.5–5.3 and **acid** foods which have a pH value in the range 3.7–4.5 since pH 4.7 is generally considered to be the lower limit for growth of *Cl. botulinum*. Foods with pH values greater than 4.5 must therefore undergo a thermal processing sufficient to ensure that endospores of *Cl. botulinum* are destroyed ('botulinum cook'). Less severe processes may be applied to foods with pH values below 4.5. In such cases it is necessary to ensure that the actual pH value of the product is below 4.5 and it is not sufficient to rely on historical experiences. In addition it is important to note that the mean product pH value is not necessarily an adequate indication of the thermal processing required for safety. A type of stuffed peppers (peppers plus rice) for example had a mean pH value of 4.4 but the pH value of the rice stuffing was 4.9. Thus the product must be considered medium-acid and should be processed accordingly.

It is also necessary to take account of heat penetration parameters and factors that alter heat penetration and circulation. A change in particle size of chopped vegetables for example may alter heat penetration, or the use of starch thickener of different gel properties may prevent heat circulation during processing. It is important that the effect of recipe changes on process efficiency are fully evaluated before new recipes are introduced.

Actual in-plant operation of the retorting process should be under the control of responsible and experienced personnel who, if not technically qualified, should be trained in the basic theory of canning. Such personnel should be required to maintain a complete record of all aspects of the retort operation and to ensure that automatic monitoring devices are functioning correctly. In many countries there is a legal requirement to monitor time and temperature relationships during retorting and to maintain written records for a minimum stated period. It is not sufficient to rely solely on thermograph records or on retort pressure but independent temperature measurements should be made using mercury in glass thermometers. It is also advisable to carry out additional checks to ensure that cans despatched from a factory have received the correct processing. Such checks include the use of heat sensitive indicator tape and post-processing laboratory examination.

Stage 7 – Cooling and Handling of Warm Cans

In recent years the major problem with canned products has been due not to underprocessing but to post-process contamination. Cans are at their most vulnerable to leakage after retorting before being fully cooled and dried. Although problems of leakage are exacerbated by faulty seaming **every** can must be considered to be a potential leaker. It is essential to ensure that cans are handled as little as possible, that cooling water, if used, is of potable quality, that cans are cooled in such a way as to reduce stresses due to pressure imbalance between the interior and exterior of the can and that cans are dried as quickly as possible.

Stage 8 – Release of Cans

Cans should not be released from the cannery until plant records and post-process tests indicate that processing has been satisfactory. Release should be the responsibility of a senior member of staff and appropriate records should be kept.

Semi-rigid or flexible retortable packs based on aluminium or aluminium/thermoplastic laminates have been developed in recent years. The example shown is a semi-rigid tray heat-sealed with an aluminium cover and, in this case, used to pack cannelloni. Heat penetration is faster than in conventional cans leading to improved product quality and reduced thermal energy costs. These advantages are offset by higher material costs and lower line speeds.

904 and **905.** **The criteria for examining thermal seals** on semi-rigid and flexible packs differ from those of a conventional double seam. Correctly made thermal seals differ from double seams in providing a permanent barrier to the entry of micro-organisms while at the same time the lack of a headspace vacuum reduces the likelihood of micro-organisms being drawn into the pack. Seals should be examined visually for obvious defects and also be subjected to objective testing for seal strength. The **burst test** by which the pressure at which a pack bursts is determined is commonly used while the tensile strength of seals from fully flexible packs may be determined by an instrument such as the **Instron.**

In this example the outer closed seam has been opened but the inner thermal seal has retained its integrity.

906. **Ring pull-opening cans** have come into use for sardines and other fish in recent years. While convenient to open, the mechanical strength of these cans is lower than that of conventional types and particular care must be taken to avoid stress during cooling.

907 and **908.** **Underprocessing of canned goods** is rare today and even where spoilage occurs it may be difficult to determine whether the cause was underprocessing or post-processing contamination. The examples illustrated are unusual in that raspberries, a high acid food, are involved and the micro-organism responsible is a mould not a bacterium. *Byssochlamys fulva* is present in raspberries before harvesting. It produces heat resistant **ascospores** which may survive the relatively light heat processing applied. The ascospores of *B. fulva* are also unusual in that their heat resistance increases as the pH value is lowered. After outgrowth and establishment of a mycelium pectinolytic enzymes are produced which cause the breakdown of the intact berry into individual drupelets. The two illustrations show intact berries (**907**); and berries infected by *B. fulva* (**908**). This form of spoilage is rare today in commercially canned and bottled fruit as shown but is more common in home bottled products. Only raspberries and their immediate relatives such as loganberries are affected.

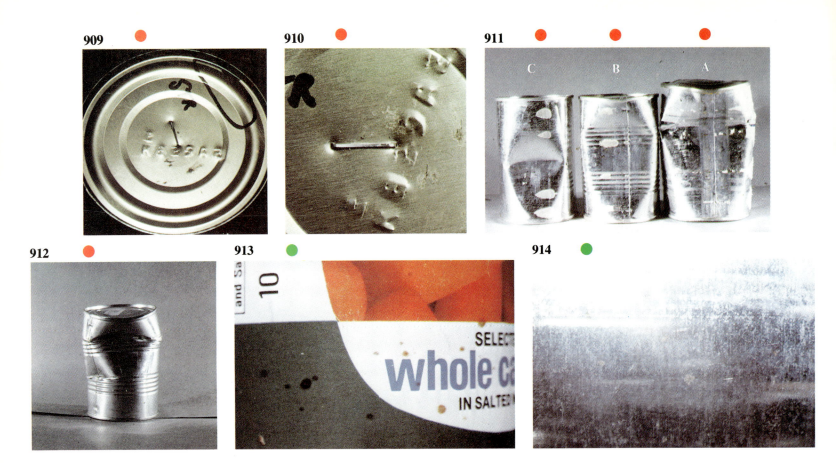

909 and **910.** **A staple placed into the lid of a can.** Staples of this gauge are not used in canning operations and its presence is likely to be due to sabotage. It is not usually possible to trace the originator of single cases of sabotage although tracing becomes possible in many instances where it is recurrent. See also Chapter 17, p. 242.

Damage to cans is also likely to occur after leaving the factory and, while cans at this stage are less susceptible to contamination, damage should be prevented as far as possible. The nature of canned goods means that conditions of handling and storage are often poor. Cans are subject not only to mechanical damage but to mishaps such as flooding. In any situation where decisions as to whether or not to accept cans have to be made experienced and senior personnel should be involved. This is not always possible in large scale retailing operations where semi-skilled or unskilled staff are employed for shelf-filling operations. In these cases, 'photoguides' showing 'unacceptable' levels of damage, as well as 'acceptable' may be of value.

Some aspects of large scale retailing such as the use of 'dump-baskets' for unstacked displays of canned goods represent a retrograde step for not only are damaged cans more difficult to detect but the possibility of damage during dumping is high.

It should be mentioned that in the U.K. at least a thriving black economy exists whereby badly damaged and suspect cans destined for destruction or animal feed find their way into the retail market via market stalls etc. It is paradoxical that in some cases such cans are only marginally cheaper than their sound equivalents.

Panel damage usually occurs during post-processing handling of cans and may happen at any stage up to the retail outlet and indeed in the home. Although damage may be extreme such cans are frequently undetected due to the relatively low level of training of retail staff and the unawareness of shoppers. Since cans are usually dry when panel damage occurs, the associated public health risks are relatively low. Illustrated guidelines are of value in assessing the suitability of damaged cans for sale.

911. **Shows varying degrees of panel damage.** The can on the left (C) is badly damaged but neither side nor top seam is affected. Can B (centre) is badly damaged, in this case also affecting the side and top seam. Can A (right) is most severely damaged with the top of the can distorted. None of these cans would be suitable for sale.

912. **Compression damage occurs** when a force is applied downward onto a can. Individually cans may be damaged in this way if dropped on end but where a large number are found it is frequently due to the collapse of stacked pallets due to overloading or overstacking. This is a typical example. Severe seam damage occurs and such cans are not suitable for sale. Superficially damage may appear to be similar to that of severe panel damage but close examination shows the two to be readily distinguishable.

913 and **914.** **Rust is also a post-process problem** with stored cans, and may be due to improper drying at the cannery or incorrect or excessively long storage. A small amount of superficial rust as shown does not preclude the can from sale although it must be

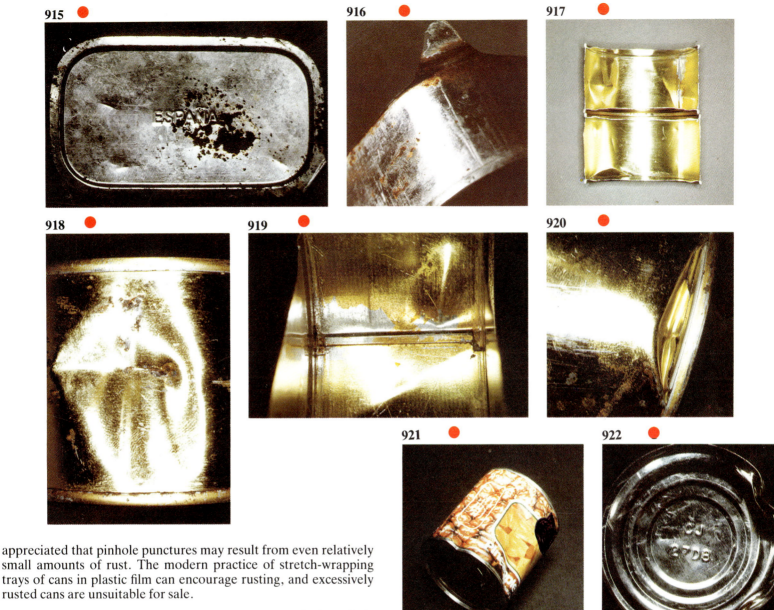

915 ● **916** ● **917** ●

918 ● **919** ● **920** ●

921 ● **922** ●

appreciated that pinhole punctures may result from even relatively small amounts of rust. The modern practice of stretch-wrapping trays of cans in plastic film can encourage rusting, and excessively rusted cans are unsuitable for sale.

915 and **916.** **Severe rusting rendering the can unfit for sale** is illustrated with this sardine can (**915**). Rusting is seen to be particularly severe around the opening tab (**916**). The can purchased in a London street market was unlabelled and apart from a factory code and year of canning (1977, purchased 1982) there is no means of tracing its history or the distribution chain. The condition of the outer carton, however, suggested that flood damage had occurred.

917 to **920.** **Multiple damage to a can of salmon.** The can has been emptied of its contents and cut in half to show the damage. Viewed from the inside it shows indentation damage due to impact with a sharp object such as another can. Small holes can be made by such damage permitting the entry of micro-organisms. There has also been some damage to the lacquer along the seam of the can. The lower half shows further damage at the rim. Here it appears that the body has been ripped away from the lid/base either by a sharp object after seaming or due to incorrect seamer operation.

It should be appreciated that lacquer is applied to protect the metal from corrosion due to (as in this case) salt or acid. Stripping of the lacquer is a common problem with high acid products particularly if storage is prolonged, and it may occur also as a consequence of mechanical damage or during seaming. Attack on the metal by acid foods may result in formation of hydrogen gas causing the can to become swollen. This type of spoilage is known as a '**hydrogen swell**'.

921 and **922.** **A further example of severe mechanical damage** affecting both body and seams of a can of pineapple. **921** shows body damage while the top view (**922**) shows the damage to the canners' seam.

923 🔴

924 🔴

925 🔴

926 🔴

927 🔴

928 🟢

929

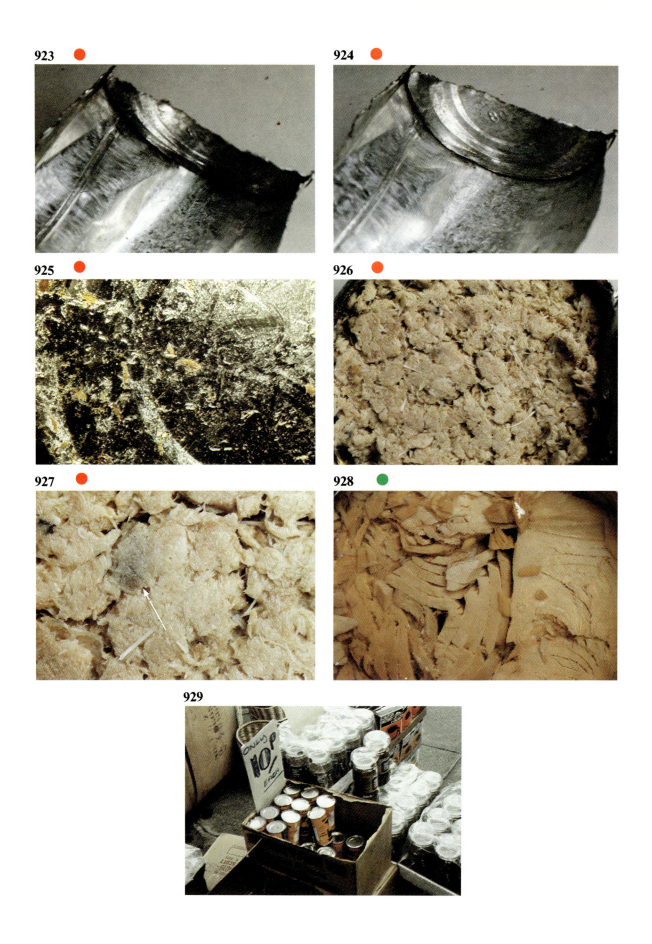

Table 23 Laboratory Tests Canned Goods

I. Microbiological
N.B. Good control at manufacturing level is the only means of ensuring the safety of canned goods. It is stressed that any laboratory examinations are at best of secondary value and at worst misleading as to the safety of the product.

A. Commercially sterile packs
Incubation tests. Although positive incubation tests are indicative of non-sterility negative tests do not necessarily mean either that cans are sterile or that thermal processing was satisfactory. When carrying out incubation tests it is not sufficient to rely on gas production blowing cans and, after incubation, cans should be opened, a visual examination made of the contents and the pH value determined. Culture of the contents is not required unless the can is blown or shows other signs of non-sterility. The contents of incubated cans should **never be tasted.** The authors do not recommend the use of rapid chemical tests such as those of Ackland *et al* (1981), Ackland and Reeder (1984). It should be noted that any laboratory examination of high acid canned foods is considered unnecessary by some authorities (ICMSF, 1980).

B. Non-commercially sterile packs
Group D Streptococci (Pasteurised cured meats only)
Osmotolerant yeasts (Sweetened condensed milk only)

A wide range of semi-preserved canned seafoods are produced. In some cases such as anchovies these may receive no heat treatment whatsoever (see p. 213). The safety of some of these products is minimal and although total viable counts give some indication of the overall microbial status their significance is difficult to assess particularly with respect to safety.

II. Chemical/physical
Choice of analyses is dependent on the nature of the product but the following precautions should be observed:

Where low pH value permits reduced thermal processing the pH value of the product must be checked. This is of particular importance with 'borderline' products such as tomatoes.

In products such as canned ham and semi-preserved fish which are wholly or partly dependent on preservatives for safety and stability the levels of preservatives should be determined. With canned smoked salmon, for example, NaCl and Aw levels are safety determinants.

Physical examination should always include seam dimensions (see pp. 217–218, 228), pack vacuum and general condition of the pack. Equivalent procedures should be adopted for flexible and semi-rigid retortable packs.

Notes:
(1) These tests are examples only and are not inclusive.
(2) Examples are for finished products only.
(3) Techniques are detailed in specialist publications such as ICMSF (1978) – Microbiology: Egan *et al* (1981) – Chemistry.

923 and **924.** **Post-process seam damage** seen from two angles. Such cans are not fit for sale.

925. **The exterior of a can of tuna fish** showing debris and exudate from other cans in the same batch which had burst due to mechanical damage when in a swollen condition. Anyone handling such a can is subject to infection and unless thoroughly cleaned and sanitised the risk of contaminating the contents on opening is considerable.

926, 927 and **928.** **The open can of tuna** showing poor quality fish and an absence of oil seen in general view (**926**) and close-up (**927**). The exterior of the can was coated with oil which had leaked through damaged seams. Note in **927** the purple patch (arrowed), the origin of which is not known but is probably a dye. **928** shows as contrast good quality canned tuna in oil. The use of poor quality fish increases greatly the likelihood of scombrotoxicosis (see Chapter 8, p. 102).

929. **The black economy in operation.** Badly damaged cans on sale at cut prices in a London street market, only a few yards from the Department of Health and Social Security Headquarters.

Summary of Requirements for Correct Double Seam

1. **Tightness** to ensure that the seam lining compound is held under compression. This is assessed by:
   (a) **Degree of wrinkling** which is measured on a scale of 0–5. Thus 0 is no wrinkling and 5 a wrinkle extending the depth of the cover hook. Ratings of 2 and 1 for the A1 and UT size covers respectively are the highest permitted.

   (b) **Free space** where –
   Free space = seam thickness – (2(tb) + 3 (te))
   (tb = body plate thickness
    te = end plate thickness)

2. **Body hook correctly embedded in lining compound** assessed by 'percent body hook butting' which should be not less than 70%

   $$\text{Percent body hook butting (PB)} = \left(\frac{BH - 1.1\,tb}{SL - 1.1\,(2te + tb)} \right) \times 100$$
   (BH = body hook
    SL = seam length)

3. **Correct body and end hook overlap** which should be as large as possible without producing wrinkling.

   Overlap = EH + BH + 1.1 te − SL
   (EH = end hook length)

4. **Freedom from obvious defects such as gross local distortions.**

Seam measurement parameters are shown in **930**.

Full details are available in The Double Seam Manual (Anon 1973).

Adapted from Hersom and Hulland (1980).

Examples of foreign bodies in canned goods may also be found in Chapter 17, Foreign Bodies and Infestations as follows:

Pp. 244–245 *Tenebrionidae* from peas
Pp. 244–245 Weevil from peaches
Pp. 246–247 Insect (unidentified) from tuna
Pp. 246–247 Spider leg in sardine and tomato paste

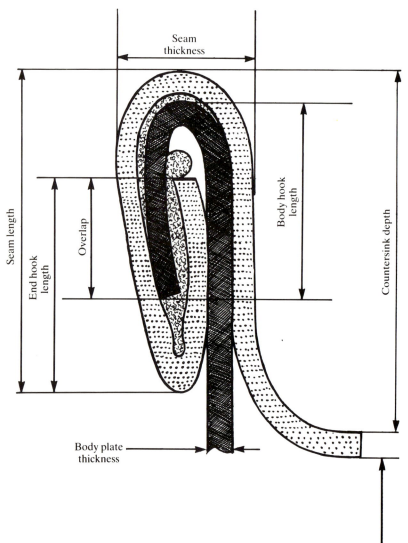

930. Can seam section.

Chapter 16

Alcoholic Beverages

Although alcoholic beverages are only marginally classed as foods they are included in this book due to their close association with food and with the food industry. On a world wide basis alcoholic beverages account for considerably more than half the gross value of fermented products with, in 1976, 72.6% of world output at a value of £25,200 million (Bronn, 1976). With the exception of the strictly Islamic nations there is no region of the world which does not produce and consume alcoholic beverages of some description. Overall *per capita* consumption of alcohol is increasing giving rise in many nations to concern over increasing alcoholism particularly amongst women and teenagers where the increase in consumption has been greatest.

The number of alcoholic beverages produced is vast. There are three basic types defined primarily by means of production.

(a) **Beers** produced by fermentation of heat extracted ingredients and produced to a standard recipe.

(b) **Wines and ciders** (including fortified wines such as sherry) produced by fermentation of whole crushed fruit or juice and thus less reproducible than beer.

(c) **Distilled** products where the alcohol content of fermented liquor is increased by distillation.

In each case definition of quality is intangible and depends largely on organoleptic assessment. Although sophisticated chemical and microbiological analyses are made as a part of the manufacturing control of alcoholic drinks such tests cannot accurately predict the quality of the finished product. Quality of alcoholic drinks is also more than usually subject to personal preference and prejudice. This aspect is worth a fuller discussion since changes in attitude towards quality (or perceivable quality) of beer in the U.K. provides one of the most striking illustrations of the power of 'consumerism': a phenomenon no one concerned with food quality can ignore.

During the 1960s the trend in U.K. beer production was towards keg beer, a filtered, pasteurised product sold from pressurised aluminium kegs. Brands were mass marketed on a national basis, a typical example being Watney's Red Barrel (later re-launched as Watney's Red). Small, independent but unprofitable breweries usually operating in rural and less populous areas became increasingly less profitable and undercapitalised and were steadily being wholly or partially taken over by the large national breweries.

In the U.K. during the early 1970s a pressure group, The Campaign for Real Ale (CAMRA), was formed with the objectives of preserving traditional beers and traditional brewing and beer handling practices and, to a lesser extent, the preservation of the traditional British public house. Although largely a volunteer organisation CAMRA rapidly developed sufficient influence to direct drinking taste (and breweries' marketing strategy) away from the bland, though consistent, keg beers back to the more traditional types. In turn the small independent breweries gained increased sales and have been able to markedly increase profitability.

Significantly, however, and possibly ironically in view of CAMRA's original objectives, many of these breweries have pursued the increasingly important supermarket and 'free' off-licence trade rather than invest further in public houses. One of the more progressive has gone so far as to dispose of all but one of its public outlets. Despite the efforts of CAMRA and the breweries overall beer consumption in the U.K. is falling while wines, light spirits such as white rum and other drinks, are becoming of increasing importance.

Examples of laboratory tests for alcoholic beverages are shown in Table 26, p. 242.

Beer (931–956)

A basic flow diagram for beer production is shown (**931**). Traditional U.K. beer (or ale) differs from that produced elsewhere by the use of a 'top' yeast as opposed to a 'bottom' yeast during fermentation. The latter gives a lager type beer. Within the U.K. beers there are many different types (Bitter, Mild, Brown etc.) but the basic technology is similar and no further details will be given unless related to quality. It will be noted that barley is referred to as the cereal source in beer but other cereals such as rice and maize are used in other parts of the world. The process is however analogous.

Beer production is discussed by stage as denoted in the flow diagram (**931**). Quality defects which may arise at each stage are described.

Malting

In preparation of malt, barley grains are soaked at 10°–16°C, germinated at 16°–21°C for 5 to 7 days and dried to ca. 5% moisture. Sprouts are removed and the malt is then milled before use. The exact malting procedure varies according to the type of beer to be brewed. Malt has an effect on the organoleptic properties of beer and good quality malt is essential for good quality beer.

Mashing

During mashing the fermentable compounds and yeast growth factors from the malt and from starchy malt adjuncts such as

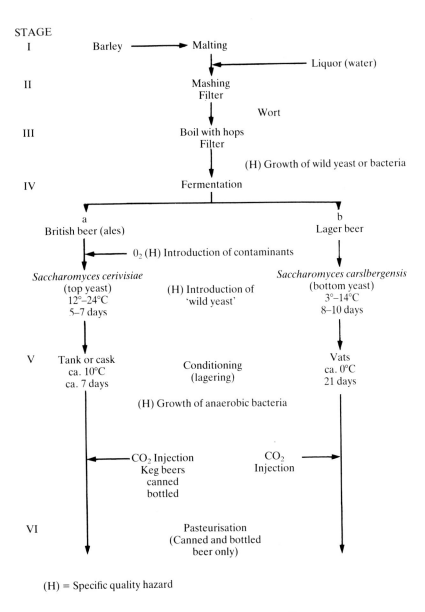

STAGE

I   Barley ⟶ Malting

⟵ Liquor (water)

II   Mashing
Filter

Wort

III   Boil with hops
Filter

(H) Growth of wild yeast or bacteria

IV   Fermentation

a — British beer (ales)   b — Lager beer

O₂ (H) Introduction of contaminants

Saccharomyces cerivisiae
(top yeast)
12°–24°C
5–7 days

(H) Introduction of 'wild yeast'

Saccharomyces carslbergensis
(bottom yeast)
3°–14°C
8–10 days

V   Tank or cask
ca. 10°C
ca. 7 days

Conditioning
(lagering)

Vats
ca. 0°C
21 days

(H) Growth of anaerobic bacteria

CO₂ Injection
Keg beers
canned
bottled

CO₂
Injection

VI   Pasteurisation
(Canned and bottled
beer only)

(H) = Specific quality hazard

931. A flow diagram for beer production. The diagram is based on traditional batch fermentations and in modern large scale breweries continuous fermenters may be used.

unmalted barley are solubilised while at the same time **amylases** and **proteases** derived from malt hydrolyse starch and other polysaccharides to simple **sugars**, and proteins to low and medium weight **polypeptides**.

The main malt mash is prepared by mixing the ground malt with liquor. The mash is carried out at 65°–70°C after which the temperature is raised to ca. 75°C to inactivate enzymes. The resultant liquid is filtered through the insoluble constituents of the mash and emerges as a relatively clear liquid – the **wort**. If the mashing procedure is extended thermophilic *Clostridium spp* may carry out a **butyric acid** fermentation which can result in off-flavours in the finished beer.

Boil with Hops

The final pre-fermentation stage is to boil the wort with hops (*Humulus lupulus*). The quantity of hops varies according to the type of beer and a boiling time of 2½ hours is required although a shorter time is required if hop extracts are used. The boiling is sufficient to kill all vegetative bacteria. Hops are important as sources of **anti-microbial agents** (humulone, cohumulone and adhumulone), which help control the growth of Gram-positive spoilage bacteria, **bitter acids** and **resins** which aid in flavour stability and head retention, and of **essential oils** which add flavour. The choice and quality of hop is therefore of importance in determining the organoleptic quality of beer.

Excess boiling results in extraction of large quantities of tannins which result in poor flavours or haziness of the finished beer.

Fermentation

Although the principles of fermentation, the conversion of sugars to ethyl alcohol and carbon dioxide are the same for all beers there are two distinct fermentation procedures. The first is used for virtually all U.K. beers (ales). In this *Saccharomyces cerevisiae* is used at a temperature of 12°–24°C for 5 to 7 days. Oxygen is usually injected into the wort to enhance yeast growth. *Sacch. cerevisiae* produces copious amounts of CO₂ and floats to the top of the fermentation vessel; it is thus referred to as a 'top yeast'. A variant of *Sacch. cerevisiae*, *Sacch. carlsbergensis*, is used for lager beers. Gas production by *Sacch. carlsbergensis* is less than by *Sacch. cerevisiae* and the yeast remains at the bottom of the fermentation vessel, thus a 'bottom yeast'. A lower fermentation temperature of 3°–14°C is used for a period of 8 to 10 days.

In either type of fermentation this stage is an important source of contamination leading possibly to beer spoilage notably by members of the *Enterobacteriaceae* such as *Obesumbacterium* (*Hafnia protea*) which can grow during the early stages of fermentation. The major source is the yeast itself, although where oxygen is injected into the wort in production of U.K. beer contamination may result from dirty oxygen lines. Traditionally the yeast for a new inoculum was taken from excess yeast produced in a previous fermentation batch, such yeast almost invariably being contaminated with 'wild' yeast and bacteria. In modern practice a master culture of yeast is obtained from a central source and then bulk cultured at the brewery. Wild yeast and bacteria may contaminate the yeast at the bulk culture stage or, more rarely, be derived from the master culture.

Conditioning

U.K. beer is conditioned either in bulk or in casks for ca. 7 days at ca. 10°C. Traditionally brewed lager beer is conditioned at ca. 0°C for a minimum of 21 days and up to several months. During conditioning proteins, resins and yeast are precipitated out clarifying the beer (**isinglass** may be added to aid this process) while chemical changes largely involving production of esters improve flavour and body (the maturation time of 'modern' lagers particu-

933

932

934

935

larly those produced in the U.K. is likely to be much reduced). Cask conditioned beer is despatched to retail outlets without further treatment whereas bulk conditioned beer is packed into kegs, barrels, bottles or cans. CO_2 is injected into keg, canned and bottled beers and a further clarification by centrifugation or filtration is often applied.

Cloudiness and off-flavours in lager type beer caused by the obligately anaerobic bacterium *Pectinatus* (Lee, 1984) have been reported in the United States, Germany and Finland. Spoilage of this type is associated with the extension of the storage life by conditioning and storage under reduced oxygen tension.

932 and **933.   The interior of a hop kettle (932).** Note the rotating stirrer and the slotted floor through which the wort is drawn by gravity out of the kettle (**933**). The majority of contamination problems occur after this stage.

934.   Haze in beer. This is not of microbiological origin but consists of starch and protein-tannin complexes. The beer is of high solute content, the haze being attributed to excessive tannin content in the hops used.

935.   Propagation of the yeast starter culture. Genetic engineering technology is being applied in attempts to both improve the fermentation performance of yeasts and to 'tailor' them to meet specific requirements. Problems with co-transfer of genes carrying undesirable properties such as the production of a phenolic taint due to 4-vinyl guaiacol have been encountered but these can be overcome by the use of more refined techniques such as **transformation** and **spheroplast fusion**. (See Pellon and Sinskey, 1983.)

Pasteurisation

The majority of canned and bottled beers are pasteurised either by hot water or by steam prior to release from the brewery. Pasteurisation of the filled, sealed container is most common although some breweries use continuous treatments either by filtration or by heat, followed by bottling. This latter procedure carries a high risk of contamination and spoilage as filling machines, even when designed for so called aseptic filling, are notorious sources of contamination.

A few bottled beers are not pasteurised and have a small quantity of fermenting wort added before bottling to produce a secondary fermentation. If the fermentation proceeds too far the beer is spoilt and would have a restricted shelf life.

In the foregoing description of the brewing process specific sources of contamination have been described. It must, however, be appreciated that contamination may occur at any point after the third stage (boiling with hops) up to, in the case of barrel and keg beer, service to the consumer. Spoilage micro-organisms are inevitably present in the environment of the brewery and any premises regularly handling beer. These will multiply rapidly in dirty equipment and utmost cleanliness is required.

In addition spoilage organisms may enter from the air and reflect the environs of the brewery. Wild yeast are particularly common aerial contaminants in rural areas while *Zymomonas* is frequently associated with building work and demolition.

Micro-organisms normally associated with spoilage may enhance the properties of certain beers when present in small numbers. Di-acetyl produced by *Pediococcus spp* for example will spoil beers at levels of 0.2 ppm and yet enhance the flavour of certain types of brown ale.

Principal types of beer spoilage micro-organisms and the nature of spoilage produced are shown in Table 24.

It should be noted that the relationship between micro-organisms isolated and spoilage is not always precise and that the nomenclature used by the brewing industry does not always correspond to that more generally used in other areas of food microbiology.

Finally, the financial aspects of a major contamination problem should be mentioned. While any major contaminant leads to loss of finished product, problems may be sufficient to prevent production for days or even weeks. The problem can be sufficiently serious to require replacement and rebuilding of plant and in at least one case total closure of a brewery.

Table 24 Micro-organisms Commonly Responsible for Spoilage of Beer

Micro-organism	Type of Spoilage
Wild yeasts	
Saccharomyces pastorianus	Cloudiness, off-tastes
Lactic acid bacteria	
Pediococcus damnosus	Sourness, turbidity, ropiness; some strains produce di-acetyl
Lactobacillus brevis	Sourness, silky turbidity
Acetic acid bacteria	
Acetobacter pasteurianus; *A. hansenii*	Vinegary taste, ropiness, turbidity
Other bacteria	
Zymomonas	Hydrogen sulphide and acetaldehyde production, turbidity
Pectinatus	Sourness, rotting egg odours, turbidity

N.B. *Zymomonas* is generally considered to be the most important contaminant in ales but the organism is not found in lager type beers (Swings and DeLey, 1984).
Pectinatus has only been reported in lager type beers.

936. Wort from the hop kettle is cooled in a plate heat exchanger (arrowed A) and injected with oxygen (arrowed B) before passing to fermentation vessels. Care must be taken to ensure that contamination does not enter the wort from the oxygen line.

937. Shows an active top fermentation of English Ale (bitter ale) with *Sacch. cerevisiae*. The foam is produced by the typical rapid CO_2 evolution. A considerable rise in temperature occurs and cooling is necessary. In these vessels cooling panels are inset in the stainless steel walls. Earlier types were fitted with cooling coils which were difficult to clean and a focal point for contamination.

938. Shows the yeast layer at the top of a vessel where active fermentation has ceased and the beer is ready for conditioning. Note (arrowed) the spray balls in the roof of the vessel for in-place cleaning (CIP).

939. Centrifuge used for clarifying beer for bottling or canning after conditioning. Centrifuges are complex pieces of equipment and if imperfectly cleaned may themselves serve as focal points for contamination.

940. Filter presses used in addition to, or in place of centrifuges for clarification of conditioned beer (and wine). The performance of filters must be carefully monitored throughout their usage.

941. A keg washer. Imperfectly cleaned kegs are a major source of contamination of cask conditioned and keg beers. The problem is particularly acute where wooden kegs are used, these being notoriously difficult to sanitise. Aluminium kegs which are now in common usage present fewer problems.

936

937

938

939

940

941

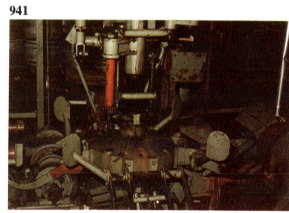

942 and **943. Any filling operation** bears the implicit risk of contamination from the environment or from the container itself. **942** shows 7 pint cans being washed before filling. **943** shows a high speed bottle filling line. Note the high standard of cleanliness and the degree of protection afforded the bottles, which in this line, are non-returnable and thus obviate the problems of contamination in the consumer's home.

944. Mould is only rarely a cause of beer spoilage. The mould pellicles shown were present in a bottle of English light ale. Such a fault is due to growth of the mould in the dregs of beer remaining in a re-usable bottle after consumption and then failure of the bottle washer to remove the mould before re-filling. The use of non-returnable (single trip) bottles obviates this possibility.

952　　**953**　●　　**954**

955　●　　**956**　●

945 and **946.**　**Beer in 7 pint cans** leaving a hot water continuous pasteuriser (**945**). The performance of the pasteuriser is checked using the instrument illustrated (**946**) which measures and records temperatures inside the can as it travels through. Similar devices are used to measure oven temperatures in bakeries etc.

Examples of various types of beer. While each example is of good quality for its type, individual tastes vary widely and, in the absence of obvious faults, judgement of quality is subjective.

947.　**English keg beer,** conditioned in bulk, pasteurised and CO_2 pressurised. Such beer is reviled by many consumers.

948.　**Traditional English cask conditioned bitter.** The flavour is considerably stronger than that of keg beer and the alcohol content higher.

949.　**English brown ale.** Almost invariably bottled or canned the colour is derived from caramel and heavily roasted malt. The sugar content is high.

950.　**German lager.** This is a traditional product with limited secondary fermentation in the bottle. Where such lagers are exported over long distances as with Polish lager the fermentation is stopped by pasteurisation to ensure a long shelf life.

951 and **952.**　**English bottled bitter (light ale).** In this particular type a strong secondary fermentation takes place in the bottle resulting in a beer of high alcohol content containing a heavy yeast sediment (**952**) from which the beer is decanted when serving.

953.　**Massive growth of wild yeast in beer.** The source of infection was an improperly cleaned and maintained dispensing pump (beer engine) at the retail outlet which permitted contaminated beer to drain back into the barrel. This growth occurred over a 3 day period.

954.　**Acetic acid bacteria** (*Acetobacter sp*) × 1,000. A cause of spoilage in beer, wine and cider. The bacteria are held in clusters in a slime matrix. The variation in size of cells is an important feature in visual identification. After Lüthi and Vetsch (1981).

955 and **956.**　**Guinness** is a famous Irish stout. The colour is derived from caramel and dark roasted malt and the beer is of high carbohydrate content. A particular feature is the large 'head' of foam (**955**). It is conditioned by a secondary fermentation in the bottle. Its shelf life is thus strictly limited and if stored for too long has a poor flavour and a change in body which affects the head (**956**).

It is of interest to note that among the legendary properties of Guinness and similar stouts was the reputation as a health food and indeed the advertising slogan 'Guinness is good for you' was used for many years. The concept of alcoholic beverages as a ready source of nutrient is common in the U.K. with the sale of tonic wines such as Wincarnis. Lactose was an ingredient of stouts such as Mackeson, while a stout containing meat extract (Mercer's Meat Stout) was produced at least until the 1930s. This latter is said to have been thoroughly unpleasant in taste and was drunk for its real or supposed medicinal properties rather than for the more usual properties of beer.

Wine and Cider (957–988)

Wine and cider (hard cider) are discussed together since the technology employed is basically similar. Wine consumption in the U.K. has increased considerably over recent years at the expense of beer and spirits while cider, traditionally the drink of social misfits, has become more fashionable.

In terms both of its production and consumption wine has acquired a folklore and mystique unrivalled in other foods, and drink quality *per se* is defined hedonistically in the taster's own unique vocabulary. It is not intended to explore any part of this mystique but to treat the winemaking process purely in a technological manner relating, where possible, defects in the finished product to the technological processes. Full discussions of the production and types of wine may be obtained in specialist books e.g., The Wine Book (Dorozynski and Bell, 1969); Alcoholic Beverages (Rose, 1977).

Wine may be made from any fruit and vegetables, even those such as parsnip, marrow and potato, as well as flowers like dandelion and rose petals. However, the vast majority of commercial wine production in the generally recognised European, U.S.A. and Australian sense is grape wine. This is discussed below and the basic process illustrated in **957**. Brief mention is also made of 'wines' which are of economic importance in other parts of the world such as sake in Japan and palm wine in East Africa.

Grapes

Grapes are the sole raw material of most wines. There are many types of wine grape, most originally associated with a particular region or type of wine but now grown in wine producing areas the world over. Harvesting grapes is a manual procedure and care must be taken to avoid damage prior to crushing at the winery to reduce the possibility of growth of wild yeasts. The presence of wild yeasts, usually of the apiculate type, is inevitable and must be controlled at all stages of the winemaking process.

Grapes are prone to fungal rots primarily caused by species of *Botrytis*. In most cases this reduces the quality and economic value of the grape. An exception is the rot caused by *B. cinerea*, the 'noble rot' which by dehydrating the grape increases the sugar content and makes the grape suitable for production of the sweet French Sauternes and the famous Hungarian Tokay wines.

Crushing and Stemming

In modern winemaking practice crushing and stemming is carried out mechanically. The resultant juice (with or without skin and pips), referred to as '**must**', is treated with sulphur dioxide to inhibit wild yeast and spoilage bacteria. Some strains of lactic acid bacteria are resistant to SO_2 and if the must is stored for significant periods prior to fermentation growth of lactic acid bacteria may produce off-flavours which are carried over to the final product.

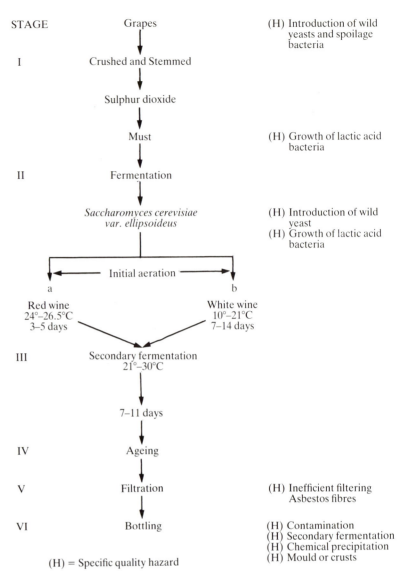

957. A flow diagram for the manufacture of red and white wine. The process varies from one type of wine to another and as a considerable craft element is involved detailed procedures also vary according to the condition of the raw material and to the practices of individual winemakers.

Fermentation

Fermentation of the must is by the wine yeast *Sacch. cerevisiae var. ellipsoideus*. This is added as an active culture at a level of 2%–5% in order to suppress any wild yeasts or spoilage bacteria present. Small producers may lack specialist yeast culture facilities and under these conditions the starter yeast itself may be an important source of wild yeast. If for any reason initial growth of the starter yeast is slow lactic acid bacteria may produce off-flavours. For this reason the must is usually aerated in the early stages of fermentation to encourage yeast growth. Red wines (where the must contains skin and pips) is fermented initially for 3 to 5 days at 24°–26.5°C whereas white wine is fermented for 7 to 14 days at 10°–21°C.

Secondary Fermentation

Secondary fermentation takes place in tanks for 7 to 11 days at 21°–30°C. No specific contamination or spoilage occurs at this stage.

Ageing

Wine is aged in wooden, concrete or stainless steel tanks (final ageing may be in the bottle). The length of ageing varies according to the type of wine and varies from a few months to years. Much of the hedonistic quality of the wine is developed at this stage. Ageing involves complex chemical changes which impart to the wine its final flavour and body.

A particularly important reaction at this stage is the malolactic fermentation. This is brought about by lactic acid bacteria notably *L. plantarum* and *Leuconostoc oenos* (*Leuc. gracile*) and involves the conversion of L-malic acid to L-lactic acid (together with formation of carbon dioxide and other compounds). A malolactic fermentation is undesirable in young, low alcohol table wines such as Californian, but is desirable in, for example the highly acid wines produced in the Italian Piedmont region.

The malolactic fermentation by lactic acid bacteria is slow and inconsistent. This may lead to quality problems in high acid wines where the malolactic fermentation is desirable to reduce acidity. Attempts are therefore being made to produce strains of wine yeast capable of carrying out the malolactic fermentation by cloning the malolactic gene from *Leuc. oenos* into a yeast (Subden, 1984).

Filtration

Wine is usually filtered before bottling to remove yeast cells, precipitated proteins, tannins etc. If filtration is inefficient yeast may pass through the filter to the bottling plant and additionally filter presses have been found to be sources of wild yeasts.

Asbestos wool filters have been used in some sectors of the wine industry. This has resulted in concern as to the possibility of carryover of asbestos fibres into wine and the risk of asbestosis to the consumer. Asbestos based filters should not therefore be used.

Bottling

Bottling of wine is an important stage both with respect to intrinsic quality and to contamination. It is essential that bottling is carried out at the correct stage. Too early bottling enhances the risk of yeast growth and an undesirable secondary fermentation or a precipitation of tartrates and protein-pigment complexes.

Equally bottling too late results in a flat, lifeless wine. The correct time for bottling varies as indicated in discussion of ageing. All filling processes carry a risk of contamination and this is particularly true of wineries which frequently have a high indigenous population of yeast and bacteria. Filling machines are notorious sources of contamination and a rigorous cleaning schedule is essential.

The cork plays an important role in maintaining the quality of wine. A porous or ill-fitting cork which permits air to enter will result in oxidation causing flavour defects, while bitter flavours may be imparted to the wine by mould growth on the cork.

It is also necessary to take stringent precautions against flying insects at the bottling stage because wine exerts a particular attraction for many species which, becoming intoxicated, fall readily into the wine. It should be noted that wine is now packaged in a variety of forms in addition to the conventional corked bottle; these include screw-capped bottles, cans and latterly the wine box (a collapsible plastic foil laminated bag in a rigid cardboard outer box).

Special Types of Wine

Sparkling Wines
Traditional sparkling wines are produced by a secondary fermentation. After the main fermentation is complete sugar and yeasts are added to produce a secondary fermentation with an excess of carbon dioxide. This may take place in the bottle, as with champagne, or in closed vats. Sparkling wines may also be produced without a secondary fermentation by injection of carbon dioxide.

Fortified Wines
Fortified wines include sherry, port, Madeira etc. The alcohol content is increased to 19%–21% by addition of grape brandy. Sherry and Madeira are matured in a solera – a continuous maturation process. The distinctive flavour of sherry is attributed to the growth in the solera of the film yeast *Mycoderma vini*.

Palm Wine
Palm wine is an alcoholic beverage produced from the sap of various palms which contains a heavy suspension of yeasts and bacteria giving the drink a milky white colour (Okafor, 1975). It is consumed throughout the tropics and is of considerable importance in the agrarian economy of producing countries. Although official production figures do not exist a conservative estimate for Nigeria alone is 91,000 to 137,000 litres per day. Palm wine differs from other alcoholic beverages in that the micro-organisms responsible for its production are present at the time of consumption. These same micro-organisms are also responsible for its spoilage since continuing metabolism replaces the characteristic sugary taste with various organic acids within 36 to 48 hours.

Sake (rice wine)
Sake is classified as a wine although some aspects of its production are more akin to beer. Sake is of considerable economic importance in Asia and is the most widely consumed alcoholic beverage although western type grape wines and spirits are becoming more widely consumed, particularly in Japan.

A starter for sake is made by growing *Aspergillus oryzae* on steamed rice to obtain amylases and proteases (analogous to the malting stage in beer production). This starter is mixed with rice mash and mixed strains of indigenous *Saccharomyces* carry out a fermentation for 10 to 14 days.

The alcohol content is 15%–17% and spoilage by yeast is rare. Lactic acid bacteria, however, and *Acetobacter* produce spoilage similar to that of beer (see Table 24). Moulds other than *A. oryzae* may grow during the starter process and produce off-flavours. In addition the possibility of mycotoxin formation during production of the starter has been noted (Saito *et al*, 1971). A similar product, Sonti, is made in India using *Rhizopus sonti* to produce the starter.

Pulque

The Mexican beverage pulque made from *Agave* sap is unique in that the alcoholic fermentation is produced by *Zymomonas spp* rather than by yeast.

Spoilage of Wine

The presence of small numbers of yeast or bacteria in wine does not necessarily indicate that the wine will spoil. Many factors are involved in determining whether or not spoilage organisms can grow, the most important being acidity, sugar content, alcohol content and level of sulphur dioxide. Common causes of wine spoilage are shown in Table 25.

958. The noble rot (*pourriture noble*) mould *Botrytis cinerea* (× 250). After Lüthi and Vetsch (1981).

Good quality examples of wine. Note that the exact colour varies from type to type.

959. Good quality red (Beaujolais).

960. Good quality white (Sauternes).

961. Good quality rosé (Anjou). Note that good quality rosé should not be made by blending red and white wine.

962. Wild yeasts infecting red wine. Note the evidence (arrowed) of gas production as well as cloudiness. The fault was attributed to a faulty filter before bottling.

963. Yeast growth in wine. The consumer had complained of the wine being 'watered down' and laboratory examination indicated a low (<5%) alcohol content. This permitted rapid yeast growth with the associated cloudy appearance. A likely cause of the additional water was a cleaning-in-place (CIP) system which had not been properly purged. Related complaints which are found with all bottled liquids are 'disinfectant' or 'detergent' taints again due to incorrect CIP practice (see p. 242). Note that watering of alcoholic drinks to disguise illicit consumption (or fraud) is a relatively common practice.

Presumptive preliminary identification of yeasts causing spoilage of wines and other alcoholic beverages is often possible on the basis of microscopic examination. Four examples are illustrated here.

964. *Saccharomyces cerevisiae var. ellipsoideus* (× 1,200), the wine starter yeast. Note the budding cells (arrowed A) and vacuoles (arrowed B).

Table 25
Micro-organisms Commonly Responsible for Spoilage of Wine

Micro-organism	Type of Spoilage
Wild yeasts	
Saccharomyces pastorianus	Cloudiness, off-taste, gas production
Brettanomyces spp	Mousy taste
Lactic acid bacteria	
Lactobacillus brevis;	
L. hilgardii	Souring, mousy taste, discoloration
Leuconostoc oenos;	
L. plantarum	Reduction in acidity*
Leuc. mesenteroides	Slime (rope) formation
Acetic acid bacteria	
Acetobacter pasteurianus	Vinegary taste, mousy, sweet-sour taste

*Acidity reduction is desirable in some wines (see p. 237). Spoilage of cider and related drinks is similar to that of wines although *Zymomonas* has been implicated in the spoilage of cider but not of wine.

965. *Kloeckera apiculata* (× 1,200). A common and fast growing yeast producing off-flavours in wines and ciders. The microscopic appearance of the yeasts is characteristic, cells varying in shape from the almost spherical (arrowed A) to pointed (arrow B). A number of cells are apiculate (lemon shaped, arrowed C).

966. *Pichia farinosa* (× 1,200). A pellicle forming yeast responsible for off-flavours in wine and fruit juice. In wine the glycerine and ethanol content are reduced. Note the constriction in the centre of some cells (arrowed A), the terminal (arrowed B) and shoulder (arrowed C) budding.

967. *Hansenula minuta* (× 1,600). A pellicle forming yeast common in fruit juices but also found in wine. The yeast forms esters of ethanoic acid and produces powerful off-flavours. After Lüthi and Vetsch (1981).

968. A *Pediococcus spp* infection in a sweet red wine. Note the cloudiness due to the bacterial cells and the colour change due to pH fall. Some strains of pediococci produce slime (rope) in wines. The source of infection is thought to have been the bottle filler.

969. *Pediococcus sp* (× 1,200) in wine. Care should be taken to avoid confusing pediococci with tannins (see **975**). After Lüthi and Vetsch (1981).

970 and 971. Vin blanc de cassis (a blend of white wine with cassis: a blackcurrant based liquor) with a medium (**970**) and close-up (**971**) view of tartrate crystals precipitated out of the wine. This fault occurs intermittently in wines particularly where fruit of high tartaric acid content such as blackcurrants have been used. The problem may also be associated with premature bottling.

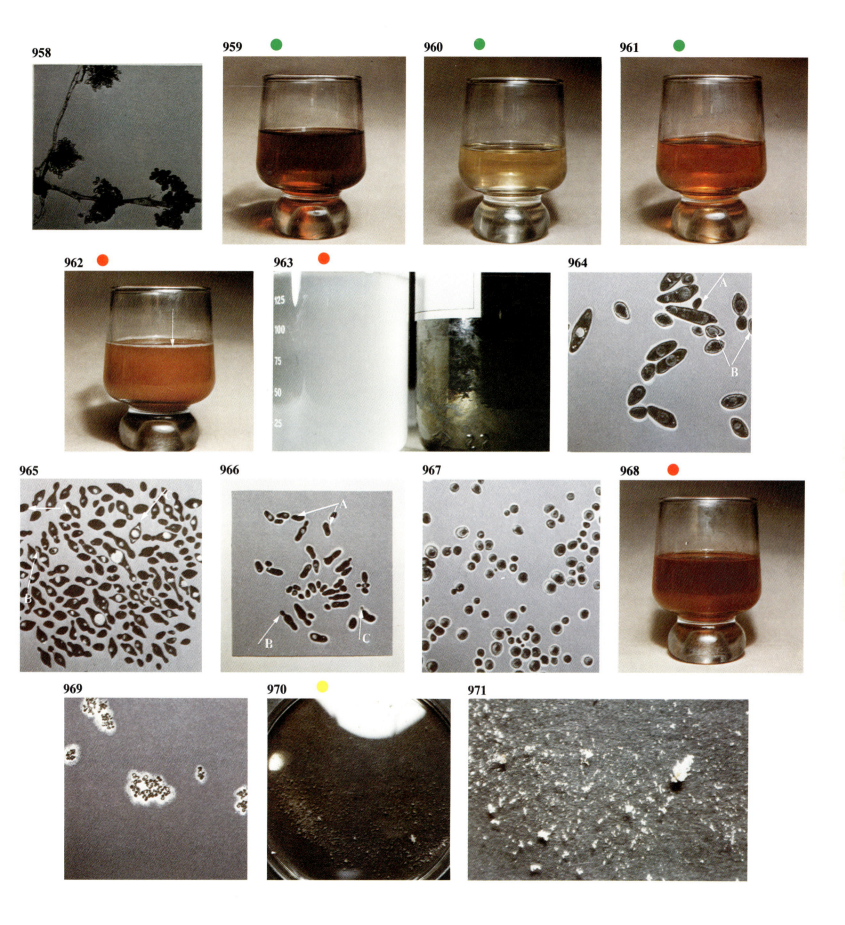

958

959 🟢

960 🟢

961 🟢

962 🔴

963 🔴

964

965

966

967

968 🔴

969

970 🟡

971

239

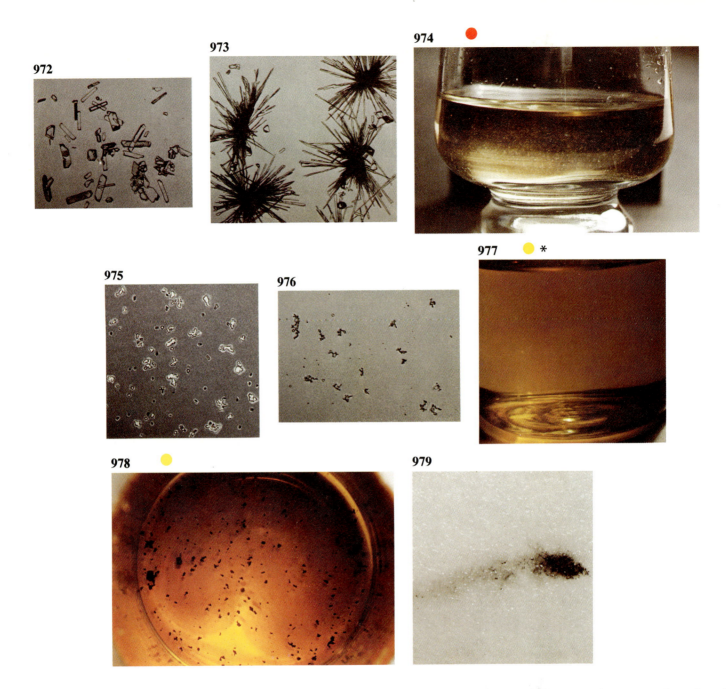

972 and **973.** **Calcium tartrate crystals** (× 100) from bottled wine (**972**). Tartrates are also found as a double salt between d-tartaric acid and L-malic. The star shaped crystals illustrated (**973**) (× 50) are readily recognised microscopically. Such crystals may be formed after de-acidication treatment with a proprietary agent such as **Acidex.** After Lüthi and Vetsch (1981).

974 to **976.** **Haze in white wine** derived from an in-bottle sediment (**974**). The sediment was amorphous vegetable material probably resulting from inefficient filtration. Other sediments form after bottling and may be identified by microscopic examination. **975** illustrates tannins (× 1,300) from red wine. The particles are relatively large and of different shape and size. They

may thus be distinguished from those of trace metals such as iron (**976**) (× 1,300) which are smaller and of uniform size. After Lüthi and Vetsch (1981).

977. **A slight cloudiness in fino sherry.** This is due to precipitation of tannins.

978 and **979.** **A black sediment in apricot wine** identified as a mixture of graphite and grease (**978**). This mixture is used to grease the moulds in which the bottles are made, thus facilitating removal of the bottle. If not fully cleaned from the bottle it is likely to contaminate the contents as shown. **979** illustrates the greasy nature of the contaminant when smeared onto filter paper.

980 and **981.** **Filter aids** added to improve the efficiency of plate filters are not uncommon contaminants of wine and are readily identified microscopically. **980** illustrates particles of diatoms (× 1,200) comprising diatomaceous earth and **981** illustrates asbestos fibres (× 50). Such fibres may appear as small 'needles' to the unaided eye.

982. **A Green lace wing (*Chrysopa carnea*)** in an intact bottle of rosé wine. Such insects appeared in large numbers in the area of the bottling plant at the time the wine was being bottled. The incidence of contamination was very high indicating both inadequate flying insect control measures and inadequate post-bottling inspection.

983. **A crane fly (*Tipula sp*)** in sherry. Many flying insects are attracted to sweet, alcoholic substances and the fly probably entered the sherry at the bottling stage.

984 and **985.** **Traditional farmhouse cider** (Scrumpy) (**984**) compared with mass produced cider which has been clarified and carbonated (**985**). Note the lighter colour, lack of haze and effervescence of the mass produced product. Farmhouse cider varies widely in colour and quality from batch to batch but is frequently characterised by a strong taste and odour of vinegar suggesting a high level of *Acetobacter* activity. It is also noted for its unpredictable and sometimes spectacular effects on the human metabolism!

986. **Yeast spoilage of cider.** The stage at which the contamination occurred is not known.

987. **A sample of cider allegedly containing oil.** Although the cider was obviously contaminated with an oily substance it would appear that contamination occurred after opening the bottle. The complainant, who was apparently acting in good faith, had consumed most of the cider during the course of a day's work on a building site. He noticed the oily appearance and a 'fruity' smell on opening the bottle during the afternoon break. The cider was strongly tainted and it is unlikely that this would not have been noticed earlier in the day. The most likely explanation is that a workmate has added something to the cider as a practical joke. It should be noted that as the bottle was not re-usable (non-returnable) the possibility of it previously having been used for storage of other material may be discounted.

988

988. The contents of a full bottle of 'cider'. The liquid had a strong smell of bleach (hypochlorite) and was obviously a type of cleaning fluid. The seal of the bottle was apparently intact before opening and the most likely source of the problem was the bottling plant. Cleaning fluid could enter bottles as a result of insufficient rinsing after cleaning-in-place (CIP) operations or due to incorrect operation of the filling line drawing fluid from a tank undergoing cleaning. In either case a large number of similar complaints would be expected.

A further possibility is that the bottle was used as a container for cleaning fluid at the bottling plant and was placed onto the line in error. The practice of using food containers for other purposes on filling lines is strongly condemned. Although in this example the bottling plant appears to be clearly implicated the bottle was re-usable (returnable) and it is possible that the cleaning fluid was placed in the bottle by a consumer, the bottle then being confused with cider containing bottles at some later time.

In the U.K. at least cider is a favourite drink of vagrants who imbibe direct from the bottle. Under such a circumstance severe injury could have been caused had any of the cleaning fluid been swallowed.

Distilled Alcoholic Drinks

Distilled alcoholic drinks include brandy, whisky, gin, rum and vodka. The alcohol content of the distillate is typically 40%–45% ($^v/_v$) and there are no problems of microbiological spoilage although growth of bacteria, especially lactic acid bacteria, occurs during the later stages of fermentation and is involved in flavour development in drinks such as whisky and dark rum. Assessment of quality is made primarily on a hedonistic basis and personal preferences play a major part.

Table 26 Laboratory Tests Alcoholic Beverages

I. Microbiological
Yeast
Acetobacter spp
Zymomonas mobilis (Forcing test)
Lactic acid bacteria (some such as *Leuc. oenos* require special media – see Garvie, 1967).

Direct microscopic examination is valuable in investigating overt spoilage. N.B. Microbiological examination is not usually required for distilled products.

II. Chemical/physical
Alcohol content
Sugar content
pH value

Direct microscopic examination is valuable in examination of chemical precipitates.

Gas liquid chromatography may be used to establish flavour profiles.

Foreign body/infestation examination not normally required.

Notes:
(1) These tests are examples only and are not inclusive.
(2) Examples are for finished products only.
(3) Techniques are detailed in specialist publications such as ICMSF (1978) – Microbiology: Egan *et al* (1981) – Chemistry.
(4) Instrumental techniques such as Impediometry may be substituted for some classical microbiological methods.

Chapter 17

Foreign Bodies and Infestations

Foreign bodies and infestations are approached in two ways in this publication. First, where these illustrate problems which are largely specific to a given product either by virtue of specific infection or by the nature of the product interfering with detection or recognition of foreign bodies, examples are included with the commodity chapter but cross referenced in this chapter. Second, where the problems illustrated pertain to food quality as a whole they are discussed in the present chapter but cross-referenced in the product chapters.

Foreign bodies and infestations may enter foods at any stage from primary production to domestic storage directly before consumption. In some cases it is possible to determine the stage at which the problem occurred but in other cases not. Some products invariably carry a low level of foreign bodies or infestations. Examples are stones in sun-dried fruit and infestants in flour. In some countries such as the U.S.A. the degree of (harmless) contamination acceptable is legally defined but in other countries, including the U.K. the situation is less clear. In theory the presence of a single foreign body may be sufficient to render the vendor liable for prosecution but in practice the nature and degree of contamination is likely to be taken into account. Consistently high levels of infestation or foreign body contamination, however, irrespective of the nature of the product, are indicative of a breakdown in hygiene and require urgent remedial action.

For the sake of clarity the chapter is divided into sections each dealing with specific types of problems. The sections are listed in Table 27.

Table 27 Categories of Foreign Bodies and Infestations

1. Insects commonly associated with production or storage of foods.
2. Insects generally present in the environment.
3. Rodents.
4. Other animals and birds.
5. Foreign material derived from food but normally removed during processing.
6. Foreign material derived from personnel.
7. Foreign material derived from plant, premises, packaging etc.
8. Foreign material derived from sabotage.
9. Miscellaneous.

Insects Commonly Associated with Production or Storage of Foods (989–1005)

Foods of plant origin are normally subject to infestations of this type, which may occur at either the pre- or post-harvest stage. In each case the potential economic loss is massive. A classic example of the former is the damage caused by locusts. The most vulnerable economies are those of the developing and underdeveloped nations where a large proportion of the population is dependent on a single crop and where effective pesticides may not be readily available. Even within developed economies the loss due to pre-harvest damage by pests such as the screw worm in the U.S.A. is considerable.

Post-harvest infestations cause considerable economic loss and this again is most serious in underdeveloped nations where storage facilities may be poor and effective pesticides are not available. Heavy loss, however, occurs elsewhere either as a result of importing heavily infested material or due to infestations previously established in food handling premises. While it may be argued that it is not possible to totally eliminate infestations from grain, dried pulses etc. control is both necessary and possible. Control must be of a continuous and preventive nature and must extend to such considerations as building construction and design. It is still too common to find food manufacturers dispensing with contract pest control services because of their success in preventing problems i.e. the continuing absence of infestations is taken to mean that control is no longer required.

It is also necessary to be aware that infestations not only occur in domestic premises but are an increasing problem, particularly as a consequence of central heating, fitted carpets and built-in kitchens. Determining the original source of the infestation may be a point of contention between the householder and the manufacturer/retailer and may require delicate negotiations.

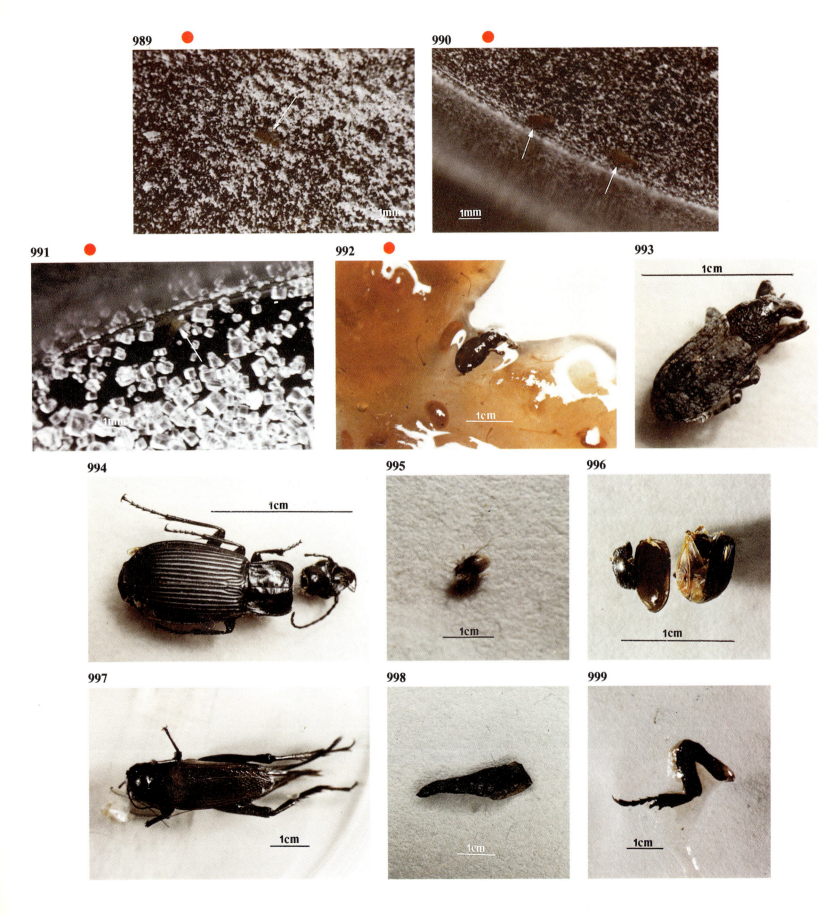

989

990

991

992

993

994

995

996

997

998

999

1000. Merchant grain beetle

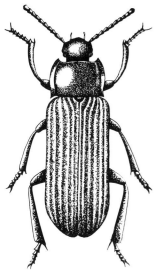

(*Tribolium mercator*) × 12.
See pages 118–119.

1001. Confused grain beetle

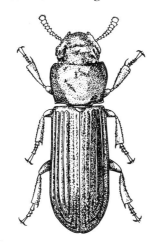

(*Tribolium confusum*) × 12.

1002. Pea weevil

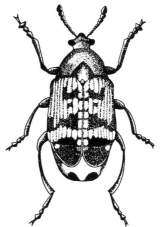

(*Bruchus pisorum*) × 30.
See pages 126-127.

1003. Drug store beetle

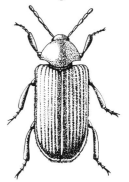

(*Stegobium paniceum*) × 12.
See pages 120–121.

989, 990 and **991.** **Psocids** (*Liposcelis* and other genera. Family *Psocidae*) were previously a problem in libraries and paper stores where they feed on microfungi growing on the paper. They are now a common pest in dry foods such as flour, sugar and cereals and again feed on microfungi rather than the product itself. Psocids may infest food at any stage from production to domestic storage and their small size means they are able to enter some types of pack without causing damage to the packaging material. Psocids in flour are illustrated (arrowed) (**989** and **990**), and in sugar (**991**). The blur in the latter example is due to the speed of the insects' movement and this with the small size means that detection of a light infestation may be difficult.

992. **A damaged insect,** a member of the family *Curculionidae* in jam. Contamination occurred prior to filling.

993. **A damaged weevil** (family *Curculionidae*) found in canned peach slices. Contamination occurred before processing.

994. **A slightly damaged beetle** (family *Tenebrionidae*). Found in canned peas. The beetle was probably present in the crop before harvesting.

995. **The drugstore beetle** (*Stegobium paniceum*) is a relatively common pest of stored foods and can also be troublesome in warehouses. It has a prediliction for spices such as cayenne pepper from which this example was isolated. (See Chapter 9, Grocery pp. 154–155.)

996. **Fragments of a leaf beetle** (*Chrysolina sp*) from dried rosemary (*Rosmarinus officinalis*). Contamination probably occurred during cultivation of the herb and was not detected until domestic usage. Classical taxonomic guides are of little use when specimens are as badly damaged as this.

1004. Grain weevil

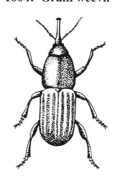

(*Sitophilus granarius*) × 15.

1005. Psocid

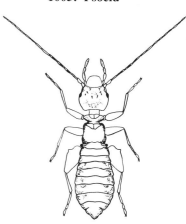

(*Liposcelis* and other genera) × 50.

997. **A locust** from North African salads.

998. **Empty larva case found in a fruit jelly.** The larva case may have been derived from the fruit used in manufacture of the jelly.

999. **An insect leg found in jam** is illustrative of the condition in which foreign bodies are often found. While contamination originated in the jam precise identification of the insect was not possible and thus detailed control procedures could not be recommended to the production unit.

It is not the intention or purpose of this volume to provide an identification manual for insects found in foodstuffs and reference should be made to specialist texts such as Hinton and Corbet (1972). For the purpose of comparison with examples illustrated line drawings of intact and magnified common food infestants are reproduced here (**1000** to **1005**). The difficulties of relating 'perfect' specimens such as these to insects found in foods may be considerable.

Examples of insects commonly associated with the production or storage of foods may also be found in the following commodity chapters.

Insects Generally Present in the Environment (1006–1022)

A number of such insects may find their way into foods. Economically they are less significant than those discussed earlier since actual attack of the food may be absent or limited. They may be a cause of a high level of consumer dissatisfaction and their presence at significant levels is indicative of poor plant housekeeping. As with infestants control should be preventive and continuous.

Personnel involved in food inspection should be aware that, unlike infestants, environmental insects may be dangerous. The best known examples are of course tropical fruit such as bananas containing tarantula spiders. In addition live scorpions have been found in citrus fruit, and other potentially dangerous insects reported in foods include tropical centipedes, hornets and ants.

1006. **The importance of electric flying insect control devices** such as **Insectocuters** is illustrated. This is the 'catch' from a single Insectocuter operating over one summer in one area with a small fly population. Such devices cannot totally obviate the presence of flies but correctly positioned (and switched on!) represent a powerful means of control.

1007. **A housefly (*Musca domestica*) in packeted suet** (arrowed). Contamination probably occurred before packing.

1008. **A fly in wholemeal flour.** In the absence of other evidence it is impossible to determine whether the fly gained entry to the flour at the mill or in the home after purchase.

1009. **Fly parts in dried lentils.** The condition of the fly suggests that it entered the lentils prior to packing.

1010. **A spider's leg** (arrowed) in a jar of sardine and tomato paste. This illustrates the extreme difficulties of investigating allegations of foreign body contamination in foods which have subsequently spoiled since, on initial examination, only mould was visible. As the paste was part consumed it is possible that the spider entered in the home. In view of the degree of damage to the spider, however, contamination during manufacture is considered more probable.

1011. **Moth larvae found in Costa Rican bananas.** Note mould growth on one (arrowed). This is fairly common and is usually indicative of the larvae being present before shipping.

1012 and **1013.** **Myriapods (centipedes or millipedes) from dried fruit.** The insects were probably present in the fruit before processing. Some tropical centipedes, when live, are poisonous and should be handled with care.

1014. **Stag beetles** may be found in food storage and handling premises, and are sometimes found as spectacular, if readily recognisable, contaminants in foods. Despite their fearsome appearance male stag beetles are harmless but their presence in food could well lead the company involved to an unwelcome appearance in local news media.

1015. **Damage has caused this earwig** (*Forficula auricularia*) from breakfast cereal to lose its characteristic appearance.

1016 and **1017.** **A moth in lettuce** illustrated in general (**1016**) and in close-up (**1017**). In this example the presence of the moth was probably adventitious.

1018. **A grossly damaged insect from canned tuna fish.** The damage is too severe to permit accurate identification but it is probably a type of flying beetle. Contamination could have occurred at the filling stage.

1006

1007 🔴

1008 🔴

1009 🔴

1010 🔴

1011

1012

1013

1014

1015

1016

1017

1018

247

1019

1020

1021

1022

1023

1019. **A damaged insect larva** from bottled spring water. The water is filtered before bottling and the source of the larva is therefore the bottling plant or the bottle itself.

1020 and **1021.** **A damaged cockroach** (*Periplaneta americana*) found in brown rice illustrated from above (**1020**) and below (**1021**). The degree of damage (note the absence of legs) suggests that contamination occurred before packing but probably after hulling where more extensive damage would be expected.

1022. **A slightly damaged spider** (Order *Araneae*) found in a cake topping (chopped nuts and sugar based). The product and spider were in a sealed pack and contamination therefore occurred before packing but probably after the main manufacturing process.

Examples of insects generally present in the environment may also be found in the following commodity chapters

Chapter 2	*Dairy*
Pp. 20–21	Silverfish in milk
P. 30	Fly eggs in Brie
Pp. 34–35	Insect in Edam cheese

Chapter 4	*Fresh Meat Products*
Pp. 60–61	Fly from beef sausage
P. 64	Spider from shepherds pie

Chapter 9	*Grocery*
Pp. 116–117	Insect webbing in flour

Chapter 10	*Bakery*
P. 167	Fur beetle damage to biscuits

Chapter 16	*Alcoholic Beverages*
Pp. 240–241	Lace wing in wine

1023. **Part of a rodent** (possibly the head) found in frozen mixed vegetables. The animal was probably present in one of the crops and passed undetected through the entire cutting, blanching, packing and freezing processes. Further confirmation of the animal as a rodent may be obtained from microscopic examination of the hairs, rodent hair having a characteristic 'bicycle chain' appearance (see **1024**).

Rodents (Mice and Rats) (1023–1024)

Despite improvements in control, rodents continue to be of importance in sectors of the food industry and a cause of considerable economic loss. Cereal products are usually thought to be the subject of rodent attack but in addition vegetable products and meat may be affected. As with insects control must be on a preventive and continuous basis. Cats and to a lesser extent dogs may be used in rodent control but while possibly acceptable at a farm level may lead to problems in their own right (see below).

Considerable care must be taken when handling rodents or rodent infected material, as they are frequently infected with one of various serovars of *Leptospira interrogans* which can infect man *via* cuts and wounds and possibly *via* the intact skin causing the sometimes fatal leptospirosis (Weil's disease).

1024. Examples of the microscopic appearance of animal hair.

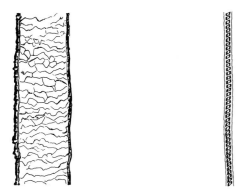

(a) Man (European). Cast of scale pattern, root region, × 340.

(b) Rat. Whole mount fine hair × 150.

(c) Cat. Whole mount fine hair × 150.

(d) Dog. Whole mount fine hair × 150.

1025. Examples of the microscopic appearance of textile fibres.

(A) Wool. Whole mount × 120.

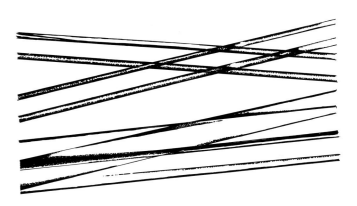

(B) Polyamide nylon (Courtaulds Type 1200) Whole mount × 120.

The nature of hairs and fibres present in food may be determined by microscopic examination of the whole fibre or of scale casts prepared from it (for details of practical techniques see Appleyard, 1960). Examples are shown (**1024**). Considerable variation occurs within species, for example between Caucasian and Oriental man or different breeds of cat and dog. Hairs from different parts of an animal body may also differ in appearance. For full details reference should be made to specialist texts such as Appleyard (1960). **1025** shows examples of microscopic preparations of textile materials. Again considerable variation may occur and reference to a specialist text such as Anon (1965) is recommended.

Examples of rodents or rodent damage may also be found in the following commodity chapter:

Other Animals (1024)

Other animals, both wild and domestic, may gain entry to food storage and processing premises. Birds for example cause problems in high roofed warehouses. In the U.K. and some other countries control of birds is complicated by the fact that protected species cannot be destroyed but must be anaesthetised and subsequently released. Breeding colonies of feral and semi-feral cats also present a considerable problem in some circumstances. Cats are acceptable for rodent control on farms and possibly in large scale grain handling facilities in countries where effective rodent control is not otherwise available. In such cases the cat population must be carefully controlled and cats should never be permitted in premises where other means of rodent control are possible or where fully processed food is handled.

Other animals tend to be a sporadic problem although it should be noted that instances of management taking pet dogs into food manufacturing premises on a regular basis exist. The taking of Guide Dogs for blind persons into food retailing premises is an emotive issue. Some retailers permit Guide Dogs partly on humanitarian grounds and partly on the premise that such dogs are likely to be more disciplined than the average pet. Other retailers prefer to exclude all dogs and provide staff assistance for the blind. Correctly and sympathetically applied this would seem to be the more consistent policy.

Apart from specific problems with cats and birds other animals may be excluded by properly maintained door and window screening and the application of good housekeeping to reduce food waste. Particular attention should be given to food refuse containers outside the food handling environment as these may serve as an attraction for a wide range of animals, insects etc.

It is necessary to be aware that domestic pets may contaminate food on domestic premises, the contamination frequently being attributed to the processing. Again cats and dogs are normally responsible but other pets such as tortoises, birds and ferrets may be involved.

Foreign Material Derived From Food But Normally Removed During Processing (1026)

Virtually all foods require removal of undesirable or inedible components during processing. It is not possible to itemise these and the majority are self-evident, examples being the shelling of peas for processing and the de-boning of meat. Problems with incorrect preparation usually involve the presence of small pieces of inedible material. Such problems are rarely of a serious nature, although a high level implies poor quality control while at the same time material may be mistaken by the consumer for more serious defects. Included with this category are bodies such as stones in sun dried fruit which, while not directly food derived, are often intimately associated with the primary processing.

Examples of foreign material derived from food but normally removed during processing may also be found in the following commodity chapters:

Chapter 9	*Grocery*
Pp. 124–125	Husk on flaked rice
Pp. 124–125	Stone in kidney beans
Pp. 158–159	Plant debris in gum

Chapter 13	*Produce*
Pp. 184–185	Deadly nightshade in spinach

Foreign Material Derived From Personnel (1025, 1027–1028)

Quite apart from public health considerations personnel may be responsible for the accidental contamination of foods. The most common items are hair, which should be adequately covered, jewellery which, with the exception of plain wedding rings and sleeper earrings should not be worn by food handlers, and adhesive plasters, wound dressings etc. The latter may be contaminated with blood or pus and are likely to be considered particularly offensive. All plasters used should be waterproof, distinctively coloured to aid visual detection and contain a metal strip to permit detection by metal detectors. Other examples include coins, handkerchiefs etc. which have fallen from pockets, and measures to obviate this possibility should be taken account of in the design of protective clothing.

Smoking constitutes a problem in its own right. It should of course be strictly forbidden on all food handling premises and breach of non-smoking rules by any member of staff (including management) justifies severe disciplinary action. The attitude of management is all important in ensuring effective control of smoking. It is still all too common particularly in small family owned companies to find management who consider themselves exempt from regulations.

Personnel may also be responsible for the introduction of taints and odours into foods as a result of contamination with cosmetics, scented hand soaps etc. For this reason cosmetics should not be worn by food handlers and soaps should be non-scented.

Foreign Material Derived From Plant, Premises, Packaging Etc. (1029–1036)

A wide variety of foreign bodies may be derived from such sources. Each hazard should be identified separately and appropriate precautions taken; e.g. correctly fitted and operated metal detectors to detect metal machine parts. In general terms, however, in plants where the standard of housekeeping is high and maintenance of both machinery and the fabric of the building carried out correctly the number of such incidents is likely to be low. Particular factors to avoid are major re-building work during production and makeshift or 'temporary' repairs (which often become 'permanent') to equipment.

1026

1027 ●

1028 ●

1029

1030 ●

1031

1032

1033

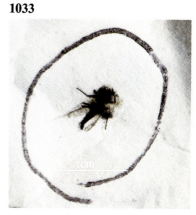

1026.   A trilobite (view of underside) found in cooked frozen prawns. This example is probably a marine species which was fished with the prawns and went through the entire cooking, freezing and packing process.

1027.   Human hair in frozen peas. The correct wearing of proper protective clothing including, where appropriate, beard bags is essential to prevent such occurrences. It should be noted that human hair can be distinguished from that of rodents and other animals such as cats by microscopic examination (see **1024**).

1028.   A cigarette filter in frozen cut green beans. The filter is intact and probably entered the beans at the packing stage. 'No smoking' regulations must be strictly enforced and observed by all members of staff. A high level of illicit smoking is usually indicative of a lack of management commitment.

1029.   General vegetable matter and dirt found in frozen peas. The vegetable matter probably derived from the pea pods and the dirt from the viner (shelling equipment).

1030.   Cobweb in dried onion. The presence of cobwebs in a factory is indicative of poor housekeeping. This example was an isolated incident and the appearance of the cobweb suggests that sabotage may have been involved.

1031.   A damaged silverfish (*Lepisma saccharina***) from flour.** Silverfish are common in domestic and commercial premises. They are able to live at low temperatures and have been known to infest the insulated lining of domestic refrigerators.

1032.   A dried up mould pellicle found in a carton of UHT orange juice. Pellicles of this nature are likely to be derived from poorly cleaned plant or utensils.

1033.   Packaging materials may be a source of foreign bodies. This example is an insect embedded in the wax coating of the coated paper used for packing 'Quick-setting gel'. The contamination occurred during the coating of the paper.

Examples of foreign material derived from plant, premises etc. may also be found in the following commodity chapters:

Chapter 2   *Dairy*
Pp. 18–19   Yellow dye in milk
Pp. 20–21   General debris in milk
Pp. 20–21   O-ring debris in milk

Chapter 4   *Fresh meat products*
P. 64   Ovenbake on pork pie

Chapter 5   *Cured meats*
Pp. 68–69   Machine dirt on bacon

Chapter 9   *Grocery*
Pp. 116–117   Grease in flour
Pp. 122–123   Mould *ex* machinery in porridge
P. 139   Caramelised material in sugar
Pp. 140–141   Machine dirt in jam
Pp. 140–141   Mould in jam jar

Chapter 10   *Bakery*
P. 162   Pan dirt on bread
P. 167   Oven bake from cake

Chapter 16   *Alcoholic beverages*
P. 234   Mould in beer
P. 240   Grease in apricot wine

Foreign Material Derived From Sabotage (1037–1038)

Sabotage is one of the most difficult sources of foreign bodies to eliminate from foods and potentially one of the most serious. Sabotage at a processing plant may be the work of a single operative with a grudge or may be a more general problem. In the latter case this is usually indicative of a low morale and dissatisfaction amongst the workforce which can be eliminated by encouraging worker participation in, for example, quality circles.

Examples of foreign bodies due to sabotage range from the relatively innocuous such as paper to the highly offensive such as contraceptives, and the dangerous such as staples. Some sabotage may be humorous in intention like plastic submarines in bottled milk but may still lead the manufacturer to a disadvantageous confrontation with the judiciary. Where possible packaging should be of such a nature as to prevent interference after sealing. This is not always possible and hence sabotage can then occur at various stages after manufacture including the home. Criminal intent has in some cases been involved in the latter either in attempting to blackmail the manufacturer or to deliberately poison other consumers.

Modern mass distribution of foods has inevitably increased the risk of mass poisoning *via* foods and there have been recent examples where this has been attempted or threatened. The most usual reasons are political as in the example of mercury injected into Israeli oranges or economic blackmail as in allegations of paraquat added to turkeys. A recent example of attempted mass poisoning in the U.S.A. pharmaceutical industry was due to a disgruntled employee and there is no doubt that the food industry is highly vulnerable in this respect.

Miscellaneous (1039–1042)

Persons responsible for investigating the presence of foreign bodies will encounter many unusual and even bizarre examples of contamination. Each must be investigated in its own right but it must be stated that the origin of many cannot be fully established and in these cases the investigator must rely on experience and intuition.

1034. **Strands of hessian found in coffee beans.** The likely origin is the sacks in which the beans have been transported to the U.K. before roasting and final packing.

material is almost certainly due to deliberate sabotage at the manufacturing stage.

1035 and **1036.** **The possible consequences** of carrying out building work while engaged in food production are illustrated. **1035** shows a masonry nail in frozen peas. Building work introduces foreign bodies such as this which are not normally present in a food processing environment while the inevitable strain on working conditions means that contamination *per se* is more likely to occur. In addition the need to use temporary sitings for production equipment may mean that all normal services, e.g. compressed air for the metal detector, are not available. **1036** shows (arrowed) plaster dust on peas produced in the same factory during the same building operations. Plaster dust would not be detected by metal detectors.

Note that the peas themselves are also of poor quality. (See Frozen Foods, Chapter 14.)

1037. **Paper on which mould growth has developed** removed from a can of lemonade with lime soft drink. The presence of such

1038. **Glass slivers** (fragments of a light bulb) embedded in sliced bread. The slivers had obviously been embedded after baking and slicing. Although not proven, the attitude of the consumer suggested the possibility of blackmail.

1039. **Fragments of plastic** found in a cake and originally mistaken for finger nails. The origin of the plastic is unknown.

1040. **A seed from a fruit** in wholemeal fruit cake mistaken by a consumer for an insect body.

1041 and **1042.** **Cellophane and a child's toy plastic cup** found in a jar of instant coffee granules are illustrated in the jar (**1041**) and removed for examination (**1042**). The foreign material may have been derived from the home but sabotage at the production stage cannot be discounted.

1034

1035 ●

1036 ●

1037

1038 ●

1039

1cm

1040

1cm

1041 ●

1042

Miscellaneous examples of foreign body contamination may also be found in the following commodity chapters:

Bibliography

Introductory Notes

Bergmeyer, H.U. (1974). *Methods of Enzymatic Analysis*, 2nd Translated Edition. Verlag Chemie: Weinheim and Academic Press: New York.

Corry, J.E.L., Roberts, D. and Skinner, F.A. (1982). *Isolation and Identification Methods for Food Poisoning Organisms*. Academic Press: London and New York.

Egan, H., Kirk, R.S. and Sawyer, R. (1981). *Pearson's Chemical Analysis of Foods*, 8th Edition. Churchill Livingstone: Edinburgh and London.

Harrigan, W.F. and McCance, M.E. (1976). *Laboratory Methods in Food and Dairy Microbiology*, Revised Edition. Academic Press: London and New York.

ICMSF (1978). *Micro-organisms in Foods. I. Their Significance and Methods of Enumeration*. 2nd Edition, University of Toronto Press: Canada.

Sharpe, A.N. (1980). *Food Microbiology – A Framework for the Future*. Charles C. Thomas Publisher: Springfield, Illinois.

Chapter 1: Food quality assurance systems

Anon (1973). *Food Safety through the Hazard Analysis and Critical Control Point System*. The Pillsbury Co.: Minneapolis.

ICMSF (1974). *Micro-organisms in Foods. 2 Sampling for Microbiological Analysis: Principles and Specific Applications*. University of Toronto Press: Canada.

Kilsby, D.C. (1982). Sampling Schemes and Limits. In *Meat Microbiology* (Brown, M.H. ed). Applied Science Publishers: London and New York.

Kilsby, D.C., Aspinall, L.J. and Baird-Parker, A.C. (1979). A system for setting numerical microbiological specifications for foods. *J. appl. Bact.*, **46**, 591.

Zottola, E.A. and Wolf, I.D. (1981). Recipe Hazard Analysis-RHAS-A systematic approach to analysing potential hazards in a recipe for food preparation/preservation. *J. Fd. Prot.*, **44**, 560.

Chapter 2: Dairy

Anon (1973). Bendenkliche Käse-Delikatesse. *Rhein-Zeitung*, August 22, 1973.

Billing, E. and Cuthbert, W.A. (1958). 'Bitty' cream: the occurrence and significance of *Bacillus cereus* spores in raw milk supplies. *J. appl. Bact.*, **48**, 297.

Bills, D.D., Morgan, M.E., Libby, L.M. and Day, E.A. (1965). Identification of compounds responsible for fruity flavour defect of experimental Cheddar cheese. *J. Dairy Sci.*, **48**, 1168.

Björck, L. and Rosen, C-G. (1976). An immobilized two-enzyme system for the activation of the lactoperoxidase antibacterial system in milk. *Biotechnology and Bioengineering*, **18**, 1463.

Blackwell, B. and Mabbit, L.A. (1965). Tyramine in cheese related to hypersensitive crises after mono-amine oxidase inhibition. *Lancet*, **1**, 938.

Chessy, B.M. (1984). Cited by Daeschel, M.A. (1984). Highlights of the 40th Annual Meeting of the Society for Industrial Microbiology, Sarasota, Fla. *Food Microbiology*, **1**, 79.

Cox, W.A., Stanley, G. and Lewis, J.E. (1978). Starters: Purpose, Production and Problems. In *Streptococci* (Skinner, F.A. and Quesnel, L.B. eds). Academic Press: London and New York.

Davies, F.L. and Wilkinson, G. (1973). *Bacillus cereus* in milk and dairy products. In *The Microbiological Safety of Food* (Hobbs, B.C. and Christian, J.H.B. eds). Academic Press: London and New York.

Davis, J.G. (1973). The Microbiology of Yoghourt. In *Lactic Acid Bacteria in Beverages and Food* (Carr, J.G., Cutting, C.V. and Whitting, G.C. eds). Academic Press: London and New York.

Davis, J.G. (1983). Personal recollections of developments in dairy bacteriology over the last 50 years. *J. appl. Bact.*, **55**, 1.

Egan, H., Kirk, R.S. and Sawyer, R. (1981). *Pearson's Chemical Analysis of Foods*, 8th Edition. Churchill Livingstone: Edinburgh and London.

Fantasia, L.D., Mestrandrea, L., Schrade, J.P. and Yager, J. (1975). Detection and growth of enteropathogenic *Escherichia coli* in soft ripened cheese. *Applied Microbiol.*, **29**, 179.

Frazier, W.C. (1967). *Food Microbiology*. 2nd Edition, McGraw-Hill: New York.

Hannington, E. (1967). Preliminary report on tyramine headache. *British Medical Journal*, **2**, 550.

Hobbs, B.C. (1974). *Food Poisoning and Food Hygiene*. 3rd Edition, Edward Arnold: London.

Hyde, K.A. and Rothwell, J. (1973). *Ice Cream*. Churchill Livingstone: Edinburgh.

ICMSF (1978). *Micro-organisms in Foods. I. Their significance and methods of enumeration*, 2nd Edition. University of Toronto Press: Canada.

Jarvis, B. (1976). Mycotoxins in Food. In *Microbiology in Agriculture, Fisheries and Food* (Skinner, F.A. and Carr, J.G. eds). Academic Press: London and New York.

Law, B.A. and Mabbitt, L.A. (1983). New methods for controlling the spoilage of milk and milk products. In *Food Microbiology: Advances and Prospects* (Roberts, T.A. and Skinner, F.A. eds). Academic Press: London and New York.

Law, B.A. and Sharpe, M.E. (1973). Lactic Acid Bacteria and Flavour in Cheese. In *Lactic Acid Bacteria in Beverages and Food* (Carr, J.G., Cutting, C.V. and Whiting, G.C. eds). Academic Press: London and New York.

Law, B.A. and Sharpe, M.E. (1978). Streptococci in the Dairy Industry. In *Streptococci* (Skinner, F.A. and Quesnel, L.B. eds). Academic Press: London and New York.

Law, B.A., Sharpe, M.E., Mabbit, L.A. and Cole, C.B. (1973). Microflora of Cheddar cheese and some of the Metabolic Products. In *Sampling – Microbiological Monitoring of Environ-*

ments (Board, R.G. and Lovelock, D.W. eds). Academic Press: London and New York.

Lowrie, R.J. and Lawrence, R.C. (1972). Cheddar cheese flavour, IV. A new hypothesis to account for the development of bitterness. *New Zealand J. Dairy Sci. and Tech.*, **7**, 51.

Reiter, B. (1981). The impact of the lactoperoxidase system on the psychrotrophic microflora in milk. In *Psychrotrophic Micro-organisms in Spoilage and Pathogenicity* Roberts, T.A., Hobbs, G., Christian, J.H.B. and Skovgaard, N. eds). Academic Press: London and New York.

Rothwell, J. (1981). Microbiology of Ice Cream and Related Products. In *Dairy Microbiology; Volume 2, The Microbiology of Milk Products* (Robinson, R.K. ed). Applied Science Publishers: London and New Jersey.

Scott, R. (1981). *Cheesemaking Practice*. Applied Science Publishers: London and New York.

Sheldon, R.M., Lindsay, R.C., Libby, L.M. and Morgan, M.E. (1971). Chemical nature of malty flavour and aroma produced by *Streptococcus lactis var maltigenes*. *Applied Microbiol.*, **22**, 263.

Wei, R.D., Still, D.E., Smalley, E.B., Schnoes, H.K. and Strong, F.M. (1973). Isolation and partial characterization of a mycotoxin from *Penicillium roquefortii*. *Applied Microbiol.*, **25**, 111.

Zall, R.R. (1980). Can cheesemaking be improved by heat treating milk on the farm? *Dairy Industries International*, **45**, 25, 48.

Chapter 3: Fresh meat

Anon (1984). *Food-borne and Water-borne disease in Canada, Annual Summary 1978*, Health Protection Branch, Health and Welfare Canada: Ottawa.

Bailey, A.J. (1983). Changes in meat handling and quality. *Proceedings: Institute of Food Science and Technology*, **16**, 16.

DeKruijf, J.M., Van Logtestijn, J.G., Franken, P. and Heder, K.A.M. (1974). Sarcosporidiosis in cattle and swine. *Tijdschr. Diergeneeskd*, **99**, 303.

Egan, H., Kirk, R.S. and Sawyer, R. (1981). *Pearson's Chemical Analysis of Foods*, 8th Edition. Churchill Livingstone: Edinburgh and London.

Forrest, J.C., Aberle, E.D., Hedrick, H.B., Judge, M.D. and Merkel, R.A. (1975). *Principles of Meat Science*. Freeman: San Francisco.

Gerrard, F. and Mallion, F.J. (1977). *The Complete Book of Meat*. Virtue Press: London.

Grimont, P.A.D. and Grimont, F. (1984) *Serratia*. In *Bergey's Manual of Systematic Bacteriology*. Volume 1 (Krieg, N.R. and Holt, J.G. eds). Williams and Wilkins: Baltimore and London.

Grossklaus, D. and Baumgarten, H.J. (1968). Die überlebensdauer von toxoplasma-cysten in schweine-fleisch I Mitteilung: Ergebnisse von lagerungsverschun bei verschiedenen temperaturen. *Fleischwirtschaft*, **48**, 930.

Heydoorn, A.O. (1977). Sarkosporidieninfiziertes. Fleisch als mögliche krankheitsursach für den menschen. *Arch. Lebensmittelhyg.*, **28**, 27.

ICMSF (1978). *Micro-organisms in Foods. I. Their significance and methods of enumeration*, 2nd Edition. University of Toronto Press: Canada.

Kotula, A.W., Murrell, K.D., Acosta-Stein, L. and Tennent, I.

(1982). Influence of rapid cooking methods on the survival of *Trichinella spiralis* in pork chops from experimentally infected pigs. *J. Fd. Sci.*, **47**, 1006.

Nottingham, P.M. (1982). Microbiology of Carcass Meats. In *Meat Microbiology* (Brown, M.H. ed). Applied Science Publishers: London and New York.

van Sprang, A.P. (1983). Parasites in meat products in the Netherlands. *Antonie von Leeuwenhoek*, **49**, 510.

Wiggins, L.S. and Wilson, A. (1976). *A Colour Atlas of Meat and Poultry Inspection*. Wolfe Medical Publications: London.

Wilson, G. (1973). Summing Up. In *The Microbiological Safety of Foods* (Hobbs, B.C. and Christian, J.H.B. eds). Academic Press: London and New York.

Chapter 4: Fresh meat products

Bull, D.C. and Solomons, G.L. (1983). The Potential for Fermentation Processes in the Food Supply. In *Food Microbiology: Advances and Prospects* (Roberts, T.A. and Skinner, F.A. eds). Academic Press: London and New York.

Egan, H., Kirk, R.S. and Sawyer, R. (1981). *Pearson's Chemical Analysis of Foods*, 8th Edition. Churchill Livingstone: Edinburgh and London.

Hobbs, B.C. (1974). *Food Poisoning and Food Hygiene*. 3rd Edition. Edward Arnold: London.

ICMSF (1978). *Micro-organisms in Foods. I. Their significance and methods of enumeration*, 2nd Edition. University of Toronto Press: Canada.

Pinegar, J.A. and Suffield, A. (1982). The Investigation of Food Poisoning Outbreaks in England and Wales. In *Isolation and Identification Methods for Food Poisoning Organisms* (Corry, J.E.L., Roberts, D. and Skinner, F.A. eds). Academic Press: London and New York.

Whittaker, R.G., Spencer, T.L. and Copeland, J.W. (1983). An enzyme-linked immunosorbent assay for species identification of raw meat, *J.Sci.Fd.Agric.*, **34**, 1143.

Chapter 5: Cured meat

Anon (1972). Nitrites, nitrates and nitrosamines in food – a dilemma. *J. Fd. Sci.*, **37**, 984.

Avagimov, C., Freidlin, Y. and Adutskevich, V. (1964). Bacon pickling by using bacterial cultures. *Meat Industry U.S.S.R.*, **35**, 55.

Buttiaux, R. (1963). Les bactéries nitrifiantes des saumures de viandes. *Revue des Fermentations et des Industries Alimentaires*, **18**, 191.

Egan, H., Kirk, R.S. and Sawyer, R. (1981). *Pearson's Chemical Analysis of Foods*, 8th Edition, Churchill Livingstone: Edinburgh and London.

Gardner, G.A. (1973). Routine microbiological examination of Wiltshire bacon curing brines. In *Sampling – Microbiological Monitoring of Environments* (Board, R.G. and Lovelock, D.W. eds). Academic Press: London and New York.

Gardner, G.A. (1982). Microbiology of Processing: Bacon and Ham. In *Meat Microbiology* (Brown, M.H. ed). Applied Science Publishers: London and New York.

Gardner, G.A. (1983). Microbial spoilage of cured meats. In *Food Microbiology: Advances and Prospects* (Roberts, T.A. and Skinner, F.A. eds). Academic Press: London and New York.

Gardner, G.A. and Patton, J. (1978). The bacteriology of bacon curing brines from multi-needle injection machines. *24th European Congress Meat Research Workers*, Kulmbach, **3**, K6.

ICMSF (1978). *Micro-organisms in Foods. I. Their significance and methods of enumeration*, 2nd Edition. University of Toronto Press: Canada.

Ingram, M. (1971). The microbiology of food pasteurization. *Memoir No. 511*, Meat Research Institute: Langford.

Ingram, M. (1976). The microbiological role of nitrite in meat products. In *Microbiology in Agriculture, Fisheries and Food* (Skinner, F.A. and Carr, J.G. eds). Academic Press: London and New York.

Jarvis, B. (1976). Mycotoxins in food. *In Microbiology in Agriculture, Fisheries and Food* (Skinner, F.A. and Carr, J.G. eds). Academic Press: London and New York.

Penner, J.L. (1984). *Providencia. In Bergey's Manual of Systemic Bacteriology*. Volume 1 (Krieg, N.R. and Holt, J.G. eds). Williams and Wilkins: Baltimore and London.

Varnam, A.H. and Grainger, J.M. (1973). Methods for the general microbiological examinations of Wiltshire bacon curing brines. In *Sampling – The Microbiological Monitoring of Environments* (Board, R.G. and Lovelock, D.W. eds). Academic Press: London and New York.

Chapter 6: Delicatessen

Bullerman, L.B. and Ayres, J.C. (1968). Aflatoxin producing potential of fungi isolated from cured and aged meats. *Appl. Microbiol.*, **16**, 1945.

Egan, H., Kirk, R.S. and Sawyer, R. (1981). *Pearson's Chemical Analysis of Foods*, 8th Edition. Churchill Livingstone: Edinburgh and London.

Gould, G.A. (1964). Effect of Food Preservatives on the Growth of Bacteria from Spores. In *Microbial Inhibitors in Food*. Almqvist and Wiksell: Stockholm.

ICMSF (1978). *Micro-organisms in Foods. I. Their significance and methods of enumeration*, 2nd Edition. University of Toronto Press: Canada.

Lancefield, R.C. (1933). A serological differentiation of human and other groups of hemolytic streptococci. *J. Exp. Med.*, **57**, 571.

Petäjä, E., Laine, J.J. and Niinivaara, F.P. (1972). Starterkulturen bei der pökelung von fleisch. *Fleischwirtschaft*, **52**, 839.

Sutherland, J.P. and Varnam, A.H. (1982). Fresh Meat Processing. In *Meat Microbiology* (Brown, M.H. ed). Applied Science Publishers: London and New York.

Chapter 7: Poultry

Barnes, E.M. and Shrimpton, D.H. (1968). The effect of processing and marketing procedures on the bacteriological condition and shelf life of eviscerated turkeys. *British Poultry Science*, **9**, 243.

Barnes, E.M., Mead, G.C. and Griffiths, N.M. (1973). The microbiology and sensory evaluation of pheasants hung at 5°, 10° and 15°C. *British Poultry Science*, **14**, 229.

Board, R.G. (1966). The course of microbial infection of the hen's egg. *J. appl. Bact.*, **29**, 319.

Brant, A.W. (1973). Controlled sanitary immersion chilling of poultry in the U.S.A., Paper No. **A5**, *Poultry Meat Symposium, Roskilde*.

Egan, H., Kirk, R.S. and Sawyer, R. (1981). *Pearson's Chemical Analysis of Foods*, 8th Edition. Churchill Livingstone: Edinburgh and London.

Grant, I.H., Richardson, N.J. and Bokkenheuser, V.D. (1980). Broiler chickens as a potential source of *Campylobacter* infections in humans. *J. Clin. Microbiol.*, **11**, 508.

ICMSF (1978). *Micro-organisms in Foods. I. Their significance and methods of enumeration*, 2nd Edition. University of Toronto Press: Canada.

Klose, A.A. (1974). Poultry scalding: present technology and alternative methods. *XV World Poultry Congress, New Orleans*, p. 541.

Mead, G.C. (1982). Microbiology of Poultry and Game Birds. In *Meat Microbiology* (Brown, M.H. ed). Applied Science Publishers: London and New York.

Parry, R.T., Haysom, L., Thomas, N.L. and Davis, R. (1982). *A Manual of Recommended Methods for the Microbiological Examination of Poultry and Poultry Products*. British Poultry Meat Assn.: London.

Patterson, J.T. (1982). Personal Communication to Mead, G.C. (1982) above.

Pivnick, H. and Nurmi, E. (1982). The Nurmi Concept in the Control of *Salmonellae* in Poultry. In *Developments in Food Microbiology-1* (Davies, R. ed). Applied Science Publishers: London and New York.

Roberts, D. (1982). Bacteria of Public Health Significance. In *Meat Microbiology* (Brown, M.H. ed). Applied Science Publishers: London and New York.

Report (1976). Evaluation of the Hygiene Problems Related to the Chilling of Poultry Carcasses. *Information on Agriculture No. 22*, Commission of the European Communities: Brussels.

Chapter 8: Fish and fish products

Arnold, S.H. and Brown, D.W. (1978). Histamine(?) toxicity from fish products. *Advances in Food Research*, **24**, 113.

Barrow, G.I. and Miller, D.C. (1976). *Vibrio parahaemolyticus* and seafoods. In *Microbiology in Agriculture, Fisheries and Food* (Skinner, F.A. and Carr, J.G. eds). Academic Press: London and New York.

Blood, R.M. (1975). Lactic Acid Bacteria in Marinated Herring. In *Lactic Acid Bacteria in Beverages and Food* (Carr, J.G., Cutting, C.V. and Whiting, G.C. eds). Academic Press: London and New York.

Connell, J.J. (1983). Trends in fish utilisation. *Proceedings: Institute of Food Science and Technology*, **16**, 22.

Deardorff, T.L., Raybourne, R.B. and Desowitz, R.S. (1984). Behaviour and viability of the third stage larvae of *Terranova spp* (Type HA) and *Anisakis simplex* (Type I) under coolant conditions. *J. Fd. Prot.*, **47**, 49.

Egan, H., Kirk, R.S. and Sawyer, R. (1981). *Pearson's Chemical*

Analysis of Foods, 8th Edition. Churchill Livingstone: Edinburgh and London.

Gilbert, R.J., Hobbs, G., Murray, C.K., Cruikshank, J.G. and Young, S.E.J. (1980). Scombrotoxic fish poisoning: features of the first 50 incidents to be reported in Britain (1976–9). *British Medical Journal*, **281**, 71.

Herbert, R.A., Hendrie, M.S., Gibson, D.M. and Shewan, J.M. (1971). Bacteria active in the spoilage of certain seafoods. *J. appl. Bact.*, **34**, 41.

Howgate, P.F. (1982). Quality assessment and quality control. In *Fish, Handling and Processing* (Aitken, A., Mackie, I.M., Merritt, J.H. and Windsor, M.L. eds). HMSO: Edinburgh.

ICMSF (1978). *Micro-organisms in Foods. 1. Their significance and methods of enumeration*, 2nd Edition, University of Toronto Press: Canada.

Shaw, S.A. (1981). An investigation of trends in the production, distribution and consumption of salmon and trout in the U.K. University of Stirling, Industrial Projects Service.

Shewan, J.M., Hobbs, G. and Hodgkiss, W. (1960). A determinative scheme for the identification of certain genera of Gram-negative bacteria with special reference to the Pseudomonadaceae. *J. appl. Bact.*, **23**, 379.

Chapter 9: Grocery

Anon (1984). *Food-borne and Water-borne disease in Canada, Annual Summary 1978*. Health Protection Branch, Health and Welfare Canada: Ottawa.

Bull, D.C. and Solomons, G.L. (1983). The Potential for Fermentation Processes in the Food Supply. In *Food Microbiology: Advances and Prospects* (Roberts, T.A. and Skinner, F.A. eds). Academic Press: London and New York.

Frazier, W.C. (1967). *Food Microbiology*. 2nd Edition. McGraw-Hill: New York.

Hulse, J.H. (1983). Food Science, for richer or for poorer. For sickness or for health. *Proceedings: Institute of Food Science and Technology*, **16**, 2.

ICMSF (1974). *Micro-organisms in Foods. 2 Sampling for Microbiological Analysis: Principles and Specific Applications*. University of Toronto Press: Canada.

ICMSF (1980). *Microbial Ecology of Foods. Vol 2, Food Commodities*. Academic Press: London and New York.

Jarvis, B. (1976). Mycotoxins in Food. In *Microbiology in Agriculture, Fisheries and Food* (Skinner, F.A. and Carr, J.G. eds). Academic Press: London and New York.

Kent, N.L. (1966). *Technology of cereals*. Pergamon Press: Oxford.

Klaushofer, H., Hollaus, F. and Pollach, G. (1971). Microbiology of beet sugar manufacture. *Process Biochemistry*, **6**, 39.

Noah, N.D., Bender, A.E., Reaidi, G.B. and Gilbert, R.J. (1980). Food poisoning from raw red kidney beans. *British Medical Journal*, **281**, 236.

Ormay, L. and Novotny, T. (1969). The Significance of *Bacillus cereus* Food Poisoning in Hungary. In *The Microbiology of Dried Foods* (Kampelmacher, E.H., Ingram, M. and Mossel, D.A.A. eds). International Association of Microbiological Societies.

Pellon, J.R. and Sinskey, A.J. (1983). Genetic Engineering for Food and Additives. In *Food Microbiology: Advances and Prospects* (Roberts, T.A. and Skinner, F.A. eds). Academic Press: London and New York.

Report (1982). The Bacteriological Examination of Water Supplies. *Reports on Public Health and Medical Subjects, No. 71*, HMSO: London.

Ross, G. (1981). A deadly oil. *British Medical Journal*, **283**, 424.

Scott, P.M. (1972). Occurrence of ochratoxin A and citrinin in cereals and patulin in apple juice. *Abstracts: IUPAC Symposium, Control of Mycotoxins*, Göteborg.

Tilbury, R.H. (1972). Post-harvest deterioration of sugarcane: An economic appraisal. Society of Chemical Industry Symposium on post-harvest deterioration.

Tilbury, R.H. (1973). Occurrence and effects of Lactic Acid Bacteria in the Sugar Industry. In *Lactic Acid Bacteria in Beverages and Food* (Carr, J.G., Cutting, C.V. and Whiting, G.C. eds). Academic Press: London and New York.

Trench, B. (1981). Spain's toxic oil: the death toll mounts. *New Scientist*, **92**, 604.

Tuynenberg-Muys, G. (1971). Microbial safety in emulsions products. *Process Biochemistry*, **6**.

Vickery, J.R. (1983). Hotpot – Comments on food legislation and research. *Focus, Journal of Professional Affairs, Institute of Food Science and Technology (U.K.)*, **16**, 6.

Chapter 10: Bakery

Joffe, A.Z. (1971). Alimentary toxic aleukia. In *Microbial Toxins* (Kadis, A. and Ciegler, A. eds). Academic Press: New York.

Kent, N.L. (1966). *Technology of Cereals*. Pergamon Press: London.

Kent-Jones, D.W. and Amos, A.J. (1957). *Modern Cereal Chemistry*. 5th Edition. Northern Publishing Co.: Liverpool.

Chapter 11: Flans and Pizzas

Egan, H., Kirk, R.S. and Sawyer, R. (1981). *Pearson's Chemical Analysis of Foods*, 8th Edition. Churchill Livingstone: Edinburgh and London.

ICMSF (1978). *Micro-organisms in Foods. 1. Their significance and methods of enumeration*, 2nd Edition. University of Toronto Press: Canada.

Chapter 12: Prepared salads

Egan, H., Kirk, R.S. and Sawyer, R. (1981). *Pearson's Chemical Analysis of Foods*, 8th Edition. Churchill Livingstone: Edinburgh and London.

ICMSF (1978). *Micro-organisms in Foods. 1. Their significance and methods of enumeration*, 2nd Edition. University of Toronto Press: Canada.

Schlech, W.F., Lavigne, P.M., Bartolussi, M.D., Allen, A.C., Haldane, E.V., Wort, A.J., Hightower, A.W., Johnson, S.E., King, S.H., Nicholls, E.S. and Broome, C.V. (1983). Epidemic Listeriosis – Evidence for transmission by food. *New England Journal of Medicine*, **308**, 203.

Chapter 13: Produce

Anon (1979). *Approved Code of Practice for Watercress*. National Farmers' Union, St Albans.

Coursey, D.G. and Brueton, A.C. (1983). Tropical fruits and vegetables. *Proceedings: Institute of Food Science and Technology*, **16**, 28.

Ho, J.L., Shands, K.N., Friedland, G., Eckind, P. and Fraser, D.W. (1981). A multihospital outbreak of Type 4b *Listeria monocytogenes* infection. *Proceedings 21st International Congress of Antimicrobial Agents and Chemotherapy*, Chicago.

Palleroni, N.J. (1984). *Pseudomonas*. In *Bergey's Manual of Systematic Bacteriology*, Volume 1 (Krieg, N.R. and Holt, J.G. eds). Williams and Wilkins: Baltimore and London.

Rymal, K.S., Chambliss, O.L., Bond, M.D. and Smith, D.A. (1984). Squash containing toxic cucurbitacin compounds occurring in California and Alabama. *J. Fd. Prot.*, **47**, 270.

Schlech, W.F., Lavigne, P.M., Bartolussi, M.D., Allen, A.C., Haldane, E.V., Wort, A.J., Hightower, A.W., Johnson, S.E., King, S.H., Nicholls, E.S. and Broome, C.V. (1983). Epidemic Listeriosis – Evidence for transmission by food. *New England Journal of Medicine*, **308**, 203.

Wilson, A.R. (1969). The development of soft rots in bulk stores. Agricultural Research Council Meeting on Bacterial Diseases of Potatoes: London.

Wong, W.C. and Preece, T.F. (1980). *Pseudomonas tolaasii* in mushroom crops: a note on primary and secondary sources of the bacterium on a commercial farm in England. *J. appl. Bact.*, **49**, 305.

Wong, W.C., Fletcher, J.T., Unsworth, B.A. and Preece, T.F. (1982). A note on ginger blotch, a new bacterial disease of the cultivated mushroom. *Agaricus bisporus. J. appl. Bact.*, **52**, 43.

Chapter 14: Frozen foods

Egan, H., Kirk, R.S. and Sawyer, R. (1981). *Pearson's Chemical Analysis of Foods*, 8th Edition. Churchill Livingstone: Edinburgh and London.

Hall, L.P. (1982). *A Manual of Methods for the Bacteriological Examination of Frozen Foods*. 3rd Edition. Campden Food Preservation Research Association: Chipping Campden.

ICMSF (1978). *Micro-organisms in Foods. I. Their significance and methods of enumeration*, 2nd Edition. University of Toronto Press: Canada.

Chapter 15: Canned foods

Ackland, M.R., Trewhella, E.R., Reeder, J.B. and Bean, P.G. (1981). The detection of microbial spoilage in canned food using thin-layer chromatography. *J. appl. Bact.* **51**, 277.

Ackland, M.R. and Reeder, J.E. (1984). A rapid chemical spot test for the detection of lactic acid as an indicator of microbial spoilage in preserved foods. *J. appl. Bact.* **56**, 415.

Anon (1973). The Double Seam Manual, The Metal Box Co. plc, Reading.

Cho, J.J., Hayward, A.C. and Rohrbach, K.G. (1980). Nutritional requirements and biochemical activities of pineapple pink bacterial strains from Hawaii, *Antonie van Leeuwenhoek*, **46**, 191.

DeLey, J., Gillis, M. and Swings, J. (1984) *Acetobacteraceae*. In *Bergey's Manual of Systematic Bacteriology*, Volume 1 (Krieg, N.R. and Holt, J.G. eds). Williams and Wilkins: Baltimore and London.

Egan, H., Kirk, R.S. and Sawyer, R. (1981). *Pearson's Chemical Analysis of Foods*, 8th Edition. Churchill Livingstone: Edinburgh and London.

Hersom, A.C. and Hulland, E.D. (1980). *Canned Foods*. Churchill Livingstone: Edinburgh.

ICMSF (1978). *Micro-organisms in Foods. I. Their significance and methods of enumeration*, 2nd Edition. University of Toronto Press: Canada.

ICMSF (1980). *Microbial ecology of foods. Vol. 2, Food commodities*. Academic Press: London and New York.

Chapter 16: Alcoholic beverages

Bronn, W.K. (1976). *Die Brantwein Wirtschaft*, **12**, 216.

Dorozynski, D. and Bell, B. (1969). *The Wine Book*. Cliveden Press: Manchester.

Garvie, E.I. (1967). *Leuconostoc oenos* sp. nov. *J. Gen. Microbiol.* **48**, 431.

Egan, H., Kirk, R.S. and Sawyer, R. (1981). *Pearson's Chemical Analysis of Foods*, 8th Edition. Churchill Livingstone: Edinburgh and London.

ICMSF (1978). *Micro-organisms in Foods. I. Their significance and methods of enumeration*, 2nd Edition. University of Toronto Press: Canada.

Lee, S.Y. (1984). *Pectinatus*. In *Bergey's Manual of Systematic Bacteriology*, Volume **1** (Krieg, N.R. and Holt, J.G. eds). Williams and Wilkins: Baltimore and London.

Lüthi, H. and Vetsch, U. (1981). *Mikroskopische Beurteilung von Weinen und Fruchtsäften in der Praxis*. 2nd Edition. Heller Chemie und Verwaltungsgesellschaft mBH: Zurich.

Okafor, N. (1975). Preliminary microbiological studies on the preservation of palm wine. *J. appl. Bact.*, **38**, 1.

Pellon, J.R. and Sinskey, A.J. (1983). Genetic Engineering for Food and Additives. In *Food Microbiology: Advances and Prospects* (Roberts, T.A. and Skinner, F.A. eds). Academic Press: London and New York.

Rose, A.H. (1977). *Economic Microbiology. Vol. 1. Alcoholic Beverages*. Academic Press: London and New York.

Saito, M., Enomoto, M., Tatsuno, T. and Uraguchi, K. (1971). Yellowed rice toxins: luteoskyrin and related compounds, chlorine containing compounds and citrinin. In *Microbial Toxins* (Ciegler, A., Kadis, S. and Ajl, S.J. eds). Academic Press: New York.

Subden, R. (1984). Wine Yeasts. In *Developments in Industrial Microbiology*, Volume **25** (Nash, C.H. and Underkofler, L.A. eds). Society for Industrial Microbiology: Arlington, Va.

Swings, J. and DeLey, J. (1984). *Zymomonas*. In *Bergey's Manual of Systematic Bacteriology*, Volume 1 (Krieg, N.R. and Holt, J.G. eds). Williams and Wilkins: Baltimore and London.

Chapter 17: Foreign Bodies and Infestations

Anon (1965). *Identification of textile materials*, 5th Edition, The Textile Institute: Manchester.

Appleyard, H.M. (1960). *Guide to the identification of animal fibres*. Wool Industries Research Assocn.: Leeds.

Hinton, H.E. and Corbet, A.S. (1972). *Common insect pests of stored food products – A guide to their identification*. The Trustees of the British Museum: London.

Index

All numbers refer to page numbers. Those in bold type refer to pages on which illustrations appear.